The Aerobic
Endospore-forming Bacteria:

Classification and Identification

Special Publications of the Society for General Microbiology

PUBLICATIONS OFFICER: COLIN RATLEDGE

The Aerobic Endospore-forming Bacteria:

Classification and Identification

Edited by

R.C.W. BERKELEY

Department of Bacteriology
University of Bristol
Bristol, U.K.

and

M. GOODFELLOW

Department of Microbiology
The University
Newcastle-upon-Tyne, U.K.

1981

Published for the
Society for General Microbiology
by
ACADEMIC PRESS
A Subsidiary of Harcourt Brace Jovanovich Publishers
London New York Toronto Sydney San Francisco

ACADEMIC PRESS INC. (LONDON) LTD.
24/28 Oval Road,
London NW1

United States Edition published by
ACADEMIC PRESS INC.
111 Fifth Avenue
New York, New York 10003

British Library Cataloguing in Publication Data
Aerobic Endospore-forming Bacteria
Conference,
University of Cambridge, 1979
The aerobic endospore-forming bacteria –
(Society for General Microbiology. Special
Publications; 4).
1. Bacteria, Aerobic – Congresses
2. Bacteria, Sporeforming – Congresses
I. Title   II. Berkeley, R. C. W.
III. Goodfellow, Michael   IV. Series
589.9'5   QR1   80-41266

ISBN 0-12-091250-3

Printed in Great Britain by
Whitstable Litho Ltd., Whitstable, Kent

CONTRIBUTORS

BERKELEY, R.C.W. *Department of Bacteriology, University of Bristol, The Medical School, University Walk, Bristol BS8 1TD.*

BONDE, G.J. *Hygiejnisk Institut, Aarhus Universitet, DK-8000 Aarhus C, Denmark.*

CLAUS, D. *Deutsche Sammlung von Mikroorganismen, Gesellschaft für Biotechnologische Forschung mbH, D-3400 Göttingen, Federal Republic of Germany.*

CROSS, T. *Postgraduate School of Studies in Biological Sciences, University of Bradford, Bradford BD7 1DP.*

de BARJAC, H. *Institut Pasteur, 25 rue du Docteur Roux, 75015 Paris, France.*

GILBERT, R.J. *Food Hygiene Laboratory, Central Public Health Laboratory, Colindale Avenue, London NW9 5HT.*

GOODFELLOW, M. *Department of Microbiology, The Medical School, University of Newcastle-upon-Tyne, Newcastle-upon-Tyne NE1 7RU.*

GORDON, R.E. *Waksman Institute of Microbiology, Rutgers University, P.O. Box 759, Piscataway, N.J. 08854, USA.*

HUNGER, W. *Deutsche Sammlung von Mikroorganismen, Gesellschaft für Biotechnologische Forschung mbH, D-3400 Göttingen, Federal Republic of Germany.*

KRAMER, J.M. *Food Hygiene Laboratory, Central Public Health Laboratory, Colindale Avenue, London NW9 5HT.*

LOGAN, N.A. *Department of Bacteriology, University of Bristol, The Medical School, University Walk, Bristol BS8 1TD.*

MINNIKIN, D.E. *Department of Organic Chemistry, University of Newcastle-upon-Tyne, Newcastle-upon-Tyne NE1 7RU.*

NORRIS, J.R. *Cadbury Schweppes Limited, Group Research, c/o Department of Chemistry, University of Reading, Whiteknights, Reading RG6 2AD.*

O'DONNELL, A.G. *Department of Microbiology, The Medical School, University of Newcastle-upon-Tyne, Newcastle-upon-Tyne NE1 7RU.*

PARRY, J.M. *Food Hygiene Laboratory, Central Public Health Laboratory, Colindale Avenue, London NW9 5HT.*

PRIEST, F.G. *Department of Brewing and Biological Sciences, Heriot-Watt University, Chambers Street, Edinburgh EH1 1HX.*

RHODES-ROBERTS, M.E. *Department of Botany and Microbiology, University College of Wales, Aberystwyth SY23 3DA.*

SHARPE, R.J. *Centre for Applied Microbiology and Research, Porton Down, Salisbury, Wiltshire SP4 0JG.*

TURNBULL, P.C.B. *Food Hygiene Laboratory, Central Public Health Laboratory, Colindale Avenue, London NW9 5HT.*

TODD, C. *Department of Microbiology, The Medical School, University of Newcastle-upon-Tyne, Newcastle-upon-Tyne NE1 7RU.*

UNSWORTH, B.A. *Department of Plant Sciences, Agricultural Science Building, University of Leeds, Leeds LS2 9JT.*

WOLF, J. *Department of Microbiology, Agricultural Science Building, University of Leeds, Leeds LS2 9JT.*

PREFACE

The aerobic endospore-forming rods constituting the genus *Bacillus* have played a prominent role in the development of microbiology since Cohn proposed the genus in 1872. Members of the genus *Bacillus* are widely distributed in soil, water and air and because their spores are so resistant, their control is of considerable importance in the food processing industry and in the preparation of all sterile products. Most *Bacillus* strains are saprophytes which, by virtue of their extensive biochemical and physiological properties, have an important part to play in the cycling of nutrients in the detritus food chain. The taxon does, however, also accommodate the well known pathogen *Bacillus anthracis*, a notable agent of food poisoning in *Bacillus cereus*, as well as a number of insect pathogens one of which, *Bacillus thuringiensis*, is produced commercially as an insecticide active against many pests of agricultural crops, forests and food. The role of *Bacillus* species as opportunistic human pathogens is increasingly being recognized. Other *Bacillus* species are exploited as a source of extracellular enzymes and antibiotics.

Bacillus systematics is in an exciting state of transition with the existing classification often unable to accommodate new discoveries. Thus, the taxonomic status and relationships of new aerobic spore-forming rods from exotic habitats as diverse as acid, hot springs and the snowfields of Antarctica is by no means clear. With the application of modern taxonomic methods such as numerical taxonomy, cell-wall analysis, lipid and nucleic acid analyses has come the realization that the genus, as at present defined, includes bacteria of widely different properties and genetic composition. The newer taxonomic methods have also helped to clarify the status of the genera *Sporolactobacillus*, *Sporosarcina* and *Thermoactinomyces* which also contain aerobic endospore-forming bacteria, and have inevitably raised the question of the relationships of these taxa to the genus *Bacillus*.

PREFACE

This book gives a comprehensive appraisal of the systematics of the aerobic, endospore-forming bacteria discovered to date.

R.C.W. Berkeley
M. Goodfellow

ACKNOWLEDGEMENTS

This book, the fourth in the series of "Special Publications of the Society for General Microbiology", arose as a result of a symposium, entitled "Aerobic Endospore-forming Bacteria", organized by the Systematics Group of the Society for General Microbiology and held at the University of Cambridge in April 1979. We would like to express our appreciation to all of the contributors for the help that they have given in the preparation of this book. Indeed in some cases, the need to master a complex and previously unreviewed literature involved unusual effort and the labours involved have slightly delayed the publication of this volume in the Special Publication Series.

We would also like to thank the staff of Academic Press, particularly Catherine Pettyfer, for their helpfulness during the preparation of this volume, as well as Niall Logan, who compiled the index, and Renate Majumdar who typed the final copy.

R.C.W.B.
M.G.

CONTENTS

Chapter 1

ONE HUNDRED AND SEVEN YEARS OF THE GENUS *BACILLUS*

R.E. GORDON

Waksman Institute of Microbiology, Rutgers,
The State University of New Jersey, USA

History of the Genus

The genus *Bacillus* has a long history and has played a major role in the development of microbiology. The genus was created in 1872 by Ferdinand Cohn at the Institute of Plant Physiology, University of Breslau. Cohn was the foremost German microbiologist when Pasteur began his work, and he encouraged Koch in his early scientific endeavours [Lechevalier and Solotorovsky, 1965]. In his publication, "Untersuchungen über Bakterien", Cohn divided the bacteria into four tribes. His new genus was assigned to Tribe III and defined as having cells, elongate or in filaments but not flexuous. For the first species of the genus, Cohn changed the generic name of Ehrenberg's *Vibrio subtilis* to *B.subtilis*. Two more species, *B.anthracis* and *B.ulna*, were assigned to the genus.

As in every development, Cohn's new genus had its antecedents. In 1835 Ehrenberg named and described *V.subtilis* as a species of the Infusoria. Although his description is generally considered unrecognizable, his was the nomenclatural precedent of the type species of the new genus. Davaine and other workers of the eighteen sixties observed microscopic parasites in the blood of animals dead of anthrax, and Davaine demonstrated clearly that animals could be experimentally infected with anthrax by blood from diseased animals only if the blood contained these characteristic bacteria, to which he gave the generic name *Bacteridium*. After his studies of the diseases of silkworms, reported in 1870, Pasteur believed that he was the discoverer of the endospores of bacteria, because he observed and illustrated clearly spore-containing bacteria in insects that had died of a disease called flacherie. Apparently Pasteur did not realize that the spores were responsible for the longevity of his bacterium of flacherie. In 1877 Cohn described the spores of *B. subtilis* and proved their heat resistance. During that same year, Koch demonstrated the developmental cycle of *B.anthracis*, from cell to resistant spore and from spore to cell [Lechevalier and Solotorovsky, 1965].

In 1925, Buchanan published a review of systematic bacteriology in which he included keys for the identification of bacteria arranged by many of the early systematists. Cohn did not use endospore formation in his keys of 1872 and 1875, and in Table 1 it can be seen that some of the other early taxonomists incorporated sporulation as a characteristic of the genus *Bacillus* in their keys for the classification of bacteria and some did not.

TABLE 1

The use of sporulation as a character in keys
for the classification of bacteria

Used	Not used
	Luerssen, 1879
	Winter, 1880
	de Lanessau, 1880
De Bary, 1884	
Van Tieghem, 1884	
Zopf, 1885	
Schroeter, 1886	Flügge, 1886
	Trevisan, 1887
	Hansgirg, 1888
De Toni and Trevisan, 1889	
Baumgarten, 1890*	
Cornil and Babes, 1890	
Migula, 1890	
Ludwig, 1892	
Sternberg, 1892*	
Freudenreich, 1894	Migula, 1894, 1900
Fischer, 1895, 1897, 1903	
Lehmann and Neumann, 1896	
Chester, 1901*	
	Kendall, 1902
	Matzuschita, 1902
Flügge, 1907	
	Conn, 1909
Heim, 1911	
	Engler, 1912
	Meyer, 1912
Löhnis, 1913	
Buchanan, 1917	
SAB Committee on Characterization and Classification, 1917	
Bergey's Manual, 1923	

* Rods with and without spores included in the genus.

Nineteen of the early keys employed sporulation to describe the genus; 3 keys placed species comprised of rods with and without spores in the genus; and 13 keys did not rely on sporulation.

In his 1907 key, Flügge was the first to introduce aerobic growth in classifying the genus, and aerobic growth appeared in the last 3 keys listed in Table 1.

Assignment of all rod-shaped genera to the genus *Bacillus* by so many of the early taxonomists resulted in a profusion of species. Although the genus was defined in the first edition of Topley and Wilson's [1929] "Principles of Bacteriology and Immunity" as "aerobic, spore-bearing rods, usually Gram positive", its index and the indices of subsequent editions furnish an idea of early nomenclature. Nearly 200 species were listed with their current generic assignments carefully designated: *B.abortus*, See *Brucella*; *B.acidophilus*, See *Lactobacillus*; *B.acnes*, See *Corynebacterium*; *B.aerogenes*, See *Bacterium*; *B.aeruginosus*, See *Pseudomonas*; *B.aerofoetidus*, See *Clostridium*; *B.avisepticus*, See *Pasteurella*. Today's microbiologists who believe that the genus *Bacillus* is too diverse in its properties (structure and composition of the cell wall, maximum temperatures for growth, nutritional requirements, products of fermentation) can take cold comfort in the fact that in the late eighteen hundreds and early nineteen hundreds it was worse.

Although the delineation of the genus *Bacillus* as encompassing the endospore-forming rods capable of aerobic growth was quite generally accepted by the nineteen twenties, a troublesome problem remained to be resolved. The identity of the type species, *B.subtilis*, was a lovely example of confusion and a topic of much discussion in the eighteen nineties and early nineteen hundreds. At that time, 2 clearly different microorganisms bore the name *B.subtilis*; one type formed large spores that germinated at the pole; the spores of the other type were smaller and germinated equatorially. The large-spored type was typified by the Michigan strain that Novy brought from Koch's laboratory in 1888 and maintained in his laboratory at the University of Michigan. The small-spored type was represented by a strain from the University of Marburg that, according to Gottheil [1901], was isolated by Meyer and used in his work published in 1899. In a report that should, in my opinion, be read by all taxonomists, Conn [1930] described assembling strains of *B.subtilis* from various culture collections [Pribram, National Collection of Type Cultures (NCTC), American Type Culture Collection (ATCC), Universities of Michigan, Marburg, and Johns Hopkins]. Five strains proved to be the large-spore type (*B.cereus*), and 5 were the small-spore type.

Then Conn turned to the literature of 1876 to 1916 and found that the early descriptions of strains of *B.subtilis*, on the whole, favoured the Marburg strain. Conn's next evidence was obtained by duplicating Cohn's and other early workers' technique of isolating their cultures of *B*.

subtilis from hay infusions boiled for 10 to 60 mins and
then incubated at 25°C or 37°C. Because methods of isolating bacteria in pure culture had not yet been developed
in Cohn's time, Conn merely inoculated agar slants from
the tubes of hay infusion. Upon first examination some of
the cultures on the agar slants were the large-spore type,
and some were the small-spore type. After a few weeks of
continued incubation, however, the small-spore type had
overgrown the other. Conn believed that his experiment
indicated that the cultures of Cohn and the other early
workers were of the Marburg type. Personal experience
confirms Conn's conclusion; in the nineteen forties, when
I examined the ATCC cultures of bacteria, I found pure
cultures of *B.subtilis*, Marburg type, bearing nearly every
species name of the various genera in the collection's
files. It seemed evident that once a culture is contaminated by *B.subtilis*, the contaminant takes over and the
original culture is lost.

In 1936, the Second International Congress for Microbiology in London adopted the Marburg strain as the neotype of *B.subtilis*, an official act considered by Gibson
[1944] as "an important advance in the taxonomy of the
genus *Bacillus*". We should not, however, become too
complaisant about this "advance". The collection upon
which the first report of Smith and his colleagues [1946]
was based contained 7 strains of *B.cereus* that we received
as *B.subtilis*. In the smaller collection examined for a
later publication [Gordon *et al.*, 1973], there was one
ATCC strain labelled *B.subtilis* that was *B.cereus*. Who
knows how many replicates of the Michigan strain remain in
culture collections as *B.subtilis* and under what circum-
stances they may again be widely distributed?

The various editions of "Bergey's Manual of Determinative Bacteriology" from the nineteen twenties to the nineteen seventies reflect the taxonomy of the genus.
Although, in some quarters, the "Manual" is regarded as a
textbook, it is not a textbook. It is a progress report
based on taxonomic studies appearing in the literature and
sometimes there is little progress. The following definitions of the family Bacillaceae and of the genus *Bacillus*
taken from the first to the eighth editions illustrate
their development over a period of 50 years:

First edition.

Family V. Bacillaceae Fischer ... Rods producing endospores,
usually Gram-positive. Flagella, when present, peritrichous.
Often decompose protein media actively through the agency of
enzymes. Key to the genera of family Bacillaceae ... Aerobic
forms, mostly saprophites (sic). Genus I. *Bacillus* ...
Aerobic forms. Mostly saprophites (sic). Generally liquefy
gelatin. Often occur in long chains and form rhizoid
colonies. Form of rod usually not greatly changed at
sporulation ...

Second, third, and fourth editions. Definitions of Bacillaceae and *Bacillus* unchanged except for the addition of "Young organisms are Gram-positive" to the definition of the genus *Bacillus*.

Fifth edition. Definition of Bacillaceae unchanged. The genus *Bacillus* described as follows:

Rod-shaped bacteria, sometimes in chains. Aerobic. Non-motile or motile by means of peritrichous flagella. Endospores formed. Generally Gram-positive. Chemo-heterotrophic, oxidizing various organic compounds.

Sixth edition.

Family XIII. Bacillaceae Fischer ... Rod-shaped cells, capable of producing spores, either with peritrichous flagella or non-motile; monotrichous flagellation has been reported but is doubtful. Endospores are cylindrical, ellipsoidal or spherical, and are located in the center of the cell, sub-terminally or terminally. Sporangia do not differ from the vegetative cells except when bulged by spores larger than the cell diameter. Such sporangia are spindle-shaped when spores are central, or wedge- or drumstick-shaped when spores are terminal. Usually Gram-positive. Pigment formation is rare. Aerobic, micro-aerophilic or anaerobic. Gelatin is frequently liquefied. Sugars are generally fermented, sometimes with the formation of visible gas. Some species are thermophilic, i.e. will grow readily at 55°C. Mostly saprophytes, commonly found in soil. A few are animal, especially insect, parasites or pathogens. Key to the genera of family Bacillaceae ... Aerobic, catalase positive. Genus I. *Bacillus* ... Rod-shaped bacteria, some-times in chains. Sporangia usually not different from the vegetative cells. Catalase present. Aerobic, sometimes show-ing rough colonies and forming a pellicle on broth. Usually oxidize carbohydrates or proteins more or less completely, often producing slight acidity, without pronounced accumulation of characteristic products ...

Seventh edition.

Family XIII. Bacillaceae Fischer, 1895 ... Rod-shaped cells capable of producing endospores which are cylindrical, ellip-soidal or spherical, and which are located in the center of the cell, sub-terminally or terminally. Sporangia do not differ from the vegetative cells except when bulged by spores larger than the cell diameter; such sporangia are spindle-shaped when spores are central and wedge- or drumstick-shaped when spores are terminal. Motile by means of peritrichous flagella or non-motile. Usually Gram-positive. Pigment formation is rare. Gelatin is frequently hydrolyzed. Sugars are generally fermented, sometimes with the production of visible gas. Aerobic, facultatively anaerobic; anaerobic; or anaerobic, aerotolerant. Some species are capable of growth at 55°C. Mostly saprophytes, commonly found in soil; a few are animal or insect parasites or pathogens. Key to the

genera of family Bacillaceae ... Aerobic or facultatively anaerobic; catalase-positive. Genus I. *Bacillus* ... Rod-shaped
cells, sometimes in chains, capable of producing endospores
... Gram-positive, some species being Gram-variable or Gram-
negative. Some species usually occur in the rough stage,
forming a pellicle on broth, whereas other species are smooth
and the rough stage is rarely seen. Usually proteins are decomposed with the production of ammonia. Carbohydrates are
generally fermented with the production of more or less
acidity; a few also produce visible gas. Catalase-positive.
Aerobic or facultatively anaerobic. Maximum temperatures for
growth vary greatly, not only between species but also between
strains of the same species. Variations in other characters
frequently occur within a species. Mostly saprophytes,
commonly found in soil; a few are animal, especially insect,
parasites or pathogens ...

Eighth edition.

Family I. Bacillaceae Fischer 1895 ... Cells rod-shaped, in
one genus spherical. Mycelium not produced. Endospores
formed. Endospores differ from vegetative cells in being more
refractive and less susceptible to staining, in having greater
resistance to heat and other destructive agents, and in containing dipicolinic acid (5-15% of dry weight). The spore contains a central cell (the core) which is enclosed by a cortex
of peptidoglycan and an outer spore coat. Majority Gram-
positive. Motile by lateral or peritrichous flagella, or non-
motile. Aerobic, facultative or anaerobic. Key to the genera
of family Bacillaceae ... Cells rod-shaped ... Aerobic or
facultative, catalase usually produced. Genus I. *Bacillus* ...
Cells rod-shaped, straight or nearly so, 0.3-2.2 by 1.2-7.0 μm.
Majority motile; flagella typically lateral. Heat-resistant
endospores formed; not more than one in a sporangial cell.
Sporulation not repressed by exposure to air. Gram reaction:
positive, or positive only in early stages of growth, or
negative. Chemoorganotrophs; metabolism strictly respiratory,
strictly fermentative or both respiratory and fermentative,
using various substrates. The terminal electron acceptor in
respiratory metabolism is molecular oxygen, replaceable in some
species by nitrate. Catalase formed by most species. Strict
aerobes or facultative anaerobes. The G + C content of the DNA
of those strains examined ranges from 32-62 moles % (T_m and
buoyant density).

The number of species assigned to the genus *Bacillus* in
the different editions of the "Manual" have also varied.
The year of publication of each edition and number of
species recognized are presented in Table 2. The number
of species included in the first to the fifth
editions make it obvious that during that time many new
species were named and described without, in Cowan's
[Cowan and Steel, 1974] words, "... the comparative work
necessary to put an organism into its rightful place in an
existing genus or species". In the most recent edition,

TABLE 2

Number of species assigned to the genus Bacillus *in different editions of Bergey's Manual*

Edition	Year	Number of species
1	1923	75
2	1925	75
3	1930	93
4	1934	95
5	1939	146
6	1948	33
7	1957	25
8	1974	22, with 26 *incertae sedis*

Gibson and Gordon [1974] recognized 22 species and placed 26 other taxa in what would have been an appendix in earlier editions. These 26 species were all represented by extant strains but, in many cases, by only one or a very few. They all need further study.

Here tribute should be paid to Dr. Robert S. Breed, who was closely associated with Dr. D.H. Bergey in developing the "Manual" and who became the guiding spirit of the contributors to the "Manual". From the first to the seventh edition he bore the main burden of the work. From 1913 to 1946 Dr. Breed was Head of the Division of Bacteriology at the New York Agricultural Experiment Station at Geneva, New York State, which was affiliated with Cornell University. I met him when I was a student at Cornell and was the fortunate recipient of his advice and encouragement for the rest of his life. He was one of my heroes, and I shall always remember him with great respect and affection.

After his retirement from the Experiment Station in 1948 until his death in 1956, Dr. Breed continued to work on the "Manual" in an office in his home. He was interested in every scrap of information about all strains of bacteria. He welcomed every comparative study of strains that was in progress and would send strains from his own collection maintained at Geneva, N.Y., or ask other bacteriologists to send their strains. Upon request, he would supply data on the history of strains and would give advice on taxonomic problems. He had a vast store of information, great enthusiasm for his work, and he was often heard to say, "Taxonomy is fun!"

A story about Dr. Breed was told at a dinner for the contributors to the sixth edition of the "Manual", where one of the guest speakers was a Vice President of The Williams and Wilkins Company, publishers of the "Manual". The speaker said that his office frequently received

letters concerning the scientific contents of the "Manual"
and that the usual reply was, "We are forwarding your
request for information to Dr. Breed. He will answer your
question in a three-page letter by return mail".

Current Major Problems

Having reviewed the history of the taxonomy of the
genus *Bacillus*, mention will be made of some current taxonomic problems; these merge into one big one: the overwhelming amount of work that remains to be done!

Knowledge of the species of the genus, as measured by
extant unclassified strains, is by no means complete. The
26 taxa mentioned earlier were not accepted by Gibson and
Gordon [1974] as distinct species because knowledge of
them was insufficient.

Another example from personal experience is provided by
the 355 strains received for examination during the last 5
years from other investigators. Of these, 71 (20%) remain
unidentified. The criteria we are using are inadequate.
The unclassified strains accumulated in this way in the
NRS Collection since 1936 plus similar strains in the
collections of other taxonomically-minded investigators
must make a sizeable total. Yet, more strains of the
poorly represented taxa are needed so that it may be established how many of the unclassified strains typify distinct
species. For this more reliable characteristics for
delineating the new and generally unknown species are
required. The need for better tools for describing and
identifying species is by no means limited to the new
species; we need them as well for currently accepted
species.

As indicated in other chapters in this book and elsewhere in the current literature, new tests for taxonomic
application are not lacking. Members of the genus
Bacillus possess many properties of varying degrees of
stability for the strain and for the species. The stability of a characteristic and its usefulness in classification can be fixed only by testing many strains of each
species of the genus. Application of a new test only to
the nomenclatural type or neotype strains of the various
species is a preliminary study. Preliminary studies are,
of course, necessary but, because of microbial variation,
they prove very little. It is difficult to state catego-
rically how many strains of a species should be tested to
determine the taxonomic value of a characteristic. One
can only say the more strains, the better. Results of the
examination of 20 strains of a species should give fairly
good proof, but results with 50 strains will provide more.

The burden of work in preparing a distinctive pattern
of dependable criteria for characterizing each species of
the genus is increased by the fact that a property
reliable in the description of some species is not

necessarily helpful in delineating other species of the same genus. Sensitivity or resistance to lysozyme, for example, was found to be a dependable characteristic for describing *B.megaterium*, *B.cereus*, and ten other species of the genus, but it was useless for *B.subtilis*, *B.pumilus*, *B.polymyxa*, *B.circulans*, *B.brevis* and *B.sphaericus* [Gordon *et al.*, 1973]. Taxonomists seek new criteria continually in the hope that the next test devised will solve all our taxonomic problems. When the new test is applied to a sufficient number of strains, however, we usually have to settle for much less and are fortunate if the new test is helpful in describing half our species. Some *Bacillus* strains seem to be capable of living in and becoming adapted to quite different environments and the result is to increase the amount of work necessary to classify and identify them.

A study of some alkalophilic strains will serve as an example. Several years ago an examination of a collection of alkalophilic strains isolated mainly from soil, dung, and water, was begun in this laboratory. The optimum pH for growth of the bacteria was 9.7, and it was quickly learned that some of the physiological tests for the identification of *Bacillus* species were useless at such a high pH value. Consequently the means of comparing the alkalophilic strains with strains of recognized species were not available. A choice was faced between (1) selecting a new battery of tests and observations (suitable to a pH range of 7.0 to 9.7) for identifying all the recognized species and any species represented by the alkalophilic strains and (2) developing variants of the alkalophilic strains that would grow at pH 7.0 and therefore could be tested by our usual methods. The latter approach was chosen and was accomplished by successive transfer (at intervals of 5 to 10 days) of the alkalophilic cultures on soil extract agar at pH 9.0, pH 8.5, pH 8.0, pH 7.5, and pH 7.0. The resulting cultures were maintained on soil extract agar (pH 7.0) with monthly transfers for a year or more before they were examined by the methods of Gordon *et al.* [1977].

Also tested were some strains which grow at pH 9.7. Cultures of 130 strains representing 15 species were transferred to nutrient broth at pH 9.7 and incubated at 28°C for 7 days. (*B.stearothermophilus*, *B.fastidiosus*, *B.larvae*, *B.popilliae* and *B.lentimorbus* were not included.) Cultures that grew in the alkaline broth were then transferred to slants of soil extract agar at pH 9.7. Of the 130 cultures, 2 of *B.megaterium*, 2 of *B.cereus*, 8 of *B. licheniformis*, 5 of *B.pumilus*, 6 of *B.firmus* and 2 of *B. circulans* (total - 25) grew on the alkaline agar (Table 3). Because all 6 strains of *B.firmus* selected grew at the high pH, 40 more strains of the *B.firmus-B.lentus* series [Gordon *et al.*, 1977] were tested, and a total of 21 cultures (46%) grew at pH 9.7.

TABLE 3

Growth of some strains of Bacillus *species at pH 9.7*

Species	Number of strains tested	Number of strains growing on agar at pH 9.7
B.*megaterium*	10	2
B.*cereus*	10	2
B.*licheniformis*	10	8
B.*subtilis*	10	0
B.*pumilus*	10	5
B.*firmus*	6	6
B.*coagulans*	10	0
B.*polymyxa*	5	0
B.*macerans*	5	0
B.*circulans*	16	2
B.*alvei*	5	0
B.*laterosporus*	6	0
B.*brevis*	10	0
B.*sphaericus*	12	0
B.*pasteurii*	5	0
Total	130	25

It was not surprising to find among the first 105 alkalophilic strains that were examined a fairly homogeneous group of 27 strains (26%) that closely resembled the 20 strains of B.*firmus* in our collection. These 20 strains of B.*firmus* include ATCC strain 14575, the nomenclatural type strain of B.*firmus*, and similar strains from a variety of sources but not strains of B.*lentus* or the B.*firmus*-B.*lentus* intermediates. Cultures of 23 of the alkalophilic strains on soil extract agar at pH 9.7 were Gram-positive and 4 were Gram-variable. Their cells were motile and resembled in appearance cells of B.*subtilis*. With the exception of one non-sporulating strain, the cultures formed ellipsoidal, sometimes cylindrical, centrally or paracentrally located spores that did not swell the sporangia appreciably. Of the cultures that sporulated at pH 9.7, only one of their counterparts failed to sporulate at pH 7.0.

The physiological characteristics of the variants of the 27 alkalophilic strains were very like those of strains of B.*firmus* [Gordon *et al.*, 1977] as typified by ATCC strain 14575 (Table 4). Their positive reactions were: production of catalase; hydrolysis of starch; growth in 7% NaCl broth and at 35°C; decomposition of casein; and acid production from glucose, mannitol, and sucrose. Negative properties of the 27 strains were: formation of acetoin and dihydroxyacetone; anaerobic

TABLE 4

Comparison of some properties of B.firmus *and*
alkalophilic strains

Property	20 strains of *B.firmus* (% positive)	27 alkalophilic strains (% positive)
Production of:		
Acetoin	0	0
Catalase	100	100
Dihydroxyacetone	0	0
Hydrolysis of starch	100	100
Anaerobic growth	0	0*
Egg-yolk reaction	0	0
Growth in:		
7% NaCl broth	95	96
Sabouraud dextrose broth	0	0
Sabouraud dextrose agar	0	0
Resistance to lysozyme	0	0
Utilization of:		
Citrate	5	0[†]
Propionate	0	0[†]
Decomposition of:		
Casein	100	93
Tyrosine	0	0
Growth at:		
35°C	100	89
50°C	0	0
Acid from:		
Arabinose	0	0
Glucose	100	100
Mannitol	90	85
Mannose	0	0
Melibiose	0	0
Raffinose	0	0
Sorbitol	0	0
Sucrose	100	100
Xylose	0	0
Sensitivity to bacteriophage (ATCC strain 14575 of *B.firmus*)	100	100

* 18% of the cultures did not grow aerobically on the test medium.

† 40% of the cultures failed to grow on the test medium.

growth; egg-yolk reaction; growth in Sabouraud dextrose
broth or on Sabouraud dextrose agar; resistance to lysozyme; utilization of citrate and propionate; decomposition of tyrosine; growth at 50°C; and acid production

from arabinose, mannose, melibiose, raffinose, sorbitol, and xylose.

The variants of the 27 alkalophilic strains capable of growth at pH 7.0 were all lysed by a bacteriophage whose host is ATCC strain 14575 of *B.firmus*. Thus far tests with this bacteriophage show that strains of *B.firmus*, some of the *B.firmus-B.lentus* intermediate strains, and some of the alkalophilic strains were sensitive whereas 162 strains of *B.licheniformis*, *B.subtilis*, and *B.pumilus*, which morphologically resemble strains of *B.firmus*, were resistant to the bacteriophage.

As indicated in Table 4, 18% of the 27 alkalophilic strains growing at pH 7.0 did not grow aerobically in the medium used for demonstrating anaerobic growth, and 40% did not grow on citrate and propionate agars. Without aerobic growth in the thioglycollate medium and on the media used for determining utilization of citrate and propionate, negative results were not acceptable.

These results necessitated some additional work. The tests listed in Table 4 were applied to 26 strains of the *B.firmus-B.lentus* complex that had been maintained in this laboratory for 8 or more years on soil extract agar at pH 7.0 and to duplicate strains maintained for 3 years on the same agar at pH 9.7. Cultures of the 26 strains cultivated at pH 7.0 grew on the test media or in the control media, for example, nutrient broth that served as a control for determining growth in Sabouraud dextrose broth. Of the duplicate cultures maintained at pH 9.7, however, 19% failed to grow aerobically in thioglycollate medium, and 30% did not grow on the citrate and propionate agars. With proof that some strains of *B.firmus* cultivated on alkaline media could lose their ability to grow on these three test media, our 27 alkalophilic strains were accepted as strains of *B.firmus*.

Another task for taxonomists is furnishing a standard procedure for the identification of strains that will give the same results when followed in different laboratories. In this context it is pleasing that work has begun at the University of Bristol on the development of an API System for the genus by Berkeley and his co-workers (see Chapter 6).

Importance of the Genus

While taxonomists continue striving for a workable classification of the genus, they will, I trust, try to avoid a return to the fifth edition of the "Manual" with its 146 species (see Table 2). They will do well also to keep in mind the facets of the importance of the genus. The genus *Bacillus* was important in the development of microbiology, for with his *B.subtilis*, Cohn put an end to the theory of spontaneous generation, and the germ theory of disease was established with *B.anthracis* by Pasteur

and Koch [Lechevalier and Solotorovsky, 1965]. Since
those early days, strains of *Bacillus* species have maintained their place in scientific investigations. They
have a major role in 2 series of conferences (Spore
Research and The International Spore Conference) whose
ultimate goal is the discovery of the reason, or reasons,
for a spore's longevity. Thermophilic strains are used as
tools for probing the secrets of life at high temperatures.
Reports of a myriad of basic studies of strains of
Bacillus species on their production of and sensitivity to
antibiotics, formation of nearly countless enzymes,
insecticidal activity, cellular composition, spores,
bacteriophages, genetics, degradation of materials such as
fuel oils, animal hair, and insecticides, resistance to
ethylene oxide, nitrogen fixation, to name a few, appear
in current journals.

Many effects and practical applications of strains of
the genus are encountered daily. Among human and animal
diseases, anthrax still exists. According to statistics
published by the World Health Organization [1977], 379
cases of anthrax were reported in 9 European countries (5
countries for 1974 and 4 for 1975). Spain, Italy and
Greece had the highest numbers of cases. In Africa, 929
cases or estimated cases were reported in 12 countries for
the same period.

Reports of infections caused by strains other than
B.anthracis appear in the literature. During the last 2
years, published reports named strains of *B.cereus* as
causal agents of abscesses, pneumonia, osteomyelitis,
bacteremia, endocarditis, and bovine mastitis. In
addition, strains of *B.cereus* were implicated in outbreaks
of food poisoning. There were also reports of infections
due to strains of *B.alvei*, *B.laterosporus*, *B.licheni-
formis*, *B.sphaericus*, and *B.subtilis*. Among strains
isolated from patients that we have received for examination (very few of which have been described in published
case reports), *B.cereus* was represented by the largest
numbers of strains followed by smaller numbers of *B.alvei*,
B.licheniformis, *B.megaterium*, *B.pumilus*, *B.sphaericus* and
B.subtilis (see also Chapter 12).

The following examples of applied microbiology's
utilization of strains of *Bacillus* species, described by
Levinson *et al.* [1978], illustrate the importance of the
genus in industry. Strains of the genus produce extracellular enzymes such as amylases, proteases, penicillin-
ases, pectinases, β-glucanases, cellulase and xylanase,
ribonuclease, deoxyribonuclease, and alkaline phosphatase.
These enzymes, of which the amylases and proteases have
the widest industrial use, are becoming increasingly
important in foods, medicine, and brewing.

Although the streptomycetes yield more industrially and
medically important antibiotics, 45 antibiotics formed by
strains of *Bacillus* species have been isolated and

described. Some have medical value and some are used in food preservation and the control of plant diseases.

The insecticidal activity of some strains of bacilli is another important property both commercially and medically. The *B.cereus* relatives, collectively grouped as *B.thuringiensis*, strains of *B.popilliae*, *B.lentimorbus*, and *B.sphaericus* have application in the control of crop-destroying and disease-carrying insects. The spores of these strains have great advantage over chemical insecticides because they kill a wide variety of insects, and are safe for humans, mammals, birds, bees, earthworms, and plants and are specific for their particular insects. They do not pollute the environment; their target insects do not become resistant to them; and they do not deteriorate in the soil.

In industry, the resistance of the bacterial spore to heat, desiccation, ultraviolet, and other agents makes the spores valuable biological indicators of the efficiency of methods of food preservation (a vast and important field) and sterilization of many different materials such as pharmaceutical products; medical, surgical, and laboratory supplies; and even the sterilization of spacecraft.

Taxonomists concerned with the genus *Bacillus* should remember that they are responsible for the names and descriptions that give meaning to the names of the strains used by other microbiologists investigating and utilizing the bacilli. Taxonomists provide the identifications of the strains of all these workers in basic science, medicine, public health, ecology, and industry. They face an enormous task and should take pride in doing it.

Acknowledgements

This study and its presentation were supported by grants from Novo Industri A/S, Bagsvaerd, Denmark; the Charles and Johanna Busch Memorial Fund; and the New Brunswick Department of Biological Sciences, Rutgers University.

References

Buchanan, R.E. (1925). "General Systematic Bacteriology", Vol. 1. Baltimore: The Williams and Wilkins Co.
Cohn, F. (1872). Untersuchungen über Bakterien. *Beiträge zur Biologie der Pflanzen 1875* 1 (Heft 2), 127-224.
Conn, H.J. (1930). The identity of *Bacillus subtilis*. *Journal of Infectious Diseases* **46**, 1-10.
Cowan, S.T. (1974). "Cowan and Steel's Manual for the Identification of Medical Bacteria". Cambridge: Cambridge University Press.
Ehrenberg, C.G. (1835). Dritter Beitrag zur Erkenntnis grosser Organisation in der Richtung des kleinsten Raumes. *Abhandlungen der Königlichen Akademie der Wissenschaften zu Berlin aus dem Jahre 1833*, 145-336.

Gibson, T. (1944). A study of *Bacillus subtilis* and related organisms. *Journal of Dairy Research* **13**, 248-260.

Gibson, T. and Gordon, R.E. (1974). *Bacillus* Cohn 1872. In "Bergey's Manual of Determinative Bacteriology 8th Edition" (eds. R.E. Buchanan and N.E. Gibbons), pp.529-550. Baltimore: The Williams and Wilkins Co.

Gordon, R.E., Haynes, W.C. and Pang, C.H-N. (1973). "The Genus *Bacillus*", Agriculture Handbook No. 427. Washington, D.C.: United States Department of Agriculture.

Gordon, R.E., Hyde, J.L. and Moore, J.A., Jr. (1977). *Bacillus firmus-Bacillus lentus*: a series or one species? *International Journal of Systematic Bacteriology* **27**, 256-262.

Gottheil, O. (1901). Botanische Beschreibung einiger Bodenbakterien. *Zentralblatt für Bakteriologie, Parasitenkunde, Infektionskrankheiten und Hygiene*, Abt. II **7**, 627-637.

Lechevalier, H.A. and Solotorovsky, M. (1965). "Three Centuries of Microbiology". New York: McGraw-Hill Book Co.

Levinson, H.S., Feeherry, F.E. and Mandels, G.R. (1978). Bases for applications of bacterial and fungal spores. In "Spores VII" (eds. G. Chambliss and J.C. Vary), pp.3-17. Washington, D.C.: American Society for Microbiology.

Smith, N.R., Gordon, R.E. and Clark, F.E. (1946). "Aerobic Mesophilic Sporeforming Bacteria", Miscellaneous Publication No. 559. Washington, D.C.: United States Department of Agriculture.

Topley, W.W.C. and Wilson, G.S. (1929). "The Principles of Bacteriology and Immunity", Vol. 1. New York: William Wood and Co.

World Health Organization (1977). "World Health Statistics. Infectious Diseases: Cases and Deaths", Vol. 2, pp.1-64. Geneva.

Chapter 2

THE TAXONOMY OF THE ENDOSPORE-FORMING ACTINOMYCETES

T. CROSS and BRIDGET A. UNSWORTH*

*Postgraduate School of Studies in Biological Sciences,
University of Bradford, Bradford, UK*

History

In 1898 Tsiklinsky isolated two actinomycetes in pure culture which grew at temperatures between 48°C and 68°C [Tsiklinsky, 1898]. One of the strains, isolated from soil, was called "Thermoactinomyces I", but the brief description has no taxonomic validity. In the following year she described two strains of "Thermoactinomyces" isolated from manure [Tsiklinsky, 1899] and named one *Thermoactinomyces vulgaris*. This then became the nomenclatural type (type species) of the genus by monotypy [Lessel, 1960]. Her brief description was illustrated with photographs and adequately defined the species. The main features given are translated and summarized below:

A filamentous organism, filaments 0.5 μm wide. Round or oval single spores (the photograph shows at least one spore on a short sporophore). Mycelium easily stained with aniline dyes or Gram's stain; spores only took up such stains when immature. Growth between 48°C and 65°C, optimum about 57°C, no growth at 70°C, 37°C or room temperature. Cultures remained viable at room temperature for one month; survived for 20 mins at 100°C in an autoclave and for 24 h in phenolic acid. Abundant growth occurred in broth after 16 h incubation, at the bottom of the tube were long, spiral, much branched filaments with terminal spores. Snow white colonies also appeared at the surface of the broth, forming a pellicle. On agar growth appeared as a white dust. Gelatin was liquefied. Milk coagulated and liquefied, with the production of acid. Starch was not hydrolyzed. Indole not formed. The species was aerobic as demonstrated by gelatin stab culture.

No local or general morbid phenomena were observed after subcutaneous or intraperitoneal injection into mice or guinea-pigs.

For present address see list of contributors.

Some actinomycetes, described prior to Tsiklinsky's study in 1899, can now be recognized as probable *Thermoactinomyces* species because the descriptions frequently contain comments on the ability of the organisms to resist heat and/or disinfectants. The earliest description of a probable thermoactinomycete is of a thermophilic *Cladothrix* isolated from "sewage water" [Kedzior, 1896]. When sewage was incubated at 60°C for 16 h, floating white flocs appeared which, when subcultured onto an agar medium at the same temperature, grew as colonies with a "chalky white deposit". This deposit (apparently Kedzior was describing the aerial mycelium) began 1 mm from the edge of the colony and appeared as concentric rings. Other, more recent workers have noticed thermoactinomycete aerial mycelium laid down in rings; Waksman and Corke [1953] described them as fairy rings. Boltanskaya *et al.* [1972] studied the effects of temperature and relative humidity on the formation of rings in the aerial mycelium of *Thermoactinomyces vulgaris* and suggested that they were due to limited autolysis of the aerial mycelium induced by minor changes in incubation conditions.

Kedzior described the spores of his *"Cladothrix"* as "terminal swellings on the hyphae on side branches of minimal length, giving the general appearance of rounded balls 0.5 to 1.5 μm in diameter sitting almost directly on the sheaths of the filaments". The spores were reported to survive steaming for 3.5 to 4.0 h and also survived 10 days in 5% carbolic acid. These properties suggest that Kedzior had isolated a member of the genus *Thermoactinomyces* and indeed, Kedzior's "thermophilic *Cladothrix*" was listed as a synonym of *Thermoactinomyces vulgaris* by Waksman in the seventh edition of Bergey's "Manual of Determinative Bacteriology" [Breed *et al.*, 1957]. However, it is not clear whether Kedzior had in fact isolated a pure culture: he described colonies growing at 35°C which became green with age. Tsiklinsky stated that her *Thermoactinomyces vulgaris* resembled Kedzior's organism in its form, its method of spore formation and staining reactions, but differed in that her actinomycete did not grow at 35°C, nor did it become green with age. Kedzior noted that his *Cladothrix* had a characteristic smell of furniture polish (probably beeswax and turpentine) whereas *Thermoactinomyces vulgaris* was odourless according to Tsiklinsky. Kedzior may have isolated a mixture of a thermophilic *Thermoactinomyces* species which formed white colonies at 60°C and a mesophilic actinomycete, possibly a *Saccharomonospora* species which would form green colonies at 35°C.

Berestnev [1897] quotes Acoste and Grande Rossi [1893] who isolated an actinomycete with spores that were killed after autoclaving at 130°C but survived prolonged treatment at 120°C. The name *Actinomyces invulnerabilis* was suggested for this organism and the ability of its spores to withstand the heat treatment quoted suggests that it may have been a thermoactinomycete. However, without more information or reference cultures it is impossible to make a positive identification. The specific name was fortunately not validly published and the literature is not burdened with the tongue-twisting combination *"Thermoactinomyces invulnerabilis"*.

In more recent years a thermophilic actinomycete was repeatedly isolated from cheeses prepared from pasteurized milk [Bernstein and Morton, 1934]. The organism was shown to survive in milk heated at 65°C to 70°C for 30 mins once or even twice during the manufacture of cheese and the spores were quoted to have a thermal death time of 2 h at 100°C. The organism had a mycelium, 0.5 µm to 0.7 µm diameter, exhibiting true branching with "lateral, terminal and intercalated chlamydospores". Bernstein and Morton compared their isolate with the published descriptions of thermophilic fungi and actinomycetes including *Thermoactinomyces vulgaris* and concluded that they had isolated a new species to which they gave the name *Actinomyces casei*. Unfortunately their comparison was incomplete for they did not make use of all the information available and published earlier by Tsiklinsky [1899]. They ignored several similarities when justifying their proposal for a new specific name.

Confusion with the genus Micromonospora

While studying the actinomycetes in soils and composts, Waksman *et al.* [1939] isolated five strains of an actinomycete which they considered was "identical, and corresponded even in minor details to the description of *Thermoactinomyces vulgaris* given by Tsiklinsky". However, they also considered that all single spored actinomycetes should be classified in one genus and consigned all such organisms to the genus *Micromonospora* [Ørskov, 1923]. Therefore, they named their isolates *Micromonospora vulgaris* and created a confusion in actinomycete taxonomy which persisted until quite recently.

In 1923, Ørskov had suggested the generic name *Micromonospora* for his Group III actinomycetes (ray fungi). This group contained only the single species *Streptothrix chalcae* isolated from the air and named by Foulerton [1905, 1910]. According to Ørskov, this strain developed deep cinnabar red colonies on nutrient peptone agar, and on other media soon assumed a brownish-black colour. The colonies were normally dome-shaped on agar but on certain media had a "humpy" or "walnut-kernel like" appearance. No aerial mycelium was produced and the spores were borne singly at the tip of short side branches on the vegetative mycelium. Ørskov regarded this species as distinct within his collection of actinomycetes being characterized by the formation of single spores and the lack of an aerial mycelium.

In 1930, Jensen isolated and briefly described 10 actinomycete strains which he placed in the genus *Micromonospora* because of their resemblance to the description given by Ørskov. In a later paper [Jensen, 1932] he reported studies on a larger group of 67 strains, isolated from soil, together with *Micromonospora (Streptothrix) chalcae* (Foulerton) Ørskov, at that time available from E. Pribram's Mikrobiologische Sammlung, Vienna. Following this study he gave the following generic definition of *Micromonospora*:

> Actinomyces-like organisms, forming a mycelium of delicate, non-septate hyphae, 0.3 to 0.8 µm thick without aerial mycelium (or traces then without spores), but producing spores singly on the distal ends of short lateral branches of the vegetative mycelium, spores spherical to oval, 1.0 - 1.2 x 1.2 - 1.5 µm, mycelium and spores Gram-positive, not acid fast. Aerobic organisms, most frequently met with in soil. The type species is *Micromonospora chalcae*.

Both Ørskov and Jensen clearly used the generic name *Micromonospora* for actinomycetes bearing single spores on the vegetative (substrate) mycelium and lacking aerial mycelium. Their strains were also mesophilic growing at temperatures in the range of 30°C - 37°C. It was extremely unfortunate that Waksman *et al.* [1939] decided to include within this genus thermophilic strains producing single spores on the aerial mycelium. It is an example of a taxonomic opinion where one character, morphology, was used almost exclusively for classification and their reasoning can be partly explained by the methods used in the study. They were studying the increase in numbers of actinomycetes in soils and composts containing stable manure which heated spontaneously. Contact slides were used to determine the frequency and the morphological types of actinomycetes present, and stained preparations were used to identify the species. All strains were determined as members of the genus *Micromonospora* but the majority would now be included in the genus *Thermomonospora* [Henssen, 1957]. The five strains isolated in pure culture were acknowledged to correspond to Tsiklinsky's description of *Thermoactinomyces vulgaris* but nomenclatural rules of priority were ignored and they named the organism "*Micromonospora vulgaris (Micr. vulgaris* Tsiklinsky n.des., syn. *Micr. coerulea* Jensen)".

Microbial taxonomists of the stature of Waksman can exert a considerable influence on the opinions of other microbiologists, usually much more than was ever intended or considered possible. Erikson, who had studied the distribution of mesophilic *Micromonospora* species in lake waters and sediments [Erikson, 1941] would appear to have been swayed by Waksman's opinion when she later turned her attention to the thermophilic actinomycetes in grass composts [Erikson, 1953]. Erikson did notice distinctive

features of her "thermophilic *Micromonospora*'s";   their
vegetative mycelium was said to differ from that of mesophilic micromonosporas in having "a greater degree of
elongation of the cells" which led to "wider spaces
between branches and a more diffuse form of growth".   She
observed spores on the vegetative mycelium of the thermophilic micromonosporas and commented that the light
scattering properties of the spores were a characteristic
of the group.   She also noticed that only the immature
spores could be stained with common bacteriological stains
and that specific stains, such as Gray's spore stain, were
required to stain mature spores.   She also reported that
spores suspended in dilute broth or sucrose withstood water
bath temperatures of 100°C for up to 45 mins and remarked
that "such powers of heat resistance bring the non acid-
fast *Micromonospora* spores in line with certain eubacterial
endospores" [Erikson, 1952].   Erikson was so near and it
now seems hard to believe that she did not draw the obvious
conclusion which would have opened up quite a novel line of
research.   Instead, she was persuaded that the formation of
an aerial mycelium by these organisms was merely associated
with their thermophilic nature and did not justify the
existence of a separate genus.   Neither Waksman nor Erikson
appear to have considered the alternative argument, that
all monosporic actinomycetes should have been included in
the genus *Thermoactinomyces*.

Waksman and Henrici [1948] perpetuated the use of the
generic name *Micromonospora* for thermophilic monosporic
actinomycetes in their chapter in the 6th edition of
"Bergey's Manual of Determinative Bacteriology".   It must
be remembered that the assumed but awesome authority of
"Bergey" was, in the past, rarely questioned by microbiologists and this is in contrast to the present day when
even immature students are encouraged by some mentors to
doubt that it contains anything of value.   But the
influential Soviet microbiologist Krassilnikov [1941, 1949,
1959, 1964] also supported Waksman's opinion by including
two thermophilic actinomycetes in the genus *Micromonospora*
namely *Micromonospora vulgaris* (syn. *Thermoactinomyces
vulgaris* Tsiklinsky, 1899) and *Micromonospora monospora*
(syn. *Actinomyces monosporus* Schütze, 1908).   The majority
of the microbiologists who trained in his department in
Moscow or used his monographs [Krassilnikov, 1941, 1949,
1959, 1964] adhered to this terminology.   The name *Micromonospora vulgaris* continued to be used for many years;
for example in Britain [Corbaz *et al.*, 1963], the USA
[Wenzel *et al.*, 1967], in the USSR [Golovina *et al.*, 1973]
and occasional papers published quite recently in other
countries [Falkowski, 1978].

Reinstatement of the genus Thermoactinomyces

Waksman and Corke [1953] revised the classification of

the thermophilic monosporic actinomycetes by considering
Thermoactinomyces and *Micromonospora* to be distinct
genera; thermoactinomycetes being distinguished by their
ability to produce a true aerial mycelium and to grow at
high temperatures. Furthermore they proposed that the
genus *Thermoactinomyces* should contain the following three
species:
- *Thermoactinomyces vulgaris* (Tsiklinsky), with white
  aerial mycelium and producing no soluble pigments.
- *Thermoactinomyces monosporus* (Schütze) comb. nov. with
  a grey green mycelium.
- *Thermoactinomyces thalpophilus* n. sp. with a white
  aerial mycelium but also forming a wine to rose
  coloured soluble pigment in certain agar media.

The new species, *Thermoactinomyces thalpophilus*, rapidly
hydrolyzed starch and formed spherical single spores on
short sporophores or were entirely sessile. Waksman, in
the seventh edition of "Bergey's Manual of Determinative
Bacteriology" [Breed *et al*., 1957] repeated the description of *Thermoactinomyces thalpophilus* and included an
amplified description of *Thermoactinomyces vulgaris* which
stated that the "spherical and ellipsoidal spores were
borne singly at the ends of short branches from which they
are easily broken". This description closely resembled
that of Tsiklinsky [1899], except that *Thermoactinomyces
vulgaris* according to Waksman was able to hydrolyze starch,
while *Thermoactinomyces vulgaris* (Tsiklinsky) did not.

Several actinomycetes isolated from farmyard manure
were identified as *Thermoactinomyces thalpophilus* by
Henssen [1957]. She compared her isolates with a type
strain supplied by Waksman and concluded that "my strains
incontestably belong to the same species". However,
Henssen amended the genus description of Waksman and Corke
[1953] in order to include organisms with chains of spores
and described the new species *Thermoactinomyces glaucus*.
Henssen was of the opinion that the genus *Thermoactinomyces* was distinguishable from all other genera of thermophilic actinomycetes by "the mode of origin of the aerial
hyphae from bow-shaped, upward branching segments of the
substrate hyphae". She was able to observe this
morphology in hanging-drop preparations of her strains of
Thermoactinomyces thalpophilus and *Tha.glaucus*, but not in
any of the other thermophilic actinomycetes she studied.
She therefore considered that *Thermoactinomyces monosporus*
[Waksman and Corke, 1953] should be reassigned to the new
genus *Thermomonospora* as strains did not show the
morphology typical of *Thermoactinomyces* species.

Waksman retained *Thermoactinomyces monosporus* in the
genus [Waksman, 1961] and increased the number of species
to six by including *Thermoactinomyces vulgaris, Tha.
thalpophilus, Tha.glaucus, Tha.thermophilus* Berestnev 1897
comb nov and *Tha.viridis* Schuurmans, Olson and San
Clemente, 1956 comb.nov. He gave no reason for the

decision to include *Thermoactinomyces thermophilus*, which appeared to have *Streptomyces*-like spore chains, or the source of the description.

One interesting point in Henssen's description of *Thermoactinomyces thalpophilus* was the mention of wide, bacteria-free zones around colonies of the actinomycete on isolation plates. This could suggest antibiotic activity but Henssen did not take this observation any further. Kosmachev [1962] described an antibiotic producing thermophilic actinomycete with white aerial mycelium and named it as a *Micromonospora* species. His isolate hydrolyzed starch, casein and gelatin, and had single spores borne directly or on very short sporophores on the aerial and substrate hyphae. Kosmachev considered this organism to be closely related to *Thermoactinomyces thalpophilus* but, working in the USSR, he presumably adhered to Krassilnikov's classification. The description of *Micromonospora vulgaris* (Tsiklinsky) Waksman 1939 in Krassilnikov's [1941] Key to the Actinomycetales states that the species had "oval and spherical conidia borne on short sporophores" and was unable to produce amylase. Kosmachev considered that he was able to distinguish his strain T12 from *Micromonospora vulgaris* just as Waksman and Corke [1953] had been able to distinguish *Thermoactinomyces thalpophilus* from *Tha.vulgaris*. Recent work has shown that these two species are indeed separate and distinct.

In the intervening years, the confused taxonomy of the genus seems to have resulted from an apparent absence of distinguishing characters that could be used for the separation of the species. Waksman and Corke considered that the production of a pink soluble pigment was the main distinguishing character of *Thermoactinomyces thalpophilus* but Henssen was unable to demonstrate the pigment on the recommended agar medium or on others that she tried. Thus pigment production appeared to be an unreliable property with which to separate the species *Thermoactinomyces vulgaris* and *Tha.thalpophilus*. Also the production of amylase does not seem to have been regarded as being important; Waksman [1961] stated that both species produced amylase while in fact *Thermoactinomyces vulgaris*, according to Tsiklinsky did not hydrolyze starch. Thus amylase positive thermoactinomycetes that could not be shown to produce a pink pigment became known as *Thermoactinomyces vulgaris*.

One exception was the strain TA/124 isolated by Craveri *et al.* [1964] which produced the antibiotic thermorubin. They compared TA/124 with the published descriptions of *Thermoactinomyces vulgaris* and *Tha.thalpophilus* and distinguished it from *Tha.vulgaris* because the latter had been reported to grow on potato and at 65°C and had not been reported to show antibiotic activity. Craveri *et al.* considered *Thermoactinomyces thalpophilus* to be characterised by its rose soluble pigment, an ability to reduce nitrate and lack of antibiotic activity (Henssen's comments and Kosmachev's strain were apparently not considered). Following a common practice in the search for new antibiotics, the strain was claimed to be a novel species for patent purposes and given the name *Thermoactinomyces antibioticus*. Properties of this species worth noting at this stage included the observation that the spores were apparently sessile and only occasionally borne on short sporophores on both aerial mycelium and substrate mycelium. The strain produced amylase and the slight production of melanoid pigments on an agar medium containing tyrosine was recorded.

The Concept of a Single Species

It is common for the proliferation of species within a bacterial genus to be followed by the publication of a paper advocating the rigorous reduction in the number of species, and it is surprising how often such sudden pruning can be seen to have been too drastic when reviewed in later years. As a result of a study of the thermophilic actinomycete population of peat and strains from culture collections, Küster and Locci [1964] proposed a revision of the genus and concluded that it should contain only one species *Thermoactinomyces vulgaris*. They regarded the formation of a red soluble pigment to be an unreliable character as they were unable to demonstrate it, either in their own isolates or in strains named *Thermoactinomyces thalpophilus* from culture collections. Consequently they proposed that *Tha.thalpophilus* be regarded as a junior synonym of *Tha.vulgaris*. *Tha.glaucus* Henssen was regarded as a *nomen dubium* because of the lack of type or reference cultures and its reported ability to produce short chains of spores (4 to 10 spores) on the aerial mycelium only and not the single spores characteristic of *Thermoactinomyces*. *Thermoactinomyces monosporus* was also considered a *nomen dubium* because of the incomplete description and lack of a type culture although they thought that it might be synonymous with *Thermomonospora viridis*. *Thermoactinomyces viridis* had already been transferred to the genus *Thermomonospora* as *Thermomonospora viridis* [Küster and Locci, 1963]. *Thermoactinomyces thermophilus* was also regarded as a *nomen dubium* [Küster and Locci, 1964] because the revised description of this organism [Waksman, 1961] indicated it to be a thermotolerant *Streptomyces* species. The strains retained in the genus by Küster and Locci all showed "a morphological appearance typical of *Thermoactinomyces vulgaris*" and there were "only a few small variations in cultural and physiological behaviour, for example growth on various media, sugar fermentation, gelatin liquefaction and amylase activity". Küster and Locci considered that these variations did not

exceed the general extent of variation that could occur
within a species.

It is from this time that the concept of *Thermoactinomyces vulgaris* as a variable species became widespread and
later studies seemed to confirm this idea. Between 1964
and 1971 the majority of taxonomists considered that there
was only one species of *Thermoactinomyces, Tha.vulgaris*,
and that this species included all thermoactinomycetes
with white aerial mycelium. Thus, Corbaz *et al.* [1963]
and Cross *et al.* [1968] applied the specific epithet
vulgaris to the many thermoactinomycetes they encountered
in mouldy hay. However, evidence had been accumulating
which indicated that there might be more than one species
in the genus. Kuo and Hartman [1966] identified 759
thermophilic strains as *Thermoactinomyces vulgaris* in a
collection of actinomycetes isolated from dung, soil and
compost. They were searching for organisms capable of
producing amylase but found that only 218 strains produced
detectable amounts of the enzyme. Further evidence of the
variability of *Thermoactinomyces vulgaris* was provided by
the studies of Flockton and Cross [1975] on 25 selected
strains; no 2 strains were identical in the series of
tests used.

Mention must also be made of another genus named *Actinobifida* which was proposed by Krassilnikov and Agre [1964]
for monosporic actinomycetes producing an aerial mycelium
and bearing single spores on dichotomously branched sporophores. They described 2 thermophilic species, *Actinobifida dichotomica* with characteristic yellow aerial and
substrate mycelia, and *Actinobifida chromogena* [Krassilni-
kov and Agre, 1965] with a light brown aerial mycelium on
dark brown colonies. Manachini *et al.* [1966] were apparently unaware of the Soviet workers proposals when they
named a yellow thermophilic actinomycete with single
spores on dichotomously branched sporophores *Thermomonospora citrina*. Neither group seems to have considered the
even earlier description of a yellow monosporic actinomycete named *Micromonospora thermolutea* [Kudrina and
Maksimova, 1963].

Circumscription of the Genus Thermoactinomyces

An extremely useful method for distinguishing the monosporic genera *Thermoactinomyces, Micromonospora, Thermomonospora* and *Actinobifida*, and supplementing the morphological distinctions is the technique of cell wall
analysis advocated by Cummins and Harris [1958] and
developed for actinomycete taxonomy by Becker *et al.* [1965]
and by Yamaguchi [1965]. *Micromonospora* walls characteristically contain the amino acids glycine and *meso*-diamino-
pimelic acid (*meso*-DAP) as characterizing constituents
(Type II wall) whereas the other genera contain only *meso*-
DAP. Strains of *Thermomonospora viridis* (now named

Saccharomonospora viridis Nonomura and Ohara, 1971) characteristically contain substantial quantities of arabinose
and galactose (Type IV wall) but the remaining genera
forming aerial mycelium, namely *Thermoactinomyces, Actinobifida* and *Thermomonospora* have a similar wall composition
but lack the additional sugars (Type III).

Other very useful characters emerged when it was found
that *Thermoactinomyces vulgaris* and *Actinobifida dichotomica* strains shared an unusual high resistance to the antibiotic novobiocin, and also that their spores were endospores containing dipicolinic acid and exhibiting extreme
heat resistance and longevity [Cross, 1968; Cross *et al.*,
1968; Dorokhova *et al.*, 1968]. As a result it was later
proposed that the genus *Thermoactinomyces* should be
restricted to species with a Type III wall and the ability
to form endospores [Cross and Goodfellow, 1973]. The
species of the genus *Actinobifida* can now be reclassified
as *Thermoactinomyces dichotomicus* and *Thermomonospora
chromogena* because the latter species forms heat sensitive
spores that lack the endospore structure.

New Thermoactinomyces Species

The stimulus for a new phase of work on the genus was
provided by the demonstration that spores of thermophilic
actinomycetes were antigens causing the hypersensitivity
disease farmer's lung [Pepys *et al.*, 1963]. These
organisms can grow profusely in overheated fodders and
cereals which later release the spores as a fine particu-
late dust able to penetrate into the alveoli of the lungs
of agricultural workers and elicit acute or even chronic
respiratory symptoms. The spores of *Micropolyspora faeni*
are chiefly responsible for farmer's lung but this species
is normally associated with high numbers of organisms that
were initially identified as *Thermoactinomyces vulgaris*
[Corbaz *et al.*, 1963; Cross *et al.*, 1968] whose spores
can also elicit a hypersensitivity reaction. One feature
of the disease is the presence of circulating antibodies
and so serological methods were introduced to determine
the causative agents in similar clinical conditions.

Studies on patients suffering from bagassosis, a hypersensitivity disease of workers handling self-heated
bagasse, showed that they had antibodies to *Thermoactinomyces vulgaris* and also to serologically distinct strains
which were later given the name *Thermoactinomyces sacchari*
[Lacey, 1971]. One other hypersensitivity disease called
humidifier fever, developed in workers continually exposed
to material emanating from heavily contaminated humidifier
units [Banaszak *et al.*, 1970]. Workers in the United
States isolated an actinomycete from the offending humidi-
fiers which showed many similarities to reference strains
of *Thermoactinomyces vulgaris* [Fink *et al.*, 1971] but
proved to be serologically distinct and was later given

the name *Thermoactinomyces candidus* [Kurup *et al.*, 1975].
A reappraisal of strains carrying the epithet *vulgaris* in
various culture collections showed that they were antigen-
ically heterogeneous [Wenzel *et al.*, 1974;   Flaherty *et
al.*, 1974;   Edwards *et al.*, 1974;   Greatorex and Pether,
1975];   some were similar to the humidifier strains later
to be named *Thermoactinomyces candidus* and others to
reference strains carrying the name *Thermoactinomyces
vulgaris*.

The nomenclatural confusion was clarified during the
course of a numerical taxonomic classification of 184
strains collected from many diverse habitats and culture
collections [Unsworth, 1978].   One very large cluster was
obtained which included strains with a white aerial
mycelium bearing mostly sessile spores, and with the
ability to hydrolyze starch and form melanin pigments on
media containing tyrosine [Cross and Unsworth, 1976].   The
cluster included strains implicated in farmer's lung
together with reference strains named *Thermoactinomyces
antibioticus* and *Tha.thalpophilus*.   The earliest applicable
specific name, *Thermoactinomyces thalpophilus*, has been
given to these strains [Cross and Unsworth, 1977].   One
other cluster contained strains with white aerial mycelium
bearing spores on short sporophores;   they were unable to
utilize starch or form melanin pigments.   The cluster included strains named *Thermoactinomyces candidus* and other
strains commonly found in mouldy hay and associated with
cases of farmer's lung.   They had properties ascribed to
the type species by Tsiklinsky in 1899 and therefore
deserve the name *Thermoactinomyces vulgaris* [Cross and
Unsworth, 1977].   Two other clusters corresponded to the
species *Thermoactinomyces sacchari* and *Tha.dichotomicus*
but a fifth cluster appeared to be a new species and has
been given the provisional name *Thermoactinomyces putidus*.

There remains one other species, placed in the genus by
Nonomura and Ohara [1971].   *Thermoactinomyces peptonophilus* is the only mesophilic species and is also exceptional in that it is sensitive to novobiocin;   however,
its spores have the structure of endospores and contain
dipicolinic acid [Attwell, 1973, 1978].   The current
classification of the genus together with synonyms is
given in Table 1.

Conclusions

This detailed history of the classification of the
genus *Thermoactinomyces* has been given in the hope that it
will warn future bacterial taxonomists of the problems and
dangers that beset them, and also of their responsibil-
ities.   It illustrates the reluctance of bacteriologists
to break with traditional dogmas, in this case to consider
that organisms far removed from the genus *Bacillus* were
capable of producing endospores.   It also shows how

TABLE 1

Species of the genus Thermoactinomyces
[Cross and Unsworth, 1977]

Thermoactinomyces vulgaris Tsiklinsky 1899
  Synonyms: *Actinomyces casei* Bernstein and Morton, 1934
             Micromonospora vulgaris Waksman *et al.*, 1939
             Thermoactinomyces albus Orlowska and Szewczuk, 1972
             Thermoactinomyces candidus Kurup *et al.*, 1975

Thermoactinomyces thalpophilus Waksman and Corke, 1953
  Synonyms: *Thermoactinomyces antibioticus* Craveri *et al.*, 1964
             Micromonospora vulgaris A64 Corbaz *et al.*, 1963
             Thermoactinomyces vulgaris CUB76 Cross *et al.*, 1968
             Micromonospora vulgaris 136 Dorokhova *et al.*, 1968
             Thermoactinomyces vulgaris M.P. Lechevalier P111 =
                                              ATCC15733

Thermoactinomyces dichotomicus [Krassilnikov and Agre, 1964] Cross
  and Goodfellow, 1973
  Synonyms: *Actinobifida dichotomica* Krassilnikov and Agre, 1964
             Micromonospora thermolutea Kudrina and Maksimova, 1963
             Thermomonospora citrina Manachini *et al.*, 1966

Thermoactinomyces sacchari Lacey, 1971

Thermoactinomyces peptonophilus Nonomura and Ohara, 1971

Nomina rejicienda
             Thermoactinomyces thermophilus (Berestnev) Waksman, 1961
             Thermoactinomyces monosporus (Lehmann and Schütze)
                                              Waksman and Corke, 1953
             Thermoactinomyces viridis Schuurmans *et al.*, 1956
             Thermoactinomyces glaucus Henssen, 1957

---

dangerous it can be for influential microbiologists to
impose their ill-conceived opinions on microbiologists who
cannot now be expected to have a working knowledge of all
the bacterial genera. The story illustrates the way in
which a classification can change as a result of an
explosion of interest in a particular group of organisms
and how the application of alternative techniques such as
serology and especially numerical taxonomy can break new
ground. Above all it must show how vital it is to be
familiar with the earlier literature on the group of
organisms being studied and to seriously consider all the
earlier workers' observations. This is stated with some
feeling because the authors have very recently been amazed
to read an account of the properties and structure of the
endospores of a monosporic, thermophilic actinomycete that
was also resistant to novobiocin. The actinomycete was
named *Actinomyces thermovulgaris* [Egorov *et al.*, 1978]!

References

Acoste, E. and Grande Rossi, F. (1893). Descripcion de un nuevo *Cladothrix (Cladothrix invulnerabilis)*. *Crônica Medica - quirurgica* **19**, 97-100.

Attwell, R.W. (1973). "Actinomycete Spores". Ph.D. Thesis. University of Bradford.

Attwell, R.W. (1978). The spores of *Thermoactinomyces peptonophilus*. *Biology of the Actinomycetes and Related Organisms* **13**, 30.

Banaszak, E.F., Thiede, W.H. and Fink, J.N. (1970). Hypersensitivity pneumonitis due to contamination of an air conditioner. *The New England Journal of Medicine* **283**, 271-276.

Becker, B., Lechevalier, M.P. and Lechevalier, H.A. (1965). Chemical composition of cell wall preparations from strains of various form-genera of aerobic actinomycetes. *Applied Microbiology* **13**, 236-243.

Berestnev, N.M. (1897). "Actinomycosis and its Causative Agents". Imperial Moscow University, Moscow (in Russian).

Bernstein, A. and Morton, H.E. (1934). A new thermophilic actinomyces. *Journal of Bacteriology* **27**, 625-628.

Boltanskaya, E.V., Agre, N.S., Sokolov, A.A. and Kalakoutskii, L.V. (1972). Effect of changes in temperature and relative humidity on rings in colonies of *Thermoactinomyces vulgaris*. *Mikrobiologiya* **41**, 675-679 (in Russian).

Breed, R.S., Murray, E.G.D. and Smith, N.R. (1957). "Bergey's Manual of Determinative Bacteriology, 7th Edition". Baltimore: Williams and Wilkins.

Corbaz, R., Gregory, P.H. and Lacey, M.E. (1963). Thermophilic and mesophilic actinomycetes in mouldy hay. *Journal of General Microbiology* **32**, 449-455.

Craveri, R., Coronelli, C., Pagani, H. and Sensi, P. (1964). Thermorubin, a new antibiotic from a thermoactinomycete. *Clinical Medicine* **71**, 511-521.

Cross, T. (1968). Thermophilic actinomycetes. *Journal of Applied Bacteriology* **31**, 36-53.

Cross, T. and Goodfellow, M. (1973). Taxonomy and classification of actinomycetes. In "Actinomycetales; Characteristics and Practical Importance" (Eds. G. Sykes and F.A. Skinner), pp.11-112. Academic Press, London and New York.

Cross, T., Maciver, A.M. and Lacey, J. (1968). The thermophilic actinomycetes in mouldy hay: *Micropolyspora faeni*. *Journal of General Microbiology* **50**, 351-359.

Cross, T. and Unsworth, B.A. (1976). Farmer's Lung: a neglected antigen. *The Lancet* (i), 958-959.

Cross, T. and Unsworth, B.A. (1977). List of actinomycetes names: alternative proposals for the genus *Thermoactinomyces*. *Biology of the Actinomycetes and Related Organisms* **12**, 6-11.

Cross, T., Walker, P.D. and Gould, G.W. (1968). Thermophilic actinomycetes producing resistant endospores. *Nature*, **London 220**, 352-354.

Cummins, C.S. and Harris, H. (1958). Studies on the cell-wall composition and taxonomy of Actinomycetales and related groups. *Journal of General Microbiology* **18**, 173-189.

Dorokhova, L.A., Agre, N.A., Kalakoutskii, L.V. and Krassilnikov, N.A.

(1968). Fine structure of spores in a thermophilic actinomycete *Micromonospora vulgaris*. *Journal of General and Applied Microbiology* **14**, 295-303.

Edwards, J.H., Baker, J.T. and Davis, B.H. (1974). Preciptin test negative farmer's lung - activation of the alternative pathway of complement by mouldy hay dust. *Clinical Allergy* **4**, 379-388.

Egorov, N.S., Vybornykh, S.N., Loriya, Zh.L. and Ermakova, L.R. (1978). Endospores of *Actinomyces thermovulgaris*. *Mikrobiologiya* **47**, 296-299 (in Russian).

Erikson, D. (1941). Studies on some lake-mud strains of *Micromonospora*. *Journal of Bacteriology* **41**, 277-300.

Erikson, D. (1952). Temperature/growth relationships of a thermophilic actinomycete, *Micromonospora vulgaris*. *Journal of General Microbiology* **6**, 286-294.

Erikson, D. (1953). The reproductive pattern of *Micromonospora vulgaris*. *Journal of General Microbiology* **8**, 449-454.

Falkowski, J. (1978). Die thermophile Streptomycetenflora in Milchpulvern und Kondensmilchprodukten. *Zentralblatt Bakteriologie Hygiene Abteilung 1. Originale B* **167**, 165-170.

Fink, J.N., Resnick, A.J. and Salvaggio, J. (1971). Presence of thermophilic actinomycetes in residential heating systems. *Applied Microbiology* **22**, 730-731.

Flaherty, D.K., Murray, H.D. and Reed, C.E. (1974). Cross reactions to antigens causing hypersensitivity pneumonitis. *Journal of Allergy and Clinical Immunology* **53**, 329-335.

Flockton, H.I. and Cross, T. (1975). Variability in *Thermoactinomyces vulgaris*. *Journal of Applied Bacteriology* **38**, 309-313.

Foulerton, A.G.R. (1905). New species of *Streptothrix* isolated from the air. *Lancet* (i), 1199-1200.

Foulerton, A.G. (1910). The streptothricoses and tuberculosis. *Lancet* (i), 551-556.

Golovina, I.G., Guzhova, E.P., Bogdanova, T.I. and Longinova, L.G. (1973). The lytic enzymes formed by the thermophilic actinomyces *Micromonospora vulgaris*. *Mikrobiologiya* **42**, 600-626 (in Russian).

Greatorex, B.D. and Pether, J.U.S. (1975). Use of a serologically distinct strain of *Thermoactinomyces vulgaris* in diagnosis of farmer's lung disease. *Journal of Clinical Pathology* **28**, 1000-1002.

Henssen, A. (1957). Beiträge zur Morphologie und Systematik der thermophilen Actinomyceten. *Archiv für Mikrobiologie* **26**, 373-414.

Jensen, H.L. (1930). The genus *Micromonospora* Ørskov, a little known group of soil micro-organisms. *Proceedings of the Linnean Society of New South Wales* **55**, 231-248.

Jensen, H.L. (1932). Contributions to our knowledge of the Actinomycetales. III. Further observations on the genus *Micromonospora*. *Proceedings of the Linnean Society of New South Wales* **57**, 173-180.

Kedzior, D. (1896). Ueber eine thermophile *Cladothrix*. *Archiv für Hygiene und Bakteriologie, Berlin* **27**, 328-338.

Kosmachev, A.E. (1962). A thermophilic *Micromonospora* and its production of antibiotic T-12 under conditions of surface and submerged fermentation at 50°C-60°C. *Mikrobiologiya* **31**, 66-71 (in Russian).

Krassilnikov, N.A. (1941). *"Keys to Actinomycetales"*. Izdatel' stvo Akademii Nauk SSSR, Moskva-Leningrad.

Krassilnikov, N.A. (1949). "Guide to the Identification of Bacteria and Actinomycetes" (English translation of section pertaining to actinomycetes. Ed. J.B. Routien. Charles Pfizer and Co., New York, 1957).

Krassilnikov, N.A. (1959). "The Determination of Bacteria and Actinomycetes". Academy of Sciences, USSA, Moscow (in Russian).

Krassilnikov, N.A. (1964). Systematic position of ray fungi among the lower organisms. *Hindustan Antibiotics Bulletin* 7, 1-17.

Krassilnikov, N.A. and Agre, N.S. (1964). A new genus of the actinomycetes *Actinobifida*. The yellow group *Actinobifida dichotomica*. *Mikrobiologiya* 33, 935-943 (in Russian).

Krassilnikov, N.A. and Agre, N.S. (1965). The brown group of *Actinobifida chromogena* n.sp. *Mikrobiologiya* 34, 284-291 (in Russian).

Kudrina, E.S. and Maksimova, T.S. (1963). Some species of thermophilic actinomycetes from the soil of China and their antibiotic properties. *Mikrobiologiya* 32, 623-631 (in Russian).

Kuo, M.J. and Hartman, P.A. (1966). Isolation of amylolytic strains of *Thermoactinomyces vulgaris* and production of thermophilic actinomycete amylases. *Journal of Bacteriology* 92, 723-726.

Kurup, V.P., Barboriak, J.J., Fink, J.N. and Lechevalier, M.P. (1975). *Thermoactinomyces candidus*, a new species of thermophilic actinomycetes. *International Journal of Systematic Bacteriology* 25, 150-154.

Küster, E. and Locci, R. (1963). Studies on peat and peat microorganisms. 1. Taxonomic studies on thermophilic actinomycetes isolated from peat. *Archiv für Mikrobiologie* 45, 188-197.

Küster, E. and Locci, R. (1964). Taxonomic studies on the genus *Thermoactinomyces*. *International Bulletin of Bacteriological Nomenclature and Taxonomy* 14, 109-114.

Lacey, J. (1971). *Thermoactinomyces sacchari* sp.nov., a thermophilic actinomycete causing bagassosis. *Journal of General Microbiology* 66, 327-338.

Lessel, E.F. (1960). The nomenclatural status of the generic names of the Actinomycetales. *International Bulletin of Bacteriological Nomenclature and Taxonomy* 10, 87-192.

Manachini, P.L., Craveri, A. and Craveri, R. (1966). *Thermomonospora citrina*, una nuova specie di Attinomicete termofilo isolato dal suolo. *Annali di Microbiologia ed Enzimologia* 16, 83-90.

Nonomura, H. and Ohara, Y. (1971). Distribution of actinomycetes in soil. (X). New genus and species of monosporic actinomycetes. *Journal of Fermentation Technology* 49, 895-903.

Orlowska, B. and Szewczuk, A. (1972). Proteases produced by *Thermoactinomyces albus*. *Archivum Immunologiae et Therapiae Experimentalis* 20, 543-554.

Ørskov, J. (1923). "Investigations into the Morphology of the Ray Fungi". Levin and Munksgaard, Copenhagen.

Pepys, J., Jenkins, P.A., Festenstein, G.N., Gregory, P.H., Lacey, M.E. and Skinner, F.A. (1963). Farmer's lung. Thermophilic actinomycetes as a source of "farmer's lung hay" antigen. *Lancet* (ii), 607-611.

Schütze, H. (1908). Beiträge zur Kenntnis der thermophilen Aktinomyceten und ihrer Sporenbildung. *Archiv für Hygiene und Bakteriologie* 67, 35-36.

Schuurmans, D.M., Olson, B.H. and San Clemente, C.L. (1956).
  Production and isolation of thermorubin, an antibiotic produced by
  Thermoactinomyces viridia n.sp. *Applied Microbiology* **4**, 61-66.
Tsiklinsky, P. (1898). Sur les microbes thermophiles. *Annales de
  Micrographie* **10**, 286-288.
Tsiklinsky, P. (1899). Sur les mucedinées thermophiles. *Annales de
  l'Institut Pasteur* **13**, 500-504.
Unsworth, B.A. (1978). "The Genus *Thermoactinomyces* Tsiklinsky".
  Ph.D. Thesis. University of Bradford.
Waksman, S.A. (1961). *"The Actinomycetes Vol. II"*. Baltimore:
  Williams and Wilkins.
Waksman, S.A. and Henrici, A.T. (1948). Family II Actinomycetaceae
  Buchanan. In "Bergey's Manual of Determinative Bacteriology,
  Sixth Edition" (Eds. R.S. Breed, E.G.D. Murray and A.P. Hitchens),
  pp.892-928. Baltimore: The Williams and Wilkins Co.
Waksman, S.A. and Corke, C.T. (1953). *Thermoactinomyces* Tsiklinsky,
  a genus of thermophilic actinomycetes. *Journal of Bacteriology*
  66, 377-378.
Waksman, S.A., Umbreit, W.W. and Cordon, T.C. (1939). Thermophilic
  actinomycetes and fungi in soils and composts. *Soil Science* **47**,
  37-61.
Wenzel, F.J., Emanuel, D.A. and Lawton, B.R. (1967). Pneumonitis due
  to *Micromonospora vulgaris*. *American Review of Respiratory
  Diseases* **95**, 652-655.
Wenzel, F.J., Gray, R.L., Roberts, R.C. and Emanuel, D.A. (1974).
  Serologic studies in farmer's lung: precipitins to the thermophilic actinomycetes. *American Review of Respiratory Diseases* **109**,
  464-468.
Yamaguchi, T. (1965). Comparison of the cell wall composition of
  morphologically distinct actinomycetes. *Journal of Bacteriology*
  89, 444-453.

Chapter 3

DNA HOMOLOGY IN THE GENUS *BACILLUS*

F.G. PRIEST

*Department of Brewing and Biological Sciences,
Heriot-Watt University, Edinburgh, UK*

Introduction

The genus *Bacillus* is generally regarded as including
those rod-shaped bacteria able to form heat-resistant
endospores in aerobic conditions. This definition encom-
passes a great variety of organisms endowed with markedly
different biochemical and physiological properties and
which apparently thrive in such diverse habitats as acid,
hot springs (pH 2 to 3, 75 to 80°C) and the snow fields of
Antarctica [Slepecky, 1975]. Originally taxonomists
emphasized this metabolic diversity and species were
defined monothetically with the result that by the mid-
1940's more than 150 species had been described. Despite
the rationalization of the genus by Smith *et al.* [1952]
and Gordon *et al.* [1973] many of these monothetic taxa
remain and new "species", *B.xerothermodurans* [Bond and
Favero, 1977] and varieties, *B.subtilis* var. *faecalis*
[Iglewski and Gerhardt, 1978], *B.alcalophilus* subsp. *halodurans* [Boyer *et al.*, 1973], of variable status continued
to be described. Such artificial classifications should
be discouraged because they cannot readily accommodate
strain variation and can consequently lead to serious mis-
identification.
   Several attempts have been made to construct polythetic
classifications using numerical taxonomic methods notably
by Bonde [1975] and Boeyé and Aerts [1976]. Such studies
are generally based on morphological, physiological and
biochemical characters but run into the problem of the
metabolic diversity of the genus. Thus, is it realistic
to compare the fermentation reactions of *Bacillus alcalophilus* growing at pH 10 with those of *B.acidocaldarius*
growing at pH 2, or the nutrient utilization pattern of
B.fastidiosus with the versatile *B.subtilis*? Numerical
phenetic classifications are indispensable for con-
structing diagnostic tables and dichotomous keys but
particularly in the case of *Bacillus*, it would seem essential that numerical analysis is substantiated with data
from other techniques.

The composition of virtually all the microbial cell
components which may be used for classification, for
example cell walls, cytoplasmic membranes and whole-cell
pyrolysis vary with environmental change [Oxborrow *et al.*,
1977; Rilfors *et al.*, 1978]. Because *Bacillus* species
are so diverse, it is impossible to design standard conditions that would allow the growth of strains of all of the
species so the taxonomist does not always know if variation between taxa is due to innate genetic differences or
to environmental effects. The quantitative estimation of
DNA homology avoids these difficulties because the genome
remains essentially unchanged whatever the environment.
Although many *Bacillus* strains harbour plasmids [Lovett
and Bramucci, 1976; Bernhard *et al.*, 1978], the presence
of which are environmentally determined, the amount of
plasmid DNA relative to genomic DNA is generally too small
to significantly affect DNA base composition determinations and reassociation assays. Thus, DNA isolated from
B.acidocaldarius can readily be compared with that from
B.psychrosaccharolyticus, *B.alcalophilus* or any other
species, in the confidence that the estimated affinity
will be based on the overall genetic capability of the
strains and not some specialized feature that may render
much of the phenotype unique. It seems likely, therefore,
that in such a heterogeneous taxon as *Bacillus*, DNA
homology will provide an ideal basis for classification
and should contribute invaluable additional data to
taxonomies based on alternative criteria.

DNA homology can be assessed at two levels. The overall composition can be quantified in terms of size (molecular weight) and mol% of guanine and cytosine (%GC).
Second, the sequence similarity between two genomes can be
estimated by reassociation experiments thereby providing a
direct measure of the affinity between two strains. The
application of these methods to the aerobic, endospore-
forming bacteria is reviewed and an attempt made to show
how these studies are providing sound and valuable information for the reclassification of *Bacillus* and related
taxa.

DNA Preparation

DNA is generally isolated from bacteria using established procedures [Marmur, 1961; Saito and Miura, 1963;
Kirby *et al.*, 1967]. The only difficulty likely to be
encountered with the *Bacillus* species, is efficient cell
lysis. This is usually effected with lysozyme and sodium
dodecyl sulphate (SDS) but the cell walls of strains in
species such as *B.cereus*, *B.alvei* and *B.laterosporus* are
resistant to lysozyme. This problem can be overcome by
the addition of vancomycin HCl [Bonde and Jackson, 1971]
or penicillin G [Seki *et al.*, 1978] to the culture during
the late exponential growth phase; such pretreated cells

are susceptible to lysozyme and SDS. It is necessary to use highly purified DNA for reassociation experiments but this is not so essential for %GC determinations from buoyant density measurements. Criteria for DNA purity and methods of RNA, protein and carbohydrate elimination have been discussed in detail by Bradley and Mordarski [1976].

Overall Composition

Genome Molecular Weight

The molecular weight of bacterial genome DNA may be conveniently estimated from the initial rate at which the sheared denatured molecule reassociates. The larger, and therefore the more complex the genome, the slower the rate of renaturation [Gillis *et al.*, 1970]. The taxonomic value of this parameter has yet to be fully evaluated but it seems reasonable to assume that the phenotypic potential of an organism is a reflection of the amount of DNA it contains and that strains with widely different genome sizes will be phenetically different.

The genome sizes of some of the *B.subtilis* group, are presented in Table 1. The *B.megaterium* genome is significantly larger than that of *B.subtilis* and, when more bacteria have been examined, it may be that such information will have some taxonomic value.

TABLE 1

Genome molecular weight for some Bacillus *species*

Organism	Number of determinations	k $\times 10^2$	Mol. weight of DNA $\times 10^{-9}$	Ratio vs. *B.subtilis* as 100
B.amyloliquefaciens F.	7	11.1±2.3	2.57	107
B.licheniformis 9945a	6	10.9±1.8	2.64	110
B.megaterium 899 thy^{-1}	-	-	3.07	128
B.pumilis NRRL B-3275	5	12.1±0.7	2.36	99
B.subtilis Marburg	9	12.0±1.7	2.39	100
B.subtilis W23	2	11.8±1.5	2.44	102

The apparent renaturation rate constant (k') for purified sheared, denatured DNA was determined at 70°C in 2 x SSC buffer. This was related to genome molecular weight according to Gillis *et al.*, [1970] using the published figure of 2.39 $\times 10^9$ for the *B.subtilis* Marburg genome.

[1]Data from Gillis *et al.* [1970].

DNA Base Composition

The mol %GC of genome DNA is particularly useful and
has received considerably more attention as an aid to
bacterial classification than molecular weight. The value
stems from a number of reasons. Bacteria differ widely in
%GC [Normore, 1973] thereby providing good discriminatory
potential. Because %GC reflects the genome composition,
it contains a high degree of information. Finally,
relatively simple physical, melting point and buoyant
density determination, and chemical techniques have been
developed for estimating %GC.

Closely related organisms generally possess a similar
nucleotide sequence and this is reflected in the %GC
values. Moreover, if 2 strains have been assigned to the
same species on numerical phenetic grounds but the %GC is
significantly different, this is strong evidence that the
phenetic classification is erroneous. Theoretical limits
to the extent of sequence homology between nucleic acids
of varying GC content have been provided by De Ley [1969]
who has shown that DNA molecules differing by 20-30% GC
can have virtually no sequences in common. The 34%
difference between *B.cereus* and *B.thermocatenulatus* (see
Fig. 1) therefore indicates that these organisms can be
only remotely related. As a general rule, differences of
less than 2% are taxonomically insignificant, over 5%
suggests different species and over 10% different genera
[Jones and Sneath, 1970; Bradley and Mordarski, 1976].
Mol %GC therefore provides a reliable estimate of heterogeneity within a taxon and a very wide range of %GC within
a genus suggests the need for taxonomic revision. On this
basis it has been recommended that *Flavobacterium* be
divided into at least two genera. Low (30-45%) GC, non-
motile strains have been retained in the genus *Flavobacterium* whilst high (55-70%) GC, non-motile and motile
rods have been reclassified *Empedobacter* [McMeekin and
Shewan, 1978]. Similarly, the genome composition of
Spirillum strains varies between 38% and 65% and it has
been proposed that the strains should be divided into
three genera according to DNA base composition and correlated characteristics; *Spirillum* (38% GC), *Oceano-
spirillum* (42-48% GC) and *Aquaspirillum* (50-65% GC) [Hyle-
mon *et al.*, 1973]. There are several other examples of
the value of DNA base composition for the improvement and
rationalization of heterogeneous taxa [Bradley and
Mordarski, 1976] and it would seem relevant to treat
Bacillus similarly.

An attempt has been made in Table 2 to arrange *Bacillus*
species into phena based on the numerical analyses
presented in Chapters 5 and 6. With the exception of
Taxon 6 (see below), strains within these groups vary by
no more than 12% GC and can therefore be considered the
equivalent of most bacterial genera. However, it is

emphasized that these taxa are tentative and will almost certainly require modification when adequate phenotypic descriptions and genome base determinations are available.

Taxon 1 (38-48% GC) contains the psychrophilic species; *B.psychrophilus*, *B.psychrosaccharolyticus* and *B.globisporus*, organisms for which DNA data are not available, interspersed with *B.firmus*, *B.lentus* and *B.circulans* strains. The *B.firmus* strains have GC ratios of 40-45% suggesting a homogeneous species. The *B.lentus* strains cluster separately from *B.firmus* in a phenon which includes several *B.circulans* strains. *B.lentus* is genetically heterogeneous (41-48% GC) as are the *B.firmus* - *B.lentus* "intermediate" strains [Gordon *et al.*, 1977] which vary between 40.5% and 48% GC. *B.aminovorans* (44% GC) and *B.nitritollens* (38-42% GC) are phenotypically related to *B.firmus* and *B.lentus*. It is now generally recognized that *B.megaterium* comprises at least two aggregates of strains, those with the higher base composition (40-42% GC) have been designated *B.carotarum*. Finally, *B.pantothenticus* is included in Taxon 1 on phenotypic grounds [Priest *et al.*, Chapter 5] but has been shown to be more closely related to *B.stearothermophilus* and *B.coagulans* using the AP1 system [Logan and Berkeley, Chapter 6] and to *B.cereus* from enzyme patterns [Baptist *et al.*, 1978]. The GC content of the genome of the type strain has been reported as 37% [Baptist *et al.*, 1978] and 45% GC [Bonde, 1975].

Taxon 2 (41-53% GC) comprises *B.stearothermophilus* and *B.thermodenitrificans*. The wide range of base composition reflects the heterogeneity of these species and indicates the requirement for more detailed study.

Taxon 3 (38-47% GC) contains the fastidious insect pathogens, *B.pulvifaciens* and *B.laterosporus*. Very little DNA base composition data has been reported for these organisms.

The alkali-forming species constitute Taxon 4 (36-47% GC). *B.macroides* is phenotypically related to *B.sphaericus* [Priest *et al.*, Chapter 5] but its genome base composition suggests that it may be distinct. Although the 7 strains of *B.brevis* examined to date are genetically homogeneous (44-46% GC) the phenotypic data does not fully support this. Nevertheless, this represents a compact taxon both phenotypically and genetically.

Taxon 5 (32-48% GC) has been divided into 2 groups because of the wide GC range [see Priest *et al.*, Chapter 5]. The strains of *B.megaterium* included in Table 2 (37-42% GC) probably represent both *B.megaterium* and *B.carotarum*. In fact Claus [Chapter 9] has shown that the base composition of 12 strains of *B.megaterium sensu stricto* varied very little (37-39% GC) and was consistently lower than *B.carotarum*. The remainder of Taxon 5A is occupied by *B.subtilis* and its relatives. Taxon 5B comprises *B. cereus*, *B.anthracis* and *B.thuringiensis*. Present evidence

TABLE 2

DNA base composition of Bacillus species

TAXON 1 (38-48% GC)	mol %GC	
B.pantothenticus	(1)	41
B.flexus	(1)	39
B.carotarum	(7)	40-42
B.aminovorans	(1)	44
B.firmus/lentus	(18)	39-48
B.nitritollens	(2)	38, 42
TAXON 2 (41-53% GC)		
B.stearothermophilus	(8)	44-53
B.thermodenitrificans	(2)	41, 52
TAXON 3 (38-47% GC)		
B.lentimorbus	(1)	38
B.popilliae	(1)	41
B.azotoformans	(17)	38-41
B.laterosporus	(4)	40-42
B.pulvifaciens	(2)	46-47
TAXON 4 (36-47% GC)		
B.sphaericus	(3)	36-39
B.aneurinolyticus	(1)	42
B.pasteurii	(1)	42
B.macroides	(1)	42
B.brevis	(7)	44-46
B.badius	(1)	47
B.freudenreichii	(1)	44

TAXON 5 A (36-48% GC)	mol %GC	
B.megaterium	(22)	36-45
B.pumilus	(22)	42-47
B.subtilis	(14)	42-48
(B.natto, 4)		41-43
B.globigii	(1)	43
B.subtilis var. aterrimus	(2)	43-44
B.subtilis var. niger	(4)	43-45
B.licheniformis	(7)	43-47
B.amyloliquefaciens	(8)	44-46
TAXON 5 B (32-40% GC)		
B.anthracis	(2)	32, 34
B.cereus	(24)	32-38
B.cereus var. mycoides	(3)	35-39
B.thuringiensis	(5)	33-40
TAXON 6 (35-55% GC)		
B.circulans	(4)	35-50
(B.aphorreus, 1)		44
B.coagulans	(5,1)	44-48, 55
B.polymyxa	(12)	44-51
B.alvei	(4)	46-48
B.amylosolvens	(1)	47
B.macerans	(5,6)	49-51, 53-54
TAXON 7 (61-67% GC)		
B.acidocaldarius (3)	(3)	61-62
B.thermocatenulatus	(1)	67

suggests that base composition will be of little value in separating these species.

Taxon 6 (35-55% GC) is genetically diverse but this disparity is a reflection of the poor classification of *B.circulans*. In the numerical analysis, *B.circulans* strains were recovered in Taxon 6 but, many others were dispersed throughout the dendrogram [Priest *et al.*, Chapter 5]. With the exception of *B.circulans*, the species of Taxon 6 vary between 44% and 55% GC and are genetically and phenotypically homogeneous. The 2 groups of *B.macerans* strains (49-51% and 53-54% GC) possibly represent 2 distinct species.

Finally Taxon 7 contains the two extreme thermophiles, *B.acidocaldarius* and *B.thermocatenulatus* at the high GC end of the spectrum. Obviously much more information, biochemical, physiological and morphological, must be evaluated before such a classification could be established, but the range of genome composition strongly indicates that a revision of the genus along the lines suggested above is very necessary.

If the 5% difference limit on the GC content of a species is accepted (see above), then there are few valid *Bacillus* species. Some are well-defined, 22 strains of *B.pumilus* cover 41.8-46.9% GC suggesting a neat, genetically homogeneous taxon. Similarly, *B.amyloliquefaciens*, *B.licheniformis*, *B.subtilis* and *B.brevis* are "good" species. However, other species are decidedly heterogeneous; 11 strains of *B.firmus* and 7 strains of *B.lentus* cover 39-50% GC. To a lesser extent *B.stearothermophilus* and *B.polymyxa* are, by this criterion, heterogeneous. But a closer examination of the data may suggest the remedy. Claus [Chapter 9] has placed 18 strains of *B.megaterium* into two taxa, 12 strains with 37-38.8% GC and 6 strains with 40.1-40.9% GC. Moreover, these groups can be distinguished both phenetically and by DNA reassociation. Table 2 shows two distinct groups of *B.macerans* at 53.6-54.1% and around 50% GC and it may be that these represent 2 distinct taxa. Perhaps when data for many more strains from different species are known similar patterns may emerge for the other heterogeneous species.

The need for great care when evaluating GC data has been indicated by Bradley and Mordarski [1976] and is

---

Footnotes to Table 2 (opposite)
The data were collected from: Welker and Campbell [1967], Bonde and Jackson [1971], Normore [1973], Gibson and Gordon [1974], Golovacheva *et al.* [1975], Gordon *et al.* [1977], Baptist *et al.* [1978], Pichinoty *et al.* [1978] and Claus (this volume).
The figures in brackets are the number of strains examined. Where a particular strain has been studied in more than one laboratory the average value has been used.

emphasized here. For example, the GC content of the type
strain of *B.alvei* has been variously reported as 33
[Marmur and Doty, 1962], 46.8 [Bonde and Jackson, 1971]
and 47.4% [Baptist *et al.*, 1978]. The DNA from *B.subtilis*
Marburg has 47.5 [Bonde and Jackson, 1971], 43.8 [Baptist
et al., 1978], 42.4 or 41.8% GC [Welker and Campbell,
1967], a range of 5.7% suggesting that it does not form a
species! It is therefore of the utmost importance that a
number of strains from the same species are examined and
that at least two methods are used for some determinations
so that any error in technique can be identified. In fact,
it is desirable that independent determinations are
performed in a second laboratory to confirm the reliability
of the methods and instrumentation being used.

DNA Reassociation

   Native duplex DNA is denatured into single strands at
temperatures above the melting point (T_m) or at a pH above
12 and renatures (reassociates) under suitable conditions,
that is at a temperature 15-30°C below T_m or neutral pH.
The dependence of the extent of reassociation on the
experimental conditions has been summarized by Brenner
[1970]. Assuming rigidly standardized conditions, the
amount of reassociation between 2 denatured DNA molecules
from different sources provides an estimate of nucleotide
sequence homology. High reassociation (>75%) normally
indicates either that the 2 genomes differ very slightly
throughout their entire length but not sufficiently to
prevent partial base pairing; or alternatively that the 2
molecules are virtually homologous but have minor
localized areas of non-identity. In contrast, low reassociation (<25%) indicates unrelated molecules with
varying degrees of conserved homologous loci. Highly
related genomes can be further analyzed by determining the
thermal stability of the hybrid duplex. It is generally
assumed that thermal instability in heterologous DNA is
due to unpaired bases and it has been estimated that each
1°C decrease in thermal stability of a DNA heteroduplex,
compared to that of a homologous DNA duplex, reflects 1%
mismatched mucleotide pairs [Steigerwalt *et al.*, 1976].
Thus, the partially base paired heteroduplex will have a
low thermal stability but the precisely paired heteroduplex with localized areas of non-identity will be relatively thermostable. The determination of heteroduplex
thermostability thus allows estimates of sequence divergence to be established in related nucleic acids.
   It is tedious to examine the thermal elution profiles
of all the heteroduplexes prepared in a systematic study
and so the concept of exacting (supraoptimal) and non-
exacting (optimal) reassociation temperatures was introduced. The extent of the reassociation reaction increases
with decrease in temperature but there is a concomitant

decrease in specificity such that at temperatures significantly below the optimum, relatively distantly related
sequences will reassociate and non-specific heteroduplexes
are formed. The optimum temperature combining specificity
with extent of reaction is about 30°C below T_m. At higher,
exacting temperatures (15-20°C below T_m) only precisely
matched sequences will reassociate. Therefore, by estimating reassociation at the 2 incubation temperatures an
indication of the structure of the heteroduplex is
obtained and evolutionary trends can be evaluated [Bradley
and Mordarksi, 1976].

DNA reassociation is assayed by incubating sheared
(200,000-500,000 molecular weight), denatured, radio-
actively labelled DNA with an excess of denatured DNA from
a different organism. The duplex DNA is then separated
from single stranded DNA by hydroxyapatite chromatography
[Brenner, 1970] or by initially immobilizing the denatured
"cold" DNA on nitrocellulose filters and measuring the
counts bound to the filter [Denhardt, 1966; Moore, 1974].
Perhaps the greatest problem in these procedures is
obtaining the highly labelled reference DNA. This can be
facilitated by using Thy⁻ auxotrophs which can be readily
isolated for most *Bacillus* species by trimethoprim
selection [Miller, 1972; Seki *et al.*, 1978].

An alternative procedure, which has the advantage of
not requiring radioactive labelling, is to measure the
initial rates of renaturation of 2 sheared, denatured DNA
samples and their mixture. The renaturation rate is
determined as decrease in absorbance at 260nm during the
first 40 to 60 mins of the reaction. The theoretical
equations and conditions of the measurement have been
presented [De Ley *et al.*, 1970] and the reliability of the
procedure is comparable to other methods [Gibbins and
Gregory, 1972]. Some typical results for *Bacillus* are
shown in Fig. 2 where a commercial strain labelled
B.subtilis B20, used for the production of α-amylase, is
shown to belong to the species *B.amyloliquefaciens*. The
reassociation of purified, sheared, denatured DNA from
B.subtilis B20, *B.subtilis* Marburg and a mixture of the 2
is shown in Fig. 2a. The initial reaction is not linear
due to temperature equilibration which is complete after
about six minutes. Thereafter the decrease in optical
density is linear for about 3 h. It can be seen that the
reaction rate of the mixture is less than the pure DNA
samples. However, in Fig. 2b, where DNA from *B.amylo-
liquefaciens* F and DNA from *B.subtilis* B20 are examined,
the reaction rates for the pure DNA samples and the
mixtures are comparable. From the equation developed by
De Ley *et al.* [1970] it can be calculated that the reassociation between *B.subtilis* B20 and *B.subtilis* Marburg DNA
is 16% and that between *B.subtilis* B20 and *B.amylolique-
faciens* F DNA is 94%. It can, therefore, be concluded
that *B.subtilis* B20 is a strain of *B.amyloliquefaciens*.

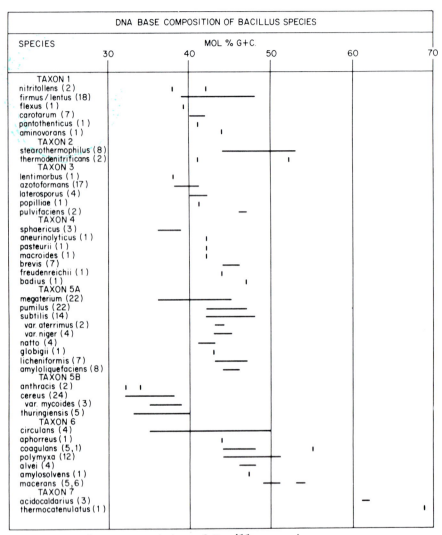

Fig. 1.   DNA base composition of *Bacillus* species

   This spectrophotometric assay has several advantages.
Optically monitored reassociation measures only matched
sequences and not open loops nor free ends. No radio-
actively labelled DNA is required so experiments are not
restricted to comparing unknown strains with a few
labelled reference strains but any combination can be
examined. Estimates of genome molecular weight can be
obtained from the renaturation rates of the pure DNA
samples. Finally, the only sophisticated equipment
required is an accurate spectrophotometer capable of monitoring the optical density of three individual samples.

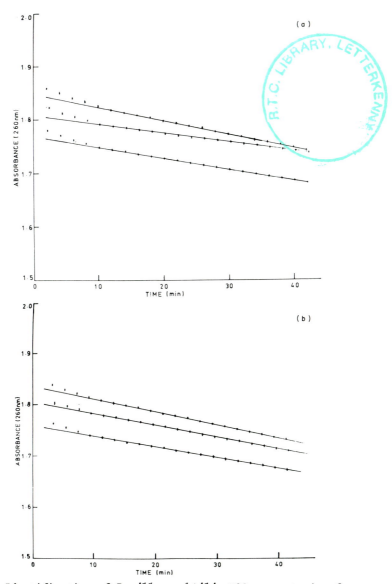

Fig. 2. Identification of *Bacillus subtilis* B20 as a strain of *Bacillus amyloliquefaciens*.
Purified, sheared (approx. 2-4 x10^5 dalton) denatured DNA from the bacteria was incubated at 70°C in a Unicam SP1800 spectrophotometer equipped with an automatic sample changer. In (a) the top curve is *B.subtilis* B20, DNA, the bottom curve *B.subtilis* Marburg DNA and the intermediate curve a mixture of the two. In (b) the top curve is *B.subtilis* B20 DNA, the bottom curve *B.amyloliquefaciens* F DNA and the intermediate curve a mixture of the two.

The method, however, does require large amounts of DNA
compared to the filter method and with strains that are
difficult to grow or to lyse this can be a severe disad-
vantage.

DNA Reassociation and *Bacillus* Taxonomy

Bacillus subtilis *Group*

   B.subtilis, *B.pumilus* and *B.licheniformis* form a highly
related aggregate phenon, and individually show few consistent phenotypic properties by which they can be distinguished [Gordon *et al.*, 1973]. A fourth member of the
group, *B.amyloliquefaciens* was isolated from soil
[Fukumoto, 1943] and given species status because of the
high levels of extracellular enzymes it secreted.
Subsequent investigators recognized this organism either
as *B.amyloliquefaciens* or as a variety of *B.subtilis* but
in many cases *B.subtilis* and *B.amyloliquefaciens* were confused. Phenotypically *B.amyloliquefaciens* differs from
B.subtilis by the production of large amounts of liquefying α-amylase. *B.subtilis* Marburg produces low levels
of saccharifying α-amylase, an enzyme that is considerably
different from the liquefying amylase [Matsuzaki *et al.*,
1974]. Welker and Campbell [1967] suggested four other
physiological tests that could be used to distinguish
these species but they have been found to be of little
value [Gordon *et al.*, 1973; Seki *et al.*, 1975].
   B.natto secretes about five times more saccharifying
α-amylase than *B.subtilis* Marburg [Yoneda *et al.*, 1974].
It was assigned to *B.subtilis* by Smith *et al.* [1946], a
recommendation that was supported by Gordon *et al.* [1973].
B.subtilis var. *amylosacchariticus* secretes about fifteen
times more saccharifying α-amylase than *B.subtilis* Marburg
[Yoneda *et al.*, 1974] and is of uncertain taxonomic
position.
   The gross genome composition of the *B.subtilis* group is
similar in molecular weight (2.35-2.64 x10^9) Table 1 and
in GC content (42-47%) Table 2. However, four homology
groups can be readily defined by DNA reassociation (Table
3).   Strains of *B.subtilis* form a homogeneous group above
67%. Included in this group are those highly amylolytic
bacilli that secrete the saccharifying α-amylase, *B.natto*
and *B.subtilis* var. *amylosacchariticus*. This supports the
recommendation that *B.natto* be classified in *B.subtilis*
and suggests that the varietal status of *B.subtilis* var.
amylosacchariticus be discontinued. Moreover, the
pigmented varieties of *B.subtilis*; var. *aterrimus* and
var. *niger* are highly related to the parent species (Table
3) and, bearing in mind the variability in production of
pigment by these strains, there is no justification for
varietal status.
   B.amyloliquefaciens comprises a homogeneous species

TABLE 3

DNA homology in the Bacillus subtilis group

DNA isolated from	B. subtilis Marburg (168) [1]	B. licheni-formis 9945a	B. pumilus NRLL B-3275	B. amylolique-faciens F
B. subtilis Marburg	100[3]	11[3], 14[2]	18[5]	25[3]
B. subtilis W23	73[2], 70[4]	12[3]	21[5]	25[2]
B. subtilis var. niger	78[3]	13[3]	12[5]	-
B. subtilis var. aterrimus	96[3]	12[3]	7[5]	-
B. subtilis var. amylosacchariticus	85[4]	-	-	-
B. natto OUT 8235	67[3]	12[3]	-	-
B. natto IFO 13169	70[3]	10[3]	-	-
B. amyloliquefaciens F	25[3], 5[2]	13[3]	-	100
B. amyloliquefaciens K	6[2]	-	-	72[2]
B. subtilis B20	16[2]	-	-	94[2]
B. amyloliquefaciens H	11[5]	-	5[5]	-
B. licheniformis 9945a	11[3]	100[5]	-	13[3]
B. licheniformis NCIB 9375	12[3]	63[2]	-	-
B. licheniformis 15 strains	9-15[3]	61-100[3]	-	-
B. licheniformis FDO-12	24[5]	-	9[5]	-
B. pumilus NRRL B-3275	18[5], 14[2]	-	100[5]	5[2]
B. pumilus NRRL B-1489	7[5]	9[5]	-	-
B. pumilus ATCC 7061	14[5], 14[2]	8[6]	58[5], 100[2]	-

[1] DNA reassociation with either Marburg strain or the trp C2 derivative, 168.
[2] Data from author's laboratory using the spectrophotometric reassociation assay of De Ley et al. [1970], at a renaturation temperature of 70°C.
[3] Data from Seki et al. [1975] obtained using the filter method incubated at 37°C for 22.5 h in the presence of 50% formamide.
[4] Data from Yoneda et al. [1974] obtained using the filter method incubated at 60°C for 3 h.
[5] Data from Lovett and Young [1969] obtained using the filter method incubated at 65°C for 16 h.
[6] Data from Seki et al. [1978]. Methodology as for (3) above.
Figures are % DNA reassociation. -, no determination.

with high (>72%) interstrain relatedness. This taxon includes the industrial strain, previously designated *B.
subtilis* B20 [Priest, 1975], and now renamed *B.amylolique-
faciens* (Fig. 1). The interspecies reassociation (6-25%)
shows that *B.amyloliquefaciens* is quite distinct from the
other taxa and should be given species status as originally proposed by Welker and Campbell [1967]. It is interesting to note that Seki *et al.* [1975] have shown that
several soil isolates were included in *B.amyloliquefaciens*
by DNA homology data but did not secrete the very high
levels of α-amylase characteristic of the industrial
strains classified in this species. Thus, the only consistent phenotypic character which separates *B.subtilis*
from *B.amyloliquefaciens* appears to be the type of amylase
produced. Sixteen strains of *B.licheniformis* displayed
61-100% DNA relatedness indicating a homogeneous species
which is quite distinct from the other species (Table 3).
Similarly DNA reassociation between 15 strains of *B.
pumilus* [Seki *et al.*, 1978] was 51-100% indicating a relatively homogeneous, discrete species (see also Table 3).
Thus, despite the scarcity of sound phenotypic data, these
four species can be readily validated by DNA reassociation
studies and it is in situations such as this that the
method is so invaluable. However, with four species
possessing such diverse genomes, it is highly probable
that an extensive numerical analysis would yield a similar
classification and allow tests to be selected for subsequent identification.

Bacillus cereus *and Related Taxa*

The low end of the GC range of the genus *Bacillus* (32-
38%) is occupied by *B.cereus* and several closely related
species whose rank is disputed, namely: *B.anthracis*, *B.
mycoides* and *B.thuringiensis*. Smith *et al.* [1952] could
find no consistent phenotypic properties that differentiated these species and proposed that *B.anthracis*, *B.
thuringiensis* and *B.mycoides* be designated varieties of
B.cereus, a scheme which was retained in a later monograph
[Gordon *et al.*, 1973]. This classification is supported
by DNA studies. The base composition of the varieties is
within the 32-38% GC range of the 24 species of *B.cereus*
examined to date (Table 2) with the sole exception of one
strain of *B.mycoides* (39% GC) [Schildkraut *et al.*, 1962]
which requires further study.
    A compilation of DNA reassociation studies performed
by Somerville and Jones [1972] and Seki *et al.* [1978] is
presented in Table 4. *B.mycoides* shares 80% homology with
B.cereus T supporting its classification as a variety of
B.cereus. In fact this varietal status has been generally
acknowledged for *B.mycoides* [Gibson and Gordon, 1974], the
DNA data are merely conclusive. However, the pathogenic
qualities of *B.anthracis* and *B.thuringiensis* have not been

so readily disregarded as the singular colonial morphology of *B.mycoides* and the recommendations for varietal status for these species [Smith *et al.*, 1952; Gordon *et al.*, 1973] have not been wholly accepted. Both "Bergey's Manual of Determinative Bacteriology" [Gibson and Gordon, 1974] and Cowan [1974] retain species status for these organisms. The available DNA reassociation data indicate a single species. The homology between DNA from *B.anthracis*, *B.thuringiensis* and *B.cereus* T is consistently within the 50-100% species range (Table 4) suggesting that organisms in these taxa should bear the same name.

　　B.thuringiensis has been divided into 14 serotypes on the basis of flagellar antigens [de Barjac, 1978; Chapter 10]. Genome relatedness between the original 11 serotypes is high (50-90%) confirming the single species. Unfortunately there are no published data of DNA homology within the serotypes but this would presumably be very high. *B. finitimus* (serotype 2) was originally given species status but the reassociation data do not support this (Table 4).

TABLE 4

DNA reassociation between Bacillus cereus *and related species*

DNA isolated from	*B.thuringiensis* serotype			*B.cereus*	*B.anthracis*
	1	3	7	T	A3
B.thuringiensis serotype 1	100	73	91	73, 80[1]	66
B.thuringiensis serotype 2	65	–	57	53	88
B.thuringiensis serotype 3	78	100	84	67	76
B.thuringiensis serotype 4	79	78	93	69	76
B.thuringiensis serotype 7	71	79	100	45, 54[1]	87
B.thuringiensis serotype 11	88	64	80	75	84
B.mycoides	–	–	–	80	74
B.anthracis A3	–	–	–	94	100
B.anthracis Sterne	–	–	–	56	100
B.cereus var. fluorescens	–	–	–	80[1]	–
B.pantothenticus ATCC 14576	–	–	–	4[1]	–
B.laterosporus	–	–	–	3[1]	–

Figures are % competition [Somerville and Jones, 1972] except where suffixed [1], -, no data available.

Those strains of *B.cereus* that produce a yellow-green
pigment have been designated var. *fluorescens* but the DNA
homology data refute this.   As with the pigmented vari-
eties of *B.subtilis*, pigmentation is too variable a
property for such status.
Bacillus pantothenticus has been included in Table 4
because comparative zone electrophoresis of enzymes
[Baptist *et al.*, 1978] revealed a high affinity with
B.cereus and the GC content (36.7%) does not exclude it.
However, it shares only limited DNA homology with *B.cereus*.
Bacillus laterosporus has some phenetic affinity with the
B.cereus group but its enzyme patterns are related to
B.amyloliquefaciens [Baptist *et al.*, 1978].   The GC
content of *B.laterosporus* DNA is 40-42.7% consistent with
its inclusion in either group.   Obviously more detailed
polyphasic data are necessary before the correct taxonomic
position of these species can be assessed.

Bacillus firmus *and* Bacillus lentus

Physiological studies suggest that these two species
represent a single "series";   that is, although strains of
B.firmus are sufficiently different from *B.lentus* to
assign them individual species status, there is a spectrum
of intermediate strains that bridges the two extremes
[Gordon *et al.*, 1977].   The GC content of 31 strains has
been determined and varies between 39-48.3% but there is
no correlation between %GC and species.
A preliminary characterization of *B.firmus* and *B.lentus*
using DNA reassociation is presented in Table 5 and it
seems likely from these results that this technique may
resolve the problem of intermediate strains.   The DNA from
those strains assigned to *B.firmus* on traditional criteria
is highly related and shares little homology with that
from *B.lentus*.   Similarly, DNA from *B.lentus* shows high
interstrain homology.   There can be little doubt that two
valid species are represented in Table 5.   Only a single
intermediate strain has been studied to date, NRS 749
which has 66% DNA homology with *B.lentus* NCIB 8773.   This
is about the minimum relatedness level normally found
amongst strains of a species in the family Enterobacteriaceae [Steigerwalt *et al.*, 1976] and is significantly greater than the 50% minimum DNA homology common
to *Bacillus* species [Seki *et al.*, 1978].   Moreover, NRS
747 is only 35% related to the *B.firmus* type strain and
19% related to *B.firmus* NRS 1131.   NRS 749 is therefore a
strain of *B.lentus*.   Perhaps when more intermediate
strains have been studied they also will cluster with one
or other of the two species.   If DNA homology can assign
intermediate strains to *B.firmus* or *B.lentus* then phenotypic data, derived from a systematic numerical taxonomic
study, should reveal the same pattern.

TABLE 5

DNA reassociation amongst some strains of
Bacillus firmus *and* Bacillus lentus

DNA isolated from	*B.firmus* NCTC 10335	*B.lentus* NClB 8773	"Intermediate" NRS 749
B.firmus NCTC 10335	100	12	35
B.firmus NRS 858	74	13	-
B.firmus NRS 1131	90	-	19
B.firmus NRS 854	101[1]	3[1]	-
B.lentus NClB 8773	12	100	66
B.lentus NRS 883	27	85	-
B.lentus NRS 769	21	-	60
B.lentus soil isolate	17	73	-

% DNA reassociation was assayed according to footnote 2 of Table 2 at a renaturation temperature of 69°C. [1], data of Seki *et al*. [1978]. -, no data available.

Miscellaneous Bacillus *Species*

The limited data available on DNA reassociation amongst *Bacillus* species other than those mentioned above, are summarized in Table 6. Although the number of strains examined is small, some generalizations can be made. *B. coagulans, B.sphaericus* and *B.circulans* appear to be discrete species and have very little homology with the other species examined. Seki *et al*. [1978] reported that 10 strains of *B.coagulans* displayed above 76% relatedness indicating a tidy homogeneous species. This is supported by the physiological properties of *B.coagulans* [Gordon *et al*., 1973] but not by the morphological characteristics. Some *B.coagulans* strains produce non-swollen sporangia (morphological group 1) whilst others possess swollen sporangia (morphological group 2). However, there appeared to be no relationship between morphological group and DNA reassociation [Seki *et al*., 1978] negating the usefulness of spore morphology for the classification of this species.

Five strains of *B.sphaericus* constitute 2 DNA homology groups (Table 6). Traditional studies indicate the boundary of *B.sphaericus* to be diffuse and it is quite likely that, when DNA reassociation between more strains has been examined, "new" species will be revealed. The 3 strains of *B.circulans* that have been examined are genetically heterogeneous. These data reinforce the original observation that this taxon represents a "complex" rather than a species [Gibson and Topping, 1938] a view that is supported by the GC data (39-48.4%). As

TABLE 6

DNA reassociation amongst some Bacillus *species*[†]

DNA from	DNA isolated from			
	B.subtilis Marburg	*B.coagulans* ATCC 7050	*B.circulans* ATCC 4513	*B.sphaericus* ATCC 14577
B.coagulans				
ATCC 7050	3	100	4	2
1FO 3887	4	113	-	-
1FO 12583	4	99	-	-
C2	4	76	-	-
B.sphaericus				
ATCC 14577	3	2	3	100
NRS 996	-	-	4	98
NRS 344	-	-	-	18
ATCC 14525	2	-	3	18
ATCC 7055	-	-	-	18
B.pantothenticus				
ATCC 14576	3	2	4	2
B.circulans				
ATCC 4513	3	2	100	3
ATCC 9966	2	-	2	2
ATCC 7049	2	-	2	2

[†]Data from Seki *et al.* [1978].   Figures represent % reassociation.
-, no data available.

with *B.sphaericus*, a polyphasic examination of more strains should reveal the internal taxonomic structure of *B.circulans*.

Finally, as described by Hunger and Claus [Chapter 9], 3 DNA homology groups have been identified in *B.megaterium* which may represent *B.megaterium* and the original species *B.carotarum* and *B.flexus*.

Wider Relations

The low interspecies DNA reassociation evident in Table 4 and Table 6 provides little information on the taxonomic structure of the genus, only that most *Bacillus* species appear to have diverged to such an extent that they are now only distantly related.   Exceptions are the *B.subtilis* group (Table 3) and *B.firmus* and *B.lentus* (Table 5) in which individual species possess a core relatedness of some 10-25%, reflecting their relatively high phenetic affinity.   However, it must be remembered that there is not a linear relationship between % DNA

homology and phenetic similarity, although it can be made
approximately linear by probit transformation [Jones and
Sneath, 1970] and organisms with about 50% similarity
(Simple matching coefficient) normally possess very few
DNA sequences in common (0-5% DNA homology). It is interesting to speculate on the nature of the residual DNA
homology between *Bacillus* species.

Genetic evidence indicates conservation within ribosomal RNA (rRNA) cistrons in bacilli [Young and Wilson,
1972]. Transformation of auxotrophic *B.subtilis* Marburg
to prototrophy with DNA from *B.amyloliquefaciens, B.lichen-
iformis* or *B.pumilus* is negligible [Coukoulis and Campbell,
1971; Young and Wilson, 1969; Seki *et al.*, 1975].
However, transformation to antibiotic resistance for
erythromycin and streptomycin using DNA from strains of
these species is detectable and approaches 80% of the
homologous reaction for *B.amyloliquefaciens* DNA. The major
barrier to interspecies transformation in *Bacillus* is
sequence heterology and not restriction of the incoming
DNA [Harris-Warwick and Lederberg, 1978] supporting the
concept of sequence homology in rRNA cistrons. The amount
of DNA devoted to rRNA synthesis is approximately equal to
the observed interspecies DNA reassociation. In *B.subtilis*
there are some 10 sets of rRNA cistrons [Potter *et al.*,
1977] each set comprising one 16S gene, one 23S gene and
spacer DNA's. This corresponds to a DNA molecular weight
of about 3 x10^7 or around 1% of the average genome.

Herndon and Bott [1969] used DNA:RNA pairing to elucidate the taxonomic position of *Sporosarcina ureae*. The
vegetative cells of *S.ureae* resemble micrococci in shape,
cell wall composition and mode of division but form endospores and are biochemically similar to *B.pasteurii*
[Norris, Chapter 14]. DNA:DNA reassociation between *S.
ureae* and bacilli (*B.subtilis, B.megaterium* and *B.pasteurii*)
is low (1-5%) but total RNA (principally rRNA) from *S.ureae*
hybridized 72 to 80% with DNA from bacilli, 44% with DNA
from *Staphylococcus aureus* and only 1% with DNA from
Sarcina lutea providing strong evidence for the inclusion
of *S.ureae* in the genus *Bacillus*. A more detailed
analysis of rRNA in *S.ureae* has been provided by Pechman
et al. [1976] using oligonucleotide mapping. In this
process, ^{32}P-labelled 16S RNA is digested with T1 nuclease
and electrophoresed. Oligonucleotide spots are then subjected to further enzymatic degradation finally providing
a sequence. The results showed that *S.ureae* belongs to
Bacillus and that *S.ureae* is more closely related to *B.
pasteurii* than either is related to *B.subtilis* or *B.
stearothermophilus*.

Fox *et al.* [1977] extended these studies to include
several bacilli and *Sporolactobacillus inulinus*. This
Gram-positive rod forms endospores but is microaerophilic
and metabolically resembles the lactobacilli. Cluster
analysis of the rRNA catalogues revealed a cluster of 3

groups;   one contained *B.pumilus*, *B.megaterium* and *B.
cereus*;   the second *B.pasteurii* and *Sporosarcina ureae* and
the third was *B.stearothermophilus*.   *Sporolactobacillus
inulinus* was outside this main cluster suggesting that
this organism may well represent an intermediate between
Bacillus and non-endospore-forming bacteria.   It is
noteworthy that the dendrogram based on rRNA sequences
closely resembled the notion of the taxonomy of *Bacillus*
species derived from traditional techniques.

It would be tedious to provide rRNA oligonucleotide
catalogues of large numbers of bacilli and related
organisms to determine taxonomic relationships.   Fortunately, analogous data can be provided by a pairing procedure recently developed by De Ley *et al.* [1978].   In this
process, the amount of ^{14}C rRNA that will hybridize with
single stranded DNA immobilized on a membrane filter is
assessed and the thermal denaturation profile of the
duplex is determined.   The depression of T_m provides an
estimate of the sequence heterology between the rRNA from
one organism and the DNA template of the corresponding
cistron from the other organisms.   There is a distinct
correlation between the thermal stability of heterologous
DNA/RNA hybrids and phenotypic similarity and this method
has been successfully used to establish the taxonomic
interrelationships between several Gram-negative genera
[De Ley *et al.*, 1978].   There is every indication that the
technique would be suitable for *Bacillus* species and
related non-sporeforming organisms.

Concluding Remarks

Perhaps the first question that arises in the evaluation of DNA reassociation data is, with such a variety of
techniques in use, are the results reliable and comparable?
Gibbins and Gregory [1972] used the filter method and the
spectrophotometric assay in their studies on *Rhizobium* and
produced congruent results.   Furthermore, the data in
Table 3 have been collated from 4 laboratories that use a
variety of methods and incubation conditions yet the
findings are consistent and meaningful.   It would seem,
therefore, that DNA reassociation can provide valid and
valuable information for the classification and identification of bacteria.

On a more speculative note it is interesting to ask if
nucleic acid studies can assist in defining taxa, particularly species;   a problem pertaining to the bacilli
because examples of "series" and "complexes" are frequent.
The DNA homology studies of the "*B.subtilis*" group (Table
3) validate the species *B.amyloliquefaciens* in the
absence, as yet, of phenetic evidence.   Moreover, DNA
reassociation and GC analyses have clarified the classification of *B.megaterium* [see Hunger and Claus, Chapter 9]
and the DNA homology data presented in Table 5 not only

show that *B.firmus* and *B.lentus* are distinct species but suggest that it may be possible to assign the intermediate strains to one of these species. Thus DNA homology is proving useful for demonstrating the affinity between strains of a species and distinguishing them from closely related species (see Fig. 2a, b). Furthermore, the information is quantitative and empirical guidelines for the rationalization of taxa can be derived. Two of these have been stated above, namely a genus should vary by no more than about 10% GC and a species by no more than about 5% GC [Jones and Sneath, 1970; Bradley and Mordarski, 1976]. From DNA reassociation studies of *Bacillus* strains it can be added that strains within a species should possess at least 50% DNA homology [Seki *et al.*, 1978]. However, from their extensive studies of the Family Enterobacteriaceae, Steigerwalt *et al.* [1976] found that strains within a species normally displayed more than 65% DNA homology. This variation merely emphasizes the inequal ranking of the Enterobacteriaceae and Baccillaceae, a feature that is also evident from rRNA studies. The relatedness of *B.pumilus* and *B.subtilis* 16S rRNA sequences is the equivalent of that found between genera within the Family Enterobacteriaceae and the relatedness between the rRNA of *B.stearothermophilus* and *B.subtilis* is comparable to that observed between the Families Enterobacteriaceae and Vibrionaceae [Woese *et al.*, 1976]. Thus, the genus *Bacillus* is the equivalent of some bacterial families and perhaps the greatest contribution that GC analysis, DNA: DNA and DNA:RNA reassociation will provide towards bacterial systematics could be the provision of guidelines such that the ranking of taxa may be standardized.

References

Baptist, J.N., Mandel, M. and Gherna, R.L. (1978). Comparative zone electrophoresis of enzymes in the genus *Bacillus*. *International Journal of Systematic Bacteriology* **28**, 229-244.

de Barjac, H. (1978). Une nouvelle variété de *Bacillus thuringiensis* très toxique pour les moustiques: *B.thuringiensis* var. *israelensis* sérotype 14. *Comptes Rendus* **286**, 797-800.

Bernhard, K., Schrempf, H. and Goebel, W. (1978). Bacteriocin and antibiotic resistance plasmids in *Bacillus cereus* and *Bacillus subtilis*. *Journal of Bacteriology* **133**, 897-903.

Boeyé, A. and Aerts, M. (1976). Numerical taxonomy of *Bacillus* isolates from North Sea sediments. *International Journal of Systematic Bacteriology* **26**, 427-441.

Bond, W.W. and Favero, M.S. (1977). *Bacillus xerothermodurans* sp. nov. a species forming endospores extremely resistant to heat. *International Journal of Systematic Bacteriology* **27**, 157-160.

Bonde, G.J. (1975). The genus *Bacillus* an experiment with cluster analysis. *Danish Medical Bulletin* **22**, 41-61.

Bonde, G.J. and Jackson, D.K. (1971). DNA-base ratios of *Bacillus* strains related to numerical and classical taxonomy. *Journal of*

General Microbiology **69**, vii.

Boyer, E.E., Ingle, M.B. and Mercer, G.D. (1973). *Bacillus alcalophilus* subsp. *halodurans* subsp. nov. an alkaline-amylase-producing alkalophilic organism. *International Journal of Systematic Bacteriology* **23**, 238-242.

Bradley, S.G. and Mordarski, M. (1976). Association of polydeoxyribonucleotides of deoxyribonucleic acids from nocardioform bacteria. In "The Biology of the Nocardiae" (Eds. M. Goodfellow, G.H. Brownell and J.A. Serrano), pp.310-336. London: Academic Press.

Brenner, D.J. (1970). Deoxyribonucleic acid divergence in Enterobacteriaceae. *Developments in Industrial Microbiology* **11**, 139-153.

Coukoulis, H. and Campbell, L.L. (1971). Transformation in *Bacillus amyloliquefaciens*. *Journal of Bacteriology* **105**, 319-322.

Cowan, S.T. (1974). "Cowan and Steel's Manual for the Identification of Medical Bacteria", 2nd edition. Cambridge: Cambridge University Press.

De Ley, J. (1969). Compositional nucleotide distribution and the theoretical prediction of homology in bacterial DNA. *Journal of Theoretical Biology* **22**, 89-116.

De Ley, J., Cattoir, H. and Reynaerts, A. (1970). The quantitative measurement of DNA hybridization from renaturation rates. *European Journal of Biochemistry* **12**, 133-142.

De Ley, J., Segers, P. and Gillis, M. (1978). Intra- and intergeneric similarities of *Chromobacterium* and *Janthinobacterium* ribosomal ribonucleic acid cistrons. *International Journal of Systematic Bacteriology* **28**, 154-168.

Denhardt, D.T. (1966). A membrane-filter technique for the detection of complementary DNA. *Biochemical Biophysical Research Communications* **23**, 641-646.

Fox, G.E., Pechman, K.R. and Woese, C.R. (1977). Comparative cataloging of 16S ribosomal ribonucleic acid: molecular approach to prokaryotic systematics. *International Journal of Systematic Bacteriology* **27**, 44-57.

Fukumoto, J. (1943). Studies on the production of bacterial amylase. 1. Isolation of bacteria secreting potent amylases and their distribution. *Journal of the Agricultural Chemical Society of Japan* **19**, 634-640 (in Japanese).

Gibbins, A.M. and Gregory, R.F. (1972). Relatedness among *Rhizobium* and *Agrobacterium* species determined by three methods of nucleic acid hybridization. *Journal of Bacteriology* **111**, 129-141.

Gibson, T. and Gordon, R.E. (1974). "*Bacillus*". In "Bergey's Manual of Determinative Bacteriology" (Eds. R.E. Buchanan and N.E. Gibbons), pp.529-550. Baltimore: The Williams and Wilkins Co.

Gibson, T. and Topping, L.E. (1938). Further studies of the aerobic spore-forming bacilli. *Abstracts of the Proceedings of the Society for Agricultural Bacteriology*, pp.43-44.

Gillis, M., De Ley, J. and De Cleene, M. (1970). The determination of molecular weight of bacterial genome DNA from renaturation rates. *European Journal of Biochemistry* **12**, 143-153.

Golovacheva, R.S., Loginova, L.G., Salikhov, T.A., Kolesnikov, A.A. and Zaitzeva, G.N. (1975). A new thermophilic species *Bacillus thermocatenulatus* nov. sp. *Mikrobiologya* **44**, 265-268.

Gordon, R.E., Haynes, W.C. and Pang, C.H-N. (1973). "The genus *Bacillus*". Washington, D.C.: United States Department of Agriculture.

Gordon, R.E., Hyde, J.L. and Moore, Jr., J.A. (1977). *Bacillus firmus* – *Bacillus lentus*: a series or one species? *International Journal of Systematic Bacteriolody* **27**, 256-262.

Harris-Warwick, R.M. and Lederberg, J. (1978). Interspecies transformation in *Bacillus*: sequence heterology as the major barrier. *Journal of Bacteriology* **133**, 1237-1245.

Heinen, U.J. and Heinen, W. (1972). Characteristics and properties of a caldo-active bacterium producing extracellular enzymes and two related strains. *Archives of Mikrobiology* **82**, 1-23.

Herndon, S.E. and Bott, K.F. (1969). Genetic relationship between *Sarcina ureae* and members of the genus *Bacillus*. *Journal of Bacteriology* **97**, 6-12.

Hylemon, P.B., Wells, Jr., J.S., Krieg, N.R. and Jannasch, H.W. (1973). The genus *Spirillum*: a taxonomic study. *International Journal of Systematic Bacteriology* **23**, 340-380.

Iglewski, W.J. and Gerhardt, N.B. (1978). Identification of an antibiotic-producing bacterium from the human intestinal tract and characterization of its antimicrobial product. *Antibiotic Agents and Chemotherapy* **13**, 81-89.

Jones, D. and Sneath, P.H.A. (1970). Genetic transfer and bacterial taxonomy. *Bacteriological Reviews* **34**, 40-81.

Kirkby, K.S., Fox-Carter, E. and Guest, M. (1967). Isolation of deoxyribonucleic acid and ribosomal ribonucleic acid from bacteria. *Biochemical Journal* **104**, 258-262.

Lovett, P.S. and Bramucci, M.G. (1976). Plasmic DNA in bacilli. In "Microbiology 1976" (Ed. D. Schlessinger), pp.388-393. Washington, D.C.: American Society for Microbiology.

Lovett, P.S. and Young, F.F. (1969). Identification of *Bacillus subtilis* NRRL B-3275 as a strain of *Bacillus pumilus*. *Journal of Bacteriology* **100**, 658-661.

Marmur, J. (1961). A procedure for the isolation of deoxyribonucleic acid from microorganisms. *Journal of Molecular Biology* **3**, 208-218.

Marmur, J. and Doty, P. (1962). Determination of the base composition of deoxyribonucleic acid from its thermal denaturation temperature. *Journal of Molecular Biology* **5**, 109-118.

Matsuzaki, H., Yamane, K., Yamaguchi, K., Nagata, T. and Maruo, B. (1974). Hybrid α-amylases produced by transformants of *Bacillus subtilis* I. Purification and characterization of extracellular α-amylases produced by the parental strains and transformants. *Biochimica et Biophysica Acta* **356**, 235-247.

McMeekin, T.A. and Shewan, J.M. (1978). Taxonomic strategies for *Flavobacterium* and related genera. *Journal of Applied Bacteriology* **45**, 321-332.

Miller, J.H. (1972). Selection of Thy⁻ strains with trimethoprim. In "Experiments in Molecular Genetics" (Ed. J.H. Miller), pp.218-220. Cold Spring Harbor Laboratory, New York.

Moore, R.L. (1974). Nucleic acid reassociation as a guide to genetic relatedness among bacteria. *Current Topic in Microbiology and Immunology* **64**, 105-128.

Normore, W.M. (1973). Guanine-plus-cytosine (GC) composition of the DNA of bacteria, fungi, algae and protozoa. In "Handbook of Micro-biology" (Eds. A.I. Laskin and H.A. Lechevalier), vol. 2, pp.285-740. CRC Press, Cleveland.

Oxborrow, G.S., Fields, N.D. and Pule, J.R. (1977). Pyrolysis gas-liquid chromatography of the genus *Bacillus*: effect of growth media on pyrochromatogram reproducibility. *Applied and Environmental Microbiology* **33**, 865-870.

Pechman, K.J., Lewis, B.J. and Woese, C.R. (1976). Phylogenetic status of *Sporosarcina ureae*. *International Journal of Systematic Bacteriology* **26**, 305-310.

Pichinoty, F., Durand, M., Job, C., Mandel, M. and Garcia, J.L. (1978). Morphological, physiological and taxonomical studies of *Bacillus azotoformans*. *Canadian Journal of Microbiology* **24**, 608-617.

Potter, S.S., Bott, K.F. and Newbold, J.E. (1977). Two-dimensional restriction analysis of the *Bacillus subtilis* genome: gene puri-fication and ribosomal ribonucleic acid gene organization. *Journal of Bacteriology* **129**, 492-500.

Priest, F.G. (1975). Effect of glucose and cyclic nucleotides on the transcription of α-amylase mRNA in *Bacillus subtilis*. *Biochemical Biophysical Research Communications* **63**, 606-610.

Rilfors, L., Wieslander, A. and Stahl, S. (1978). Lipid and protein composition of membranes of *Bacillus megaterium* variants in the temperature range 5 to 70°C. *Journal of Bacteriology* **135**, 1043-1052.

Saito, H. and Miura, K.I. (1963). Preparation of transforming deoxyribonucleic acid by phenol. *Biochimica et Biophysica Acta* **72**, 619-629.

Schildkraut, C.L., Marmur, J. and Doty, P. (1962). Determination of the base composition of deoxy-ribonucleic acid from its buoyant density in CsCl. *Journal of Molecular Biology* **4**, 430-443.

Seki, T., Chung, C-K., Mikami, H. and Oshima, Y. (1978). Deoxyribo-nucleic acid homology and taxonomy of the genus *Bacillus*. *International Journal of Systematic Bacteriology* **28**, 182-189.

Seki, T., Oshima, T. and Oshima, Y. (1975). Taxonomic study of *Bacillus* by deoxyribonucleic acid - deoxyribonucleic acid hybridization and inter-specific transformation. *International Journal of Systematic Bacteriology* **25**, 258-270.

Slepecky, R.A. (1975). Ecology of bacterial sporeformers. In "Spores VI" (Eds. P. Gerhardt, H.L. Sadoff and R.N. Costilow), pp.297-313. American Society for Microbiology, Washington.

Smith, N.R., Gordon, R.E. and Clark, F.E. (1952). "Aerobic Spore-forming Bacteria". United States Department of Agriculture, Washington, D.C.

Somerville, H.J. and Jones, M.L. (1972). DNA competition experiments within the *Bacillus cereus* group of bacilli. *Journal of General Microbiology* **73**, 257-265.

Steigerwalt, A.G., Fanning, G.R., Fife-Abury, M.A. and Brenner, D.J. (1976). DNA relatedness among species of *Enterobacter* and *Serratia*. *Canadian Journal of Microbiology* **22**, 121-137.

Welker, N.E. and Campbell, L.L. (1967). Unrelatedness of *Bacillus amyloliquefaciens* and *Bacillus subtilis*. *Journal of Bacteriology* **94**, 1124-1130.

Woese, C., Sogin, M., Stahl, D., Lewis, B.J. and Bonen, L. (1976). A comparison of the 16S ribosomal RNAs from mesophilic and thermophilic bacilli: some modifications in the Sanger method for RNA sequencing. *Journal of Molecular Evolution* **7**, 197-213.

Yoneda, Y., Yamane, K., Yamaguchi, K., Nagata, Y. and Maruo, B. (1974). Transformation of *Bacillus subtilis* in α-amylase productivity by deoxyribonucleic acid from *Bacillus subtilis* var. *amylosacchariticus*. *Journal of Bacteriology* **120**, 1144-1150.

Young, F.E. and Wilson, G.A. (1972). Genetics of *Bacillus subtilis* and other Gram-positive sporulating bacilli. In "Spores V" (Eds. H.O. Halvorson, R. Hanson and L.L. Campbell), pp.77-106, American Society for Microbiology, Washington D.C.

Chapter 4

LIPIDS IN THE CLASSIFICATION OF *BACILLUS* AND RELATED TAXA

D.E. MINNIKIN and M. GOODFELLOW

Departments of Organic Chemistry and Microbiology, The University, Newcastle upon Tyne, UK

Introduction

Traditional morphological and biochemical studies have not always provided reliable features for bacterial classification and attention has turned increasingly to chemical analyses for the provision of good taxonomic characters. Analyses of deoxyribonucleic acid, peptidoglycan, polysaccharide, protein and lipid composition have made valuable contributions to the classification of representatives of diverse bacterial taxa [see Goodfellow and Board, 1980]. The present paper is concerned with assessing the potential of lipid analyses in the taxonomy of the aerobic endospore-forming bacteria currently classified in the genera *Bacillus, Sporolactobacillus* and *Sporosarcina*.

The lipids of members of the genus *Bacillus* and related bacteria are considered to be predominantly free lipids located in the plasma membranes. The structural lipid components of plasma membranes are amphipathic polar lipids such as phospholipids and glycolipids but the membrane matrix also accommodates lipid-soluble isoprenoid quinones and pigments. Three main categories of lipids have been found to be potentially useful in the classification of aerobic endospore-forming bacteria. These three categories, to which the present review will be limited, comprise fatty acids and other non-saponifiable components of polar lipids, intact polar lipids and isoprenoid quinones. The contributions of fatty acid and polar lipid analyses to general bacterial classification have been the subject of recent reviews [Goldfine, 1972; Shaw, 1974; Lechevalier, 1977] and isoprenoid quinone composition has also been reviewed [Thomson, 1971; Threlfall and Whistance, 1971].

Lipid Composition

Fatty Acids and Other Non-saponifiable Lipids

The predominant non-saponifiable lipids of the aerobic spore-forming bacteria are fatty acids of the methyl-

branched iso and anteiso types, examples of which are 13-
methyltetradecanoic (iso-C_{15}) (I) and 12-methyltetra-
decanoic (anteiso-C_{15}) (II) acids. Normal straight-chain
acids are also usually encountered as minor components and
monounsaturated fatty acids such as *cis*-octadec-9-enoic
(Δ^9-C_{18}, oleic) (III) are found in certain strains. The
most unusual fatty acids are the cyclohexyl acids (IV)
occurring as major components in *B.acidocaldarius* [De Rosa
et al., 1971b, 1972]. Pentacyclic triterpenes of the
hopane family have also been characterized from *B.acido-
caldarius*; De Rosa *et al.* [1971a, 1973] isolated hop-22
(29)-ene (V) and Langworthy and Mayberry [1976] reported
the presence of the tetrol (VI, R = H) of undetermined
stereochemistry.

The fatty acids of the genus *Bacillus* have been
extensively analysed and reviewed by Kaneda [1977] who
suggested that, on the basis of fatty acid analysis,
Bacillus species could be separated into 6 groups (Table
1). Detailed analysis of the fatty acid patterns allowed
further possible subdivisions to be made; *B.lentimorbus*
was distinguished from *B.larvae* and *B.popilliae* [Kaneda,
1969] and *B.anthracis*, *B.cereus* and *B.thuringiensis* each
gave distinct patterns [Kaneda, 1968]. Cyclohexyl fatty
acids (IV) are only found in *B.acidocaldarius* and closely-
related organisms [De Rosa *et al.*, 1971b, 1972, 1974;
Oshima and Ariga, 1975; Langworthy *et al.*, 1976] though
a mutant of *B.subtilis* will synthesize these and other
ω-alicyclic fatty acids if the appropriate precursors are
provided [Dreher *et al.*, 1976; Blume *et al.*, 1978].

The ability of *Bacillus* species to synthesize fatty
acids from a variety of branched-chain precursors makes it
very important to compare profiles of organisms cultivated
under standard conditions [Kaneda, 1977]. It was shown,
for example, that addition of isoleucine increased the
proportion of anteiso acids and valine and leucine
enhanced the production of iso fatty acids in *B.subtilis*
[Kaneda, 1966] and a thermophilic species of *Bacillus*
[Daron, 1973]. Indeed, if unnatural ethyl and dimethyl
branched precursors were fed to *B.subtilis*, corresponding
branched fatty acids were produced in substantial proportions [Kaneda, 1971b].

A particularly informative demonstration of the effect
of substrate on fatty acid composition was provided by
Willecke and Pardee [1971] using a mutant of *Bacillus
subtilis* defective in branched-chain α-keto acid dehydrogenase thereby requiring short branched fatty acids for
growth (Table 2). The wild type of the strain studied had
anteiso-C_{15} as the predominant acid with substantial
amounts of anteiso-C_{17} and smaller amounts of C_{14} to C_{17}
iso and straight-chain acids. Supplementation of the
mutant with 2-methylbutyrate, the precursor of odd numbered
anteiso acids, gave a very simple pattern of comparable

I $CH_3 . \overset{\overset{\displaystyle CH_3}{|}}{CH} . (CH_2)_{11} . COOH$

II $CH_3 . CH_2 . \overset{\overset{\displaystyle CH_3}{|}}{CH} . (CH_2)_{10} . COOH$

III $CH_3 . (CH_2)_7 . CH = CH . (CH_2)_7 . COOH$
                    cis

IV 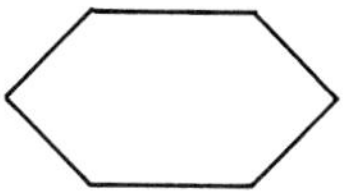 $(CH_2)_n . COOH$

n = 10 and 12

V

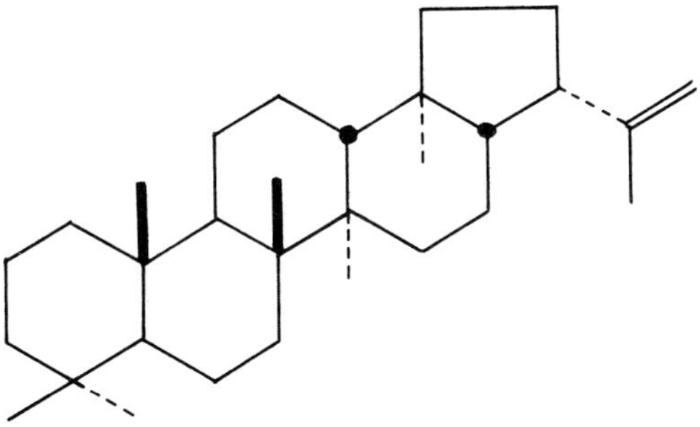

VI

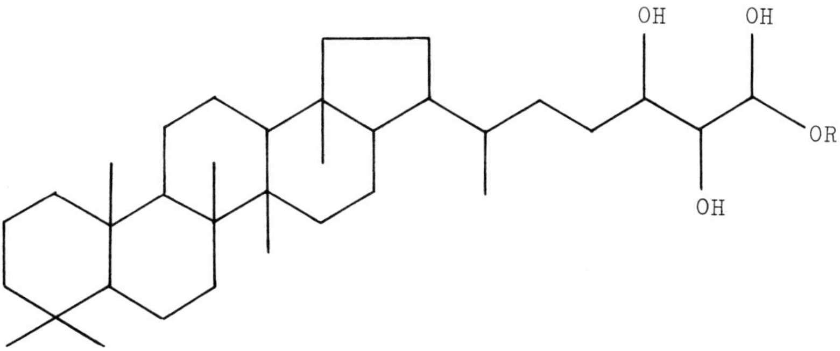

TABLE 1

Grouping of Bacillus species based on fatty acid patterns

Unsaturated acids	Predominant acids	Chain length range	Species	Kaneda group
Very small or insignificant proportion (<3%)	anteiso C_{15} (26-60%) + iso C_{15} (13-30%)	C_{14} - C_{17}	*B.alvei, B.brevis, B.circulans, B.licheniformis, B.macerans, B.megaterium, B.pumilus, B.subtilis*	A
	anteiso C_{15} (39-62%)	C_{14} - C_{17}	*B.polymyxa, B.larvae, B.lentimorbus, B.popilliae*	B
	iso-C_{15}	C_{14} - C_{17}	*B.caldolyticus, B.caldotenax, B.stearothermophilus*	C
	cyclohexyl acids	C_{17} - C_{19}	*B.acidocaldarius*	D
Small proportion (7-12%)	iso-C_{15} (19-31%)	C_{12} - C_{17}	*B.anthracis, B.cereus, B.thuringiensis*	E
Large proportion (17-28%)	anteiso-C_{15}	C_{14} - C_{17}	*B.insolitus, B.psychrophilus, B.globisporus*	F

Data from Kaneda [1977].

amounts of anteiso-C_{15} and C_{17} acids and small proportions (<10%) of straight-chain C_{14} and C_{16} acids. Growth of the mutant on the iso acid precursors isobutyrate and iso-valerate gave simple patterns consisting mainly of iso-C_{14} and iso-C_{16} acids and iso-C_{15} acids, respectively (Table 2). Long-chain iso and anteiso acids were also utilized by the mutant though less efficiently than the short-chain acids. Certain natural methyl, dimethyl and ethyl branched short-chain acids were converted into the corresponding branched long-chain acids.

The study of Willecke and Pardee [1971], outlined in the previous paragraph, also provided an insight into the role of fatty acids in *Bacillus* species. As noted above, utilization of isovalerate by the mutant gave predominantly iso-C_{15} but isobutyrate gave a mixture of iso-C_{14} and C_{16} acids suggesting that the average fatty acid size, to support growth of this mutant under the conditions employed, should be around 15 carbons for iso acids. This is in accordance with the general principle that the fatty acid composition of membranes should be composed of a balanced mixture of fatty acids to provide a suitable degree of fluidity [for reviews see McElhaney, 1976; Esser and Souza, 1976]. The correct fluidity balance is usually achieved by mixing relatively solid straight-chain or iso acids (melting points >50°C) with relatively liquid *cis*-unsaturated or anteiso acids (melting points <40°C) or by variations in chain length, the shorter chains having greater fluidity [McElhaney, 1976; Russell, 1971]. The point concerning the increased fluidity of anteiso acids with respect to the corresponding iso acids is also illustrated by a comparison of the uptake of 2-methyl-butyrate and isovalerate by the mutant studied by Willecke and Pardee [1971]. Isovalerate is converted predominantly into an iso-C_{15} acid but the more fluid anteiso-C_{15} acid produced from 2-methylbutyrate must be balanced by a substantial proportion of an anteiso-C_{17} acid (Table 2). Considerations of the role of fatty acids in membrane fluidity also offer an explanation of the predominance of anteiso-C_{15} and unsaturated acids in the psychrophilic strains, *B.insolitus*, *B.psychrophilus* and *B.globisporus* (Table 1).

Wide variations in the fatty acid compositions of representatives of *Bacillus stearothermophilus* and related thermophiles have been reported. Kaneda [1977] suggested that thermophilic bacilli could be placed in a group having predominantly iso-C_{15} fatty acids (Table 1). This assign-ment, however, was based on 2 reports of the fatty acid composition at the lowest growth temperatures of *B.stearo-thermophilus* [Yao *et al.*, 1970] and *B.caldolyticus* and *B.caldotenax* [Weerkamp and Heinen, 1972a]. In both the cases quoted the fatty acid profiles were not determined under the conditions used by Kaneda [1977] to obtain the data for most of the other strains listed in Table 1.

TABLE 2

Fatty acid composition of supplemented branched-chain α-keto acid dehydrogenase mutants of B.subtilis

	$i\text{-}C_{14}$	14:0	$i\text{-}C_{15}$	$ai\text{-}C_{15}$	$i\text{-}C_{16}$	16:0	$i\text{-}C_{17}$	$ai\text{-}C_{17}$
B.subtilis 61141 (wild type)	6		7	50	9	5	2	20
Mutant strain 626 + isobutyrate ($i\text{-}C_4$)	33	2			51	13		
+ 2-methylbutyrate ($ai\text{-}C_5$)		2	56	51		8	12	39
+ isovalerate		3		7*		16		2*

*Possibly due to 1% 2-methylbutyrate in commercial isovalerate.

Data from Willecke and Pardee [1971].

Indeed Weerkamp and Heinen [1972b] in another study, not
quoted by Kaneda [1977], of the fatty acids of *B.caldo-
lyticus* and *B.caldotenax* showed that changes in carbon
source could result in iso-C_{17} becoming the major acid as
was also the case with increasing temperature [Weerkamp
and Heinen, 1972a]. An increase in temperature from 45 to
55°C changed the predominant fatty acid in *B.stearothermo-
philus* from iso-C_{15} to normal 16:0 [Yao *et al.*, 1970] and
in another thermophilic strain iso-C_{15} was the major acid
at 55°C but anteiso-C_{15} was the main component at 37°C
[Chan *et al.*, 1973].

A comparison of the fatty acid composition of thermophilic and mesophilic *Bacillus* strains [Shen *et al.*, 1970]
showed that iso-C_{15} was the predominant acid in all cases,
including representatives of *B.licheniformis* and *B.pumilus*
found in Kaneda's [1977] studies (Table 1) to have mainly
anteiso-C_{15}. Oshima and Miyagawa [1974] found comparable
proportions of iso-C_{15} and iso-C_{17} in *B.stearothermophilus*
and in an organism labelled *B.thermoruber* [Aragozzini *et
al.*, 1976] iso-C_{15} predominated. A thermophile, unable to
hydrolyze starch, studied by Daron [1970a] and another,
starch-positive, organism [Esser and Souza, 1974; Souza
et al., 1974; McElhaney and Souza, 1976] had, in contrast,
iso-C_{16} as the major fatty acid, though with increasing
temperature the latter organism produced predominant
amounts of normal 16:0 fatty acid. A thermophilic
bacterium PS3 had iso-C_{15} as over half its fatty acid
content [Kagawa *et al.*, 1976]. It is apparent that
standardization of the growth media and temperature is
essential before productive comparisons of the fatty acid
compositions of thermophilic strains can be made.

In addition to the studies cited by Kaneda [1977] and
the analyses of the fatty acids of thermophiles mentioned
in the previous section, a number of other investigations
are worthy of consideration. The fatty acids of *B.polymyxa*
were found by Viviani *et al.* [1965] to have anteiso-C_{15} as
major component (see Table 1) though it was not certain
that iso-C_{15}, if any, was clearly separated under the
analytical conditions used. In several studies the fatty
acids of *B.anthracis*, *B.cereus*, *B.licheniformis*, *B.pumilus*,
B.subtilis and *B.thuringiensis* were compared [Fabio and
Vivoli, 1966; Vivoli *et al.*, 1966; Vivoli and Fabio,
1967]. The results, however, were rather obscured by the
variable resolution of the C_{15} acids but similar patterns
were found for 6 strains each of *B.cereus* and *B.thuringi-
ensis* (see Table 1) [Vivoli and Fabio, 1967]. The fatty
acids of *B.cereus* and related strains were compared by
Niskanen *et al.* [1975] but branched chain isomers were not
resolved; this inadequacy was met in a subsequent study by
the use of capillary gas chromatography [Niskanen *et al.*,
1978]. The effect of growth temperature on the fatty acid
composition of four temperature range variants of *Bacillus
megaterium* was studied by Rilfors *et al.* [1978]. In

general, with increasing growth temperature the relative proportions of iso acids increased, anteiso acids decreased, and the average chain length became greater. An obligatory thermophilic strain always had much more iso-C_{15} acid than anteiso-C_{15} throughout its growth range of 40 to 70°C but for a facultative thermophile the same iso acids predominated at higher temperatures (35 to 55°C) but anteiso acids were more abundant between 17 and 35°C. A mesophilic strain and a facultative thermophile always had more anteiso-C_{15} acid than iso-C_{15} acid throughout their respective growth ranges of 10 to 44°C and 5 to 35°C. The proportions of the C_{17} branched chain acids varied in a similar manner to those of the C_{15} acids for all variants of *B.megaterium*.

Although only low proportions of unsaturated fatty acids are found in *Bacillus* species, detailed analysis of their exact structure may be of value in classification since the position of the double bond is not constant [Fulco, 1974]. Monounsaturated fatty acids usually have the double bond in the middle of the chain in Δ^9 or Δ^{11} positions but unusual Δ^5, Δ^8 and Δ^{10} isomers are found in certain *Bacillus* species (Table 3). Further studies are obviously necessary in order to judge whether the analysis of unsaturated fatty acids has potential for the classification of these organisms especially since these acids are usually very minor components (Table 1).

The fatty acids of sporolactobacilli have been compared in a detailed study with possibly related strains of *Bacillus* and *Lactobacillus* [Uchida and Mogi, 1973]. Representatives of *Sporolactobacillus inulinus*, *S.laevus*, *S.racemicus*, *B.laevolacticus*, *B.myxolactis* and *B.racemilacticus*, all spore-forming lactic acid bacteria, had closely similar patterns having iso-C_{15}, anteiso-C_{15} and anteiso-C_{17} acids as major components. These profiles were similar to those recorded for *B.subtilis* and *B.coagulans* but were clearly different to those of *Lactobacillus casei*, *L.plantarum* and *L.yamanashiensis* [*L.mali*, Carr *et al.*, 1977] which contained major amounts of straight-chain, unsaturated and cyclopropane acids. In a separate study [Whiteside *et al.*, 1971], the fatty acids of *Sporosarcina ureae* were found to be mainly branched-chain C_{15} and C_{17} acids but the iso and anteiso isomers were not resolved.

Polar Lipids

Polar lipids are amphipathic molecules having a balance between hydrophilic and hydrophobic portions. Phospholipids based on phosphatidic acid (VII, Y = H) are the most common types of polar lipids but polar glycolipids, such as the diglucosyldiacylglycerol (VIII), are also widely encountered [Shaw, 1974, 1975; Lechevalier, 1977].

The polar lipid composition of representatives of the genus *Bacillus* is summarized in Table 4. The most commonly encountered phospholipids are phosphatidylglycerol (PG)

TABLE 3

Distribution of unsaturated fatty acids in Bacillus species

Kaneda group (see Table 1)		Double bond position in fatty acid	Reference
A	B.alvei		
	B.megaterium		
	B.pumilus		
	B.subtilis	Δ-5	
	B.licheniformis		
	B.licheniformis	Δ-5 and/or Δ-10	Fulco, 1969a, b; 1974;
	B.licheniformis		Fujii and Fulco, 1977
	B.brevis		
	B.macerans	Δ-8, 9, 10	
C	B.stearothermophilus		
E	B.cereus	Δ-5 and/or Δ-10	Dart and Kaneda, 1970; Kaneda, 1972a
F	B.globisporus		
	B.insolitus	Δ-5	Kaneda, 1971a
	B.psychrophilus		
-	B.stearothermophilus	Δ-5	Daron, 1970a, b

VII

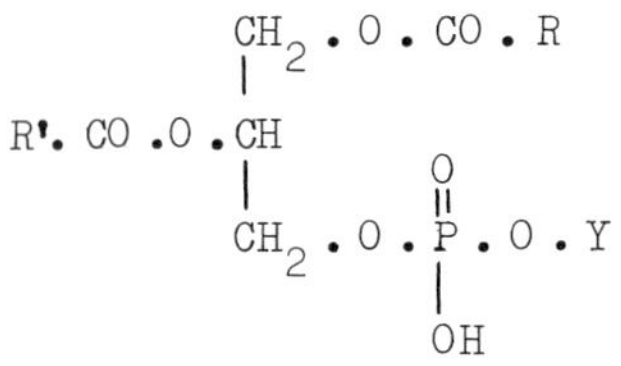

$$CH_2 . O . CO . R$$
$$R' . CO . O . CH$$
$$CH_2 . O . \overset{O}{\underset{OH}{P}} . O . Y$$

R, R' = long alkyl chains

VIII

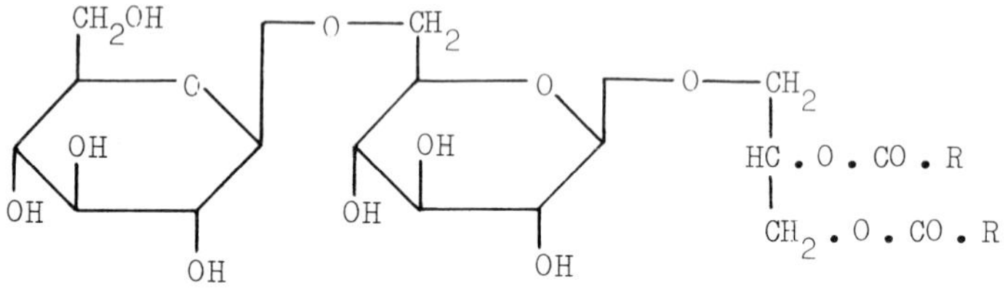

(VII, Y = glycerol), diphosphatidylglycerol (DPG) (VII, Y = PG) and phosphatidylethanolamine (PE) (VII, Y = ethanolamine). The occasional absence of DPG (Table 4) is not considered to be of particular significance since this lipid is formed directly from PG [Hirschberg and Kennedy, 1972] and their relative proportions may vary with growth conditions. PE is also found in most representatives of *Bacillus* and in *Sporosarcina* (Table 4) but its definite absence from the lipids of *B.acidocaldarius* [Langworthy *et al.*, 1976; Minnikin *et al.*, 1977] and starch-negative strains of *B.stearothermophilus* [Minnikin *et al.*, 1974, 1977] is of particular significance.

Lysine esters of PG are well-established as components of the lipids of *B.subtilis* and are probably present in *B.licheniformis* and certain strains of *B.megaterium* (Table 4). Certain representatives of *B.megaterium* contain characteristic isomers of glucosaminyl PG (IX) [MacDougall and Phizackerley, 1969; Op den Kamp *et al.*, 1971] and glucosaminyldiacylglycerol had been detected in one strain [Phizackerley *et al.*, 1972]. It is notable that with the exception of a single unconfirmed report [Singh *et al.*, 1978] diglucosyldiacylglycerol (VIII) is absent from *B.megaterium* while its presence is well-documented (Table 4) in *B.subtilis, B.licheniformis, B.cereus* and starch-negative strains of *B.stearothermophilus* which also contained substantial proportions of monoglucosyldiacyl-glycerol [Minnikin *et al.*, 1974, 1977]. Detailed invest-igations by Fischer *et al.* [1978] revealed the presence of small amounts of glycerophosphoglycolipids (X), mono-glucosyldiacylglycerol and triglucosyldiacylglycerol in

B.subtilis and *B.licheniformis*. A small amount of a polar glycolipid observed in a study of the lipids of *B.subtilis* var. *niger* [Minnikin and Abdolrahimzadeh, 1974a] had the chromatographic properties of a trihexosyldiacylglycerol.

IX

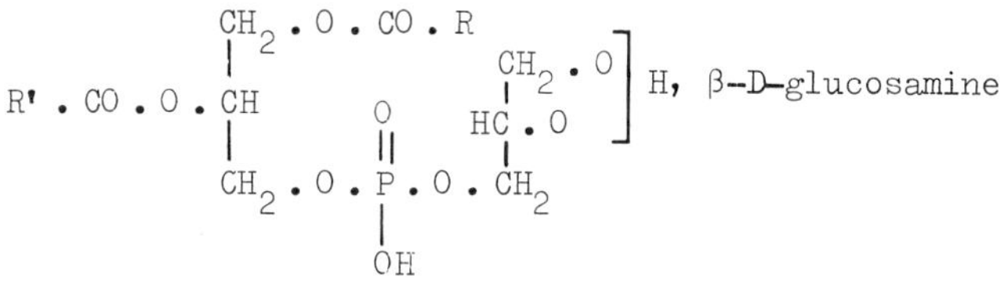

$$R' . CO . O . CH$$

CH$_2$. O . CO . R

CH$_2$. O

HC . O

H, β-D-glucosamine

CH$_2$. O . P . O . CH$_2$

OH

X

CH$_2$OH

HC . OH

CH$_2$. O — P — OH

O

CH$_2$

O — CH$_2$

OH

OH

OH

OH

OH

OH

O — CH$_2$

HC . O . CO . R

CH$_2$. O . CO . R

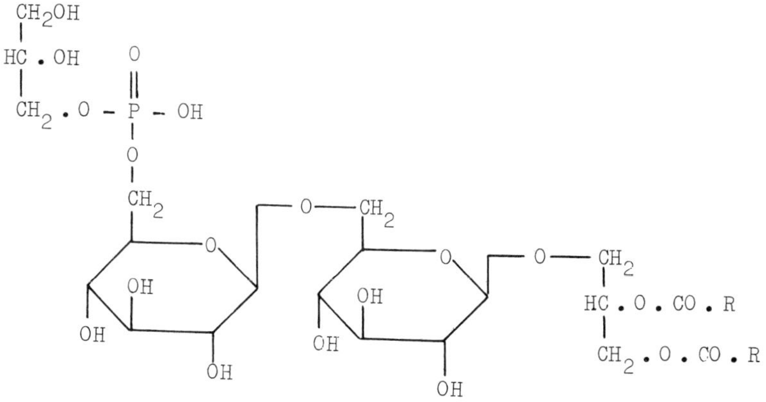

The polar lipid composition of *B.acidocaldarius* is particularly distinctive for in addition to the main phospholipids DPG and PG the other lipids are all phosphorus-free complex glycolipids [Langworthy *et al.*, 1976; Minnikin *et al.*, 1977], the essential nature of the lipids being established by the former authors. The acidic component of the glycolipids was a sulphonoglycosyldiacylglycerol (XI) but the other components contained N-acylglucosamine. The most conventional in structure of the glucosamine-containing lipids were glycosyl-N-acylglucosaminylacylglycerols (XII) but a unique component was an N-acylglucosaminyl pentacyclic tetrol (VI, R = N-acylglucosamine) [Langworthy *et al.*, 1976; Langworthy and Mayberry, 1976].   The relationship of these unconventional lipids to the unknown components described by Minnikin *et al.* [1977] remains to be established. It has been argued by Rohmer *et al.* [1979] that pentacyclic triterpene tetrols (VI, R = H) may play a membrane reinforcing role in some prokaryotes similar to that performed by sterols in eukaryotes but further studies are necessary to clarify this relationship.

Certain *Bacillus* species have been reported to contain phospholipids less commonly found in bacterial lipids (Table 4). In an early study Kates *et al.* [1962] reported

TABLE 4

Polar lipids of some Bacillus species and related taxa

Taxon	DPG[1]	PG	PE	LPG	DGDG	Other	Reference
B.acidocaldarius	+	+	-	-	-	+2 +3	Langworthy et al., 1976 Minnikin et al., 1977
B.amyloliquefaciens	+	+	+	?	+		Glenn and Gould, 1973; Paton et al., 1978
	±	ND	+	-	- +	- +4 +5	Kaneda, 1972a; Beaman et al., 1974 Kates et al., 1962 Houtsmuller and van Deenen, 1963; Mastroeni et al., 1971
B.cereus		+	+	-	+	+6 +7	Minnikin et al., 1971a, b Lang and Lundgren, 1970; Felix and Lundgren, 1972
	ND				+	+4,5,8 ND	Singh et al., 1978 Saito and Mukoyama, 1971
B.coagulans	+	+	-	-	- +	+9 +10	Frade and Guenan-Siberil, 1975 Fischer et al., 1978
B.licheniformis	+	+	+	+ -	ND +	- -	Morman and White, 1970 Button and Hemmings, 1976a, b
				+	-	- +11 +11 +11,12	Ochi et al., 1971 Mastroeni et al., 1971 Op den Kamp et al., 1967; 1969a; 1971 MacDougall and Phizackerley, 1969; Phizackerley et al., 1972
B.megaterium	+	+	+				

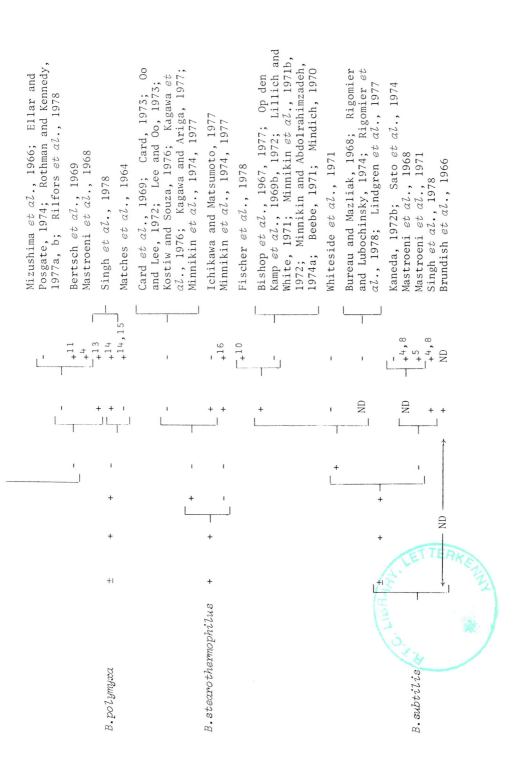

TABLE 4 (Cont'd)

Taxon	DPG	PG	PE	LPG	DGDG	Other	Reference
B.natto	+	+	+	-	ND	-	Urakami and Umetani, 1968
B.alvei	+	+	±	-	-	-	Bulla *et al.*, 1970; Bulla and St.
B.larvae							Julian, 1972
B.lentimorbus							
B.popilliae	+	+	+	-	-	-	
B.thuringiensis							
Sporosarcina ureae	+	+	+	-	-	-	Whiteside *et al.*, 1971
	+	+	+	-	-	+17	Komura *et al.*, 1975

1. Abbreviations: DPG, diphosphatidylglycerol; PG, phosphatidylglycerol; PE, phosphatidylethanolamine; LPG, lysyl PG; DGDG, diglucosyldiacylglycerol (VIII); ±, lipid absent in certain strains.
2. Sulphonoglycosyldiacylglycerol (XI), glucosyl-N-acylglucosaminylacylglycerols (XII), N-acylglucosaminyltriterpenetetrol (VI, R = N-acylglucosamine).
3. Uncharacterized glycolipids and other polyols.
4. Phosphatidylcholine.   5. Ornithine ester of PG.
6. Glucuronosyldiacylglycerol.   7. Alanine ester of PG.
8. Phosphatidylinositol.   9. Phosphatidylserine.
10. Glycerophosphoglycolipid (X) and monoglucosyl and triglucosyldiacylglycerols.
11. Glucosaminyl PG's (IX).
12. Glucosaminyldiacylglycerol.
13. Unknown glycolipid.
14. Lysophosphatidylcholine and lysophosphatidylserine.
15. Bisphosphatidic acid.
16. Monoglucosyldiacylglycerol.
17. Unidentified phospholipid.

XI

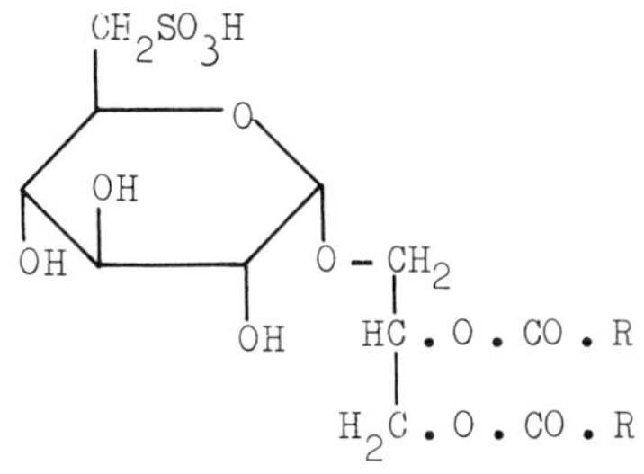

CH_2SO_3H

O

OH

OH

$O - CH_2$

OH $HC.O.CO.R$

$H_2C.O.CO.R$

XII

CH_2OH CH_2OH

O O O ── CH_2

OH OH $HC.O.CO.R$

OH $H_2C.O.CO.R$

OH $NH.CO.R$

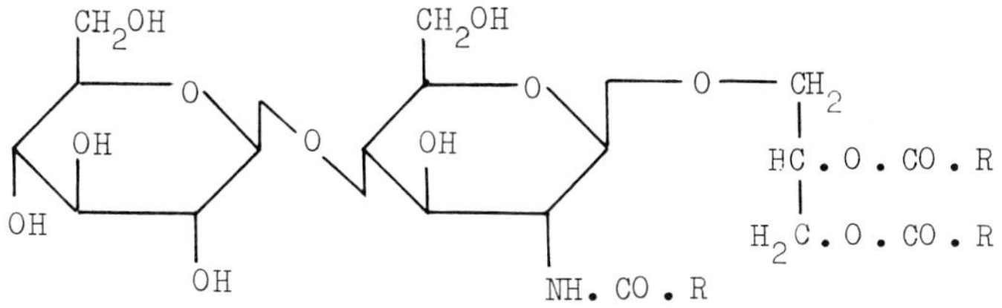

the presence of phosphatidylcholine in *B.cereus* and Singh
et al. [1978] lent support although other workers in the
interim period did not isolate this lipid from this
organism (Table 4). Singh *et al.* [1978] and Mastroeni *et
al.* [1968] identified phosphatidylcholine and phosphatidylinositol in the lipids of *B.subtilis* though, without
reference to their previous paper, Mastroeni *et al.* [1971]
reported neither of these lipids present in *B.subtilis*.
Similarly the report of phosphatidylcholine in *B.megaterium*
by Mastroeni *et al.* [1968] was not clarified by a subsequent study [Mastroeni *et al.*, 1971] in which this lipid
was apparently absent. The reported presence of phosphatidylserine in *B.coagulans* [Frade and Guenan-Siberil,
1975] and lysophosphatidylserine, lysophosphatidylcholine
and bisphosphatidic acid in *B.polymyxa* [Matches *et al.*,
1964] should be confirmed by the analysis of further
representatives of these organisms. The presence of lysyl
PG in *B.subtilis*, *B.licheniformis* and certain strains of
B.megaterium is well-documented (Table 4) but reports of
ornithine [Houtsmuller and van Deenen, 1963; Singh *et al.*,
1978] and alanine [Lang and Lundgren, 1970] esters of PG
in *B.cereus* and ornithinyl PG in *B.subtilis* [Singh *et al.*,
1978] require verification.
     Considerations of polar lipid composition should take
into account the possible interchangeability of polar
lipid types and their variation in proportion with
changing growth environment. Studies on the lipid composition of *B.cereus* T and *B.subtilis* W23 during their
growth cycle showed that under conditions of apparent
phosphate starvation diglucosyldiacylglycerols accumulated
while the proportion of phosphatidylethanolamine, an
amphoteric neutral phospholipid, was reduced [Minnikin *et*

al., 1971b]. In the same study it was found that in
B.cereus T an acidic glucuronosyldiacylglycerol increased
at the expense of the acidic phospholipids DPG and PG and
it was suggested that a degree of interchangeability in
function was possible within acidic and neutral polar
lipid classes. This hypothesis was stimulated by
Wilkinson's [1968] original suggestion that acidic glycolipids in certain pseudomonads may replace acidic phospholipids in function. Similar proposals have also been
advanced for substitutions, in higher organisms, of lipid
types within acidic and neutral polar lipid groups [Rouser
et al., 1971].

Support for the proposed interrelation of PE and
diglucosyldiacylglycerols (DGDG) in *Bacillus* species has
been obtained from a number of diverse experiments. Beebe
[1971] isolated a PE-deficient mutant of *B.subtilis*
containing increased proportions of DGDG and conversely
Button and Hemmings [1976a] found that a mutant of *B.liche-
niformis* lacking DGDG was enriched in PE. PE was present
and DGDG absent in starch-positive representatives of
B.stearothermophilus but the reverse was found in starch-
negative thermophiles [Minnikin *et al.*, 1974, 1977]. It
is apparent therefore that these bacteria require either
PE or DGDG in their membranes in addition to the acidic
polar lipids.

In other bacterial membranes PE may be replaced in
function by zwitterionic ornithine amide lipids (XIII) as
originally suggested by Wilkinson [1972]. Magnesium-
limited chemostat cultures of *Pseudomonas fluorescens*
contained PE as the sole non-acidic polar lipid but in
certain phosphate-limited cultures PE was completely replaced by an ornithine amide lipid (XIII) [Minnikin and
Abdolrahimzadeh, 1974b]. A similar interchangeability of
PE and functionally similar ornithine amide lipids has
been demonstrated for representatives of *Streptomyces*
[Batrakov and Bergelson, 1978].

$$\text{XIII} \qquad \begin{array}{c} R \cdot \underset{|}{CH} \cdot CH_2 \cdot CO \cdot NH \cdot \underset{|}{CH} \cdot COOH \\ R \cdot CO \cdot O \qquad\qquad (CH_2)_3 \cdot NH_2 \end{array}$$

Experiments involving continuous cultures of *Bacillus*
species have also shown how their polar lipid composition
may vary with changing growth environment. Magnesium and
phosphate-limited cultures of *B.subtilis* NC1B 3610 prepared
at pH 7 both lacked PE but contained substantial proportions of DGDG [Minnikin *et al.*, 1972]. In the magnesium-
limited cultures PG was the only other major lipid
occurring with DGDG lending support to a probable role for
PG in the uptake of essential metal ions such as magnesium.
The phosphate-limited cells of *B.subtilis* 3610 had
increased amounts of DGDG and DPG but greatly decreased

proportions of PG [Minnikin *et al.*, 1972]. Reduction of
the pH of phosphate-limited cultures of *B.subtilis* 3610 to
pH 5.1 resulted in the additional synthesis of PE and lysyl
PG [Minnikin and Abdolrahimzadeh, 1974a]. The production
of lysyl PG, a positively charged polar lipid, is considered to be a means whereby the bacterial membrane may be
protected against the influx of protons at low pH [Haest *et
al.*, 1972]. Analyses of the lipids of phosphate and
magnesium-limited continuous cultures of *B.subtilis* var.
niger showed that with decreasing pH the proportions of
lysyl PG increased but did not reach as high proportions
as in batch cultures [Minnikin and Abdolrahimzadeh, 1974a].

$$\text{XIV} \qquad (CH_3)_2CH \cdot (CH_2)_9 \cdot \underset{\underset{\displaystyle CO \cdot Leu \cdot Leu \cdot Asp \cdot Val\ L}{|}}{\overset{\overset{\displaystyle L \qquad\quad L}{|}}{CH}} \cdot CH_2 \cdot CO \cdot \underset{}{Glu} \quad \underset{\underset{\displaystyle Leu\ D}{|}}{Leu}$$

Certain cultures of *B.subtilis* 3610 and *B.subtilis* var.
niger studied by Minnikin *et al.* [1972] and Minnikin and
Abdolrahimzadeh [1974a] contained small amounts of an
acidic peptidolipid (XIV) previously detected in the
culture fluids of several strains of *B.subtilis* [Kakinuma
et al., 1969]. This acidic peptidolipid, named 'surfactin',
although amphipathic in character is structurally distinct
from the usual bacterial membrane polar lipids. Esperin,
a peptidolipid having the same amino acids as surfactin,
has been characterized previously from *B.subtilis* (neé
B.mesentericus) [Ito and Ogawa, 1959; Ovchinnikov *et al.*,
1966]. The distribution of these and related peptidolipids
which have antibiotic properties should be studied more
systematically and their relationship to other peptide
antibiotics elaborated by *Bacillus* species [Katz and
Demain, 1977] should also be considered. The function of
'surfactin' is unknown but it is interesting to speculate
that it may be involved in the uptake of metal ions since
it is present in phosphate-limited cultures of *B.subtilis*
when the proportions of acidic phospholipids are reduced
[Minnikin *et al.*, 1972; Minnikin and Abdolrahimzadeh,
1974a].

Menaquinones

Isoprenoid quinones are characteristic components of the
membranes of all aerobic bacteria and those from *Bacillus*
species and related taxa are exclusively menaquinones (XV).
The chemotaxonomic potential of menaquinones lies in the
variation of the numbers of isoprene units and hydrogenated
double bonds [Thomson, 1971; Threlfall and Whistance,
1971; Yamada *et al.*, 1976; Minnikin *et al.*, 1978a, b;
Minnikin and Goodfellow, 1980]. Only two systematic

studies [Watanuki and Aida, 1972; Collins and Jones, 1979]
of the menaquinones of aerobic spore-forming bacteria have
been published and the results of these and other studies
are summarized in Table 5. It is immediately apparent
that, with the exception of B.acidocaldarius, the menaquinone composition of Bacillus and related aerobic spore-
forming taxa is remarkably homogeneous.

XV

TABLE 5

Major isoprenoid quinones of some Bacillus *species and related taxa*

Major menaquinone	
MK-9	*B.acidocaldarius*[1]
MK-7	*B.alvei*[2], *B.brevis*[2,3], *B.cereus*[2,4], *B.cereus* var. *mycoides*[2], *B.circulans*[2], *B.coagulans*[2,5], *B.dextrolacticus*[5], *B.firmus*[2], *B.laevolacticus*[5], *B.lentus*[2], *B.licheniformis*[6,7], *B.macerans*[2], *B.macroides*[5], *B.myxolactis*[5], *B.megaterium*[2,6,8,9,10], *B.mesentericus*[11], *B.pantothenticus*[2], *B.polymyxa*[2], *B.pumilus*[2], *B.racemilacticus*[5], *B.sphaericus*[2,5,12], *B.stearothermophilus*[2,13], *B.subtilis*[2,4,6,8,14,15], *Desulfotomaculum nigrificans*[16], *Sporolactobacillus inulinus*[5], *Sporolact.laevus*[5], *Sporolact.racemicus*[5], *Sporosarcina ureae*[17]
Uncharacterized	*B.brevis*[18], *B.cereus*[19], *B.megaterium*[20], *B.mesentericus*[21]

1. De Rosa *et al.*, 1973; 2. Watanuki and Aida, 1972;
3. Fynn *et al.*, 1972; 4. Jacobsen and Dam, 1960;
5. Collins and Jones, 1979; 6. Salton and Schmitt, 1967;
7. Goodman *et al.*, 1976; 8. Bishop *et al.*, 1962;
9. Brown *et al.*, 1968; 10. Leistner *et al.*, 1967;
11. Egorov *et al.*, 1974; 12. Gale *et al.*, 1962;
13. Dunphy *et al.*, 1971; 14. Downey, 1964; 15. Farrand and Taber,
1974; 16. Unpublished result cited by Threlfall and Whistance, 1971;
17. Yamada *et al.*, 1976; 18. Tishler and Sampson, 1948;
19. Hollander *et al.*, 1977; 20. Kröger and Dadák, 1969;
21. Lester and Crane, 1959.

Classification

The aerobic endospore-forming rods constitute a metabolically diverse group of bacteria that are currently
classified into species recognized primarily on the basis
of a small number of morphological and physiological
properties [Gordon *et al*., 1973; Gibson and Gordon, 1974].
Although the classification of Gordon and her colleagues
provides a workable framework for the study of *Bacillus*
strains, it casts little light on the subgeneric structure
of the genus. Gordon [1977] recognized the need for
further comparative work on representative strains and
also noted that knowledge of the various species had
developed unevenly. Thus, more attention has been placed
on representatives of medically and industrially important
species and on those readily isolated from, and abundant
in, natural habitats. Given the problems inherent in
selecting strains to represent the whole range of variation
encompassed by the genus *Bacillus* it is hardly surprising
that there is no coherent picture of the overall lipid
composition of these bacteria.

There are a number of additional problems that make it
difficult to determine the value of lipid composition in
the classification of *Bacillus* and related taxa. Thus,
many investigations were aimed at determining particular
lipid structures and biochemical phenomena while the
application of a variety of analytical techniques in other
analyses made it difficult to compare the data obtained.
A more fundamental obstacle in developing a coherent
picture of the overall lipid composition of members of this
genus may be the interrelation of lipid types as already
noted for fatty acids and polar lipids. As a consequence
of such interchangeabilities, growth environments may have
a profound effect on the proportions of the individual
lipids of *Bacillus* and related aerobic endospore-forming
bacteria. It should be noted, however, that such variability of lipids is not characteristic of all bacteria,
Escherichia coli, for example, has a very stable polar
lipid composition, major amounts of PE occurring with
lesser proportions of PG and DPG [Cronan and Vagelos, 1972].

The incomplete data obtained so far on fatty acid
(Tables 1 to 3), polar lipid (Table 4) and isoprenoid
quinone (Table 5) composition do not enable all representatives of *Bacillus* and related taxa to be readily
separated into discrete groups. It is, however, immediately apparent that *B.acidocaldarius* can be characterized by its lipid composition, the presence of MK-9,
cyclohexyl fatty acids, triterpenes and complex polar
lipids being particularly distinctive. *Bacillus stearothermophilus* has been shown to be heterogeneous on the
basis of serological and physiological properties [Walker
and Wolf, 1971]. It is, therefore, interesting that
strains of *B.stearothermophilus* unable to hydrolyze starch

TABLE 6

Lipids in the classification of Bacillus species and related taxa

Group	Fatty acids[1] Unsaturated acid	Taxon	Polar lipids[2]	Menaquinone[3]
D	-	B.acidocaldarius	Complex	MK-9
		B.alvei	PE	
		B.megaterium	PE (LPG, GNPG)	
	Δ-5	B.pumilus	ND	
		B.subtilis		MK-7
A	Δ-5 and/or Δ-10	B.licheniformis	PE, LPG, DGDG	
	Δ-8, 9, 10	B.brevis, B.macerans		
	ND	B.circulans	ND	
		B.polymyxa	LPC, LPS, BPA	
B	ND	B.larvae, B.lentimorbus,	PE	ND
		B.popilliae		
		B.thuringiensis		
E	ND	B.anthracis	ND	ND
		B.cereus		MK-7
	Δ-5 and/or Δ-10	B.caldolyticus, B.caldotenax	PE (DGDG)	ND
		B.stearothermophilus	ND	MK-7
C	Δ-8, 9, 10	B.stearothermophilus	PE	ND
IA	Δ-5	B.globisporus, B.insolitus	MGDG, DGDG	ND
F	Δ-5	B.psychrophilus	ND	ND
		B.amyloliquefaciens	PE	ND
		B.coagulans	PS	MK-7
ND	ND	B.firmus, B.lentus, B.macroides,	ND	
		B.pantothenticus, B.sphaericus		MK-7
		B.thermoruber		ND
IA	ND	Desulfotomaculum nigrificans	ND	MK-7
ND	ND	Sporolactobacillus spp.[4]	ND	MK-7
IA	-	Sporosarcina ureae	ND	MK-7
IA	-		PE	MK-7

can be distinguished from starch-positive strains, particularly by their distinctly different polar lipid patterns
(Tables 4 and 6). It has been pointed out that fatty acid
composition of *Bacillus* strains may be substantially
altered by changes in growth environment and it should be
noted that of the groups proposed by Kaneda [1977] only
organisms placed in groups A, B and E (Tables 1 and 6) were
cultivated strictly according to the standard conditions
recommended by the same author [Kaneda, 1977]. The full
potential of fatty acid analyses in the classification of
Bacillus may not be realized until systematic studies
involving greater numbers of representative strains and
more rigorously standardized growth conditions have been
completed.

The available data on polar lipid composition (Tables 4
and 6) indicate possible areas where systematic analyses
could be valuable in classification. Representatives of
B.megaterium may apparently be divided into two groups
depending on whether or not they contain characteristic
glucosaminyl lipids (Table 4) and isolated reports suggesting that *B.coagulans* and *B.polymyxa* contain unusual polar
lipids (Table 4) should be substantiated. The separation
of *B.megaterium* into two groups is of particular interest
in view of numerical phenetic [Priest *et al.*, Chapter 5;
Logan and Berkeley, Chapter 6] and DNA homology [Claus and
Hunger, Chapter 9] data which show this taxon to be heterogeneous. Again, the presence or absence of glucosyldiacyl-
glycerols should also be systematically investigated since
in a number of polar lipid studies glycolipids were not
considered (Table 4).

Considering that *Bacillus* species form such a heterogeneous group with respect to their deoxyribonucleic acid
guanine plus cytosine content [Priest, Chapter 3] it is
remarkable that, with the exception of *B.acidocaldarius*,

Footnotes to Table 6 (opposite)

1. Groups A-F according to Kaneda [1977] (see Table 1); IA, fatty
    acids having iso and anteiso major components (see text) but not
    correlated with groups A-F in systematic studies. Unsaturated
    fatty acid data taken from Table 3.
2. Characteristic polar lipids occurring in addition to phosphatidyl-
    glycerol and diphosphatidylglycerol; data from Table 4.
    Abbreviations: PE, phosphatidylethanolamine; LPG, lysyl-
    phosphatidylglycerol; GNPG, glucosaminylphosphatidylglycerol;
    DGDG, diglucosyldiacylglycerol; LPC, lysophosphatidylcholine;
    LPS, lysophosphatidylserine; BPA, bisphosphatidic acid;
    MGDG, monoglucosyldiacylglycerol; PS, phosphatidylserine.
3. Data from Table 5.
4. *Sporolactobacillus inulinus*, *S.laevus*, *S.racemicus* and closely
    related *B.laevolacticus*, *B.myxolactis* and *B.racemilacticus*.

so much homogeneity is found in the lipids of these organisms. The menaquinone composition appears to be very similar in all strains, the fatty acid profiles present subtle variations on a branched-chain theme and PE and glycosyl diacylglycerols are commonly found in addition to PG and DPG. In certain cases, two groups of *B.stearo-thermophilus* for example, PE and glycosyl diacylglycerols do not co-occur but mutually exclude each other. The limited lipid analyses performed on representatives of *Sporolactobacillus* and *Sporosarcina* (Table 6) and on the anaerobe *Desulfotomaculum*, support a close relationship between these taxa and *Bacillus*. Representatives of the genus *Thermoactinomyces*, currently classified in the Order Actinomycetales, also appear to have fatty acid profiles rich in branched-chain acids similar to those of *Bacillus* [Ballio and Barcellona, 1968; Guzeva *et al.*, 1973; Kroppenstedt and Kutzner, 1978].

It is clear that extensive carefully planned systematic studies are necessary before the full value of lipid analysis in the classification of *Bacillus* and related aerobic spore-forming strains can be realized. The prospects for such studies are, however, good for it can be confidently anticipated that the application of methods such as numerical [see Logan and Berkeley, Chapter 6] and genetical [see Priest, Chapter 3] taxonomy will contribute towards significant improvements in the classification of *Bacillus* and related taxa. Improved classification of aerobic endospore-forming bacteria will make it possible to use good representative strains in the search for diagnostic lipids.

Acknowledgements

Thanks are due to M.D. Collins for data on menaquinone composition and to the Science Research Council for support (Grant No. GR/A/55479).

References

Aragozzini, F., Toppino, P., Manachini, P.L. and Craveri, R. (1976). Fatty acid composition of *Bacillus thermoruber*. *Annali di Microbiologia ed Enzimologia* **26**, 9-13.

Ballio, A. and Barcellona, S. (1968). Relations chimiques et immunologiques chez les actinomycétales. 1. Les acides gras de 43 souches d'actinomycetes aérobies. *Annales de l'Institut Pasteur* **114**, 121-137.

Batrakov, S.G. and Bergelson, L.D. (1978). Lipids of the streptomycetes. Structural investigation and biological interrelation. *Chemistry and Physics of Lipids* **21**, 1-29.

Beaman, T.C., Pankratz, H.S. and Gerhardt, P. (1974). Chemical composition and ultrastructure of native and reaggregated membranes from protoplasts of *Bacillus cereus*. *Journal of Bacteriology* **117**, 1335-1340.

Beebe, J.L. (1971). Isolation and characterisation of a phosphatidyl-ethanolamine-deficient mutant of *Bacillus subtilis*. *Journal of Bacteriology* **107**, 704-711.

Bertsch, L.L., Bonsen, P.P.M. and Kornberg, A. (1969). Biochemical studies of bacterial sporulation and germination. XIV. Phospholipids in *Bacillus megaterium*. *Journal of Bacteriology* **98**, 75-81.

Bishop, D.G., Op den Kamp, J.A.F. and van Deenen, L.L.M. (1977). The distribution of lipids in the protoplast membranes of *Bacillus subtilis*. *European Journal of Biochemistry* **80**, 381-391.

Bishop, D.G., Rutberg, L. and Samuelsson, B. (1967). The chemical composition of the cytoplasmic membrane of *Bacillus subtilis*. *European Journal of Biochemistry* **2**, 448-453.

Bishop, D.H.L., Pandya, L.P. and King, H.K. (1962). Ubiquinone and vitamin K in bacteria. *Biochemical Journal* **83**, 606-614.

Blume, A., Dreher, R. and Poralla, K. (1978). The influence of branched-chain and ω-alicyclic fatty acids on the transition temperature of *Bacillus subtilis* lipids. *Biochimica et Biophysica Acta* **512**, 489-494.

Brown, B.S., Whistance, G.R. and Threlfall, D.R. (1968). Studies on α-naphthol as a precursor of microbial menaquinone. *FEBS Letters* **1**, 323-325.

Brundish, D.E., Shaw, N. and Baddiley, J. (1966). Bacterial glycolipids. *Biochemical Journal* **99**, 546-549.

Bulla, L.Λ. and St. Julian, G. (1972). Lipid metabolism during bacterial growth and sporulation: Phospholipid pattern in *Bacillus thuringiensis* and *Bacillus popilliae*. In "Spores V" (eds. H.O. Halvorson, R. Hanson and L.L. Campbell), pp.191-196. American Society for Microbiology, Washington, D.C.

Bulla, L.A., Bennett, G.A. and Shotwell, O.L. (1970). Physiology of sporeforming bacteria associated with insects. II. Lipids of vegetative cells. *Journal of Bacteriology* **104**, 1246-1253.

Bureau, G. and Mazliak, P. (1968). Métabolisme des phospholipides chez *Bacillus subtilis* var. *niger*, lors des sporulations induites par pénurie de phosphore. *Compte rendu hebdomadaire des séances de l'Académie des sciences* **266D**, 2510-2512.

Button, D. and Hemmings, N.L. (1976a). Teichoic acids and lipids associated with the membrane of a *Bacillus licheniformis* mutant and the membrane lipids of the parental strain. *Journal of Bacteriology* **128**, 149-156.

Button, D. and Hemmings, N.L. (1976b). Lipoteichoic acid from *Bacillus licheniformis* 6346 MH-1. Comparative studies on the lipid portion of the lipoteichoic acid and the membrane glycolipid. *Biochemistry* **15**, 989-995.

Card, G.L. (1973). Metabolism of phosphatidylglycerol, phosphatidyl-ethanolamine and cardiolipin of *Bacillus stearothermophilus*. *Journal of Bacteriology* **114**, 1125-1137.

Card, G.L., Georgi, C.E. and Militzer, W.E. (1969). Phospholipids from *Bacillus stearothermophilus*. *Journal of Bacteriology* **97**, 186-192.

Carr, J.G., Davies, P.A., Dellaglio, F., Vescovo, M. and Williams, R.A.D. (1977). The relationship between *Lactobacillus mali* from cider and *Lactobacillus yamanashiensis* from wine. *Journal of Applied Bacteriology* **42**, 219-228.

Chan, M., Virmani, Y.P., Himes, R.H. and Akagi, J.M. (1973). Spin-labelling studies on the membrane of a facultative thermophilic bacillus. *Journal of Bacteriology* **113**, 322-328.

Collins, M.D. and Jones, D. (1979). Isoprenoid quinone composition as a guide to the classification of *Sporolactobacillus* and possibly related bacteria. *Journal of Applied Bacteriology* **47**, 293-297.

Cronan, J.E. and Vagelos, P.R. (1972). Metabolism and function of the membrane phospholipids of *Escherichia coli*. *Biochimica et Biophysica Acta* **265**, 25-60.

Daron, H.H. (1970a). Fatty acid composition of lipid extracts of a thermophilic *Bacillus* species. *Journal of Bacteriology* **101**, 145-151.

Daron, H.H. (1970b). The position of the double bond in unsaturated fatty acids of a thermophilic *Bacillus* species. *Biochemical and Biophysical Research Communications* **41**, 334-338.

Daron, H.H. (1973). Nutritional alteration of the fatty acid composition of a thermophilic *Bacillus* species. *Journal of Bacteriology* **116**, 1096-1099.

Dart, R.K. and Kaneda, T. (1970). The production of Δ^{10}-mono-unsaturated fatty acids by *Bacillus cereus*. *Biochimica et Biophysica Acta* **218**, 189-194.

De Rosa, M., Gambacorta, A. and Bu'lock, J.D. (1974). Effects of temperature on the fatty acid composition of *Bacillus acidocaldarius*. *Journal of Bacteriology* **117**, 212-214.

De Rosa, M., Gambacorta, A., Minale, L. and Bu'lock, J.D. (1971a). Bacterial triterpenes. *Chemical Communications* 619-620.

De Rosa, M., Gambacorta, A., Minale, L. and Bu'lock, J.D. (1971b). Cyclohexane fatty acids from a thermophilic bacterium. *Chemical Communications* 1334.

De Rosa, M., Gambacorta, A., Minale, L. and Bu'lock, J.D. (1972). The formation of ω-cyclohexyl-fatty acids from shikimate in an acidophilic thermophilic bacillus. *Biochemical Journal* **128**, 751-754.

De Rosa, M., Gambacorta, A., Minale, L. and Bu'lock, J.D. (1973). Isoprenoids of *Bacillus acidocaldarius*. *Phytochemistry* **12**, 1117-1123.

Downey, R.J. (1964). Vitamin K-mediated electron transfer in *Bacillus subtilis*. *Journal of Bacteriology* **88**, 904-911.

Dreher, R., Poralla, K. and König, W.A. (1976). Synthesis of ω-alicyclic fatty acids from cyclic precursors in *Bacillus subtilis*. *Journal of Bacteriology* **127**, 1136-1140.

Dunphy, P.J., Phillips, P.G. and Brodie, A.F. (1971). Separation and identification of menaquinones from microorganisms. *Journal of Lipid Research* **12**, 442-449.

Egorov, N.S., Ushakova, V.I. and Kalistratov, G.A. (1974). Isolation and identification of vitamin K_2 synthesized by *Bacillus mesentericus*. *Prikladnaya Biokhimiya i Mikrobiologiya* **10**, 64-67.

Ellar, D.J. and Posgate, J.A. (1974). Characterisation of forespores isolated from *Bacillus megaterium* at different stages of development into mature spores. In "Spore Research 1973" (eds. A.N. Barker, G.W. Gould and J. Wolf), pp.21-40. London: Academic Press.

Esser, A.F. and Souza, K.A. (1974). Correlation between thermal death and membrane fluidity in *Bacillus stearothermophilus*. *Proceedings of the National Academy of Sciences USA* **71**, 4111-4115.

Esser, A.F. and Souza, K.A. (1976). Growth temperature and the structure of the lipid phase in biological membranes. In "Extreme Environments. Mechanisms of Microbial Adaptation" (ed. M.R. Heinrich), pp.283-293. New York: Academic Press.

Fabio, U. and Vivoli, G. (1966). Composizione in acidi grassi di due stipiti di *Bacillus anthracis*. *Gionale di Batteriologia Virologia ed Immunologia* **59**, 552-557.

Farrand, S.K. and Taber, H.W. (1974). Changes in menaquinone concentration during growth and early sporulation in *Bacillus subtilis*. *Journal of Bacteriology* **117**, 324-326.

Felix, J. and Lundgren, D.G. (1972). Some membrane characteristics of *Bacillus cereus* during growth and sporulation. In "Spores V" (eds. H.O. Halverson, R. Hanson and L.L. Campbell), pp.35-43. Washington, D.C.: American Society for Microbiology.

Fischer, W., Nakano, M., Laine, R.A. and Bohrer, W. (1978). On the relationship between glycerophosphoglycolipids and lipoteichoic acids in gram-positive bacteria. I. The occurrence of phosphoglycolipids. *Biochimica et Biophysica Acta* **528**, 288-297.

Frade, R. and Guenan-Siberil, M.T. (1975). Effet du pH du milieu de culture sur la teneur en phospholipides des membranes de *Bacillus coagulans*. *Biochimie* **57**, 1397-1400.

Fujii, D.K. and Fulco, A.J. (1977). Biosynthesis of unsaturated fatty acids by bacilli. Hyperinduction and modulation of desaturase synthesis. *Journal of Biological Chemistry* **252**, 3660-3670.

Fulco, A.J. (1969a). The biosynthesis of unsaturated fatty acids by bacilli. I. Temperature induction of the desaturation reaction. *Journal of Biological Chemistry* **244**, 889-895.

Fulco, A.J. (1969b). Bacterial biosynthesis of polyunsaturated fatty acids. *Biochimica et Biophysica Acta* **187**, 169-171.

Fulco, A.J. (1974). Metabolic alterations of fatty acids. *Annual Review of Biochemistry* **43**, 215-241.

Fynn, G.H., Thomas, D.V. and Seddon, B. (1972). On the role of menaquinone in the reduced nicotinamide adenine dinucleotide oxidative pathway of *Bacillus brevis*. *Journal of General Microbiology* **70**, 271-275.

Gale, P.H., Page, A.C., Stoudt, T.H. and Folkers, K. (1962). Identification of vitamin $K_{2(35)}$, an apparent cofactor of a steroidal-1-dehydrogenase of *Bacillus sphaericus*. *Biochemistry* **1**, 788-792.

Gibson, T. and Gordon, R.E. (1974). *Bacillus*. In "Bergey's Manual of Determinative Bacteriology, Eighth Edition" (ed. R.E. Buchanan and N.E. Gibbons), pp.529-550. Baltimore: The Williams and Wilkins Co.

Glenn, A.R. and Gould, A.R. (1973). Inhibition of lipid synthesis in *Bacillus amyloliquefaciens* by inhibitors of protein synthesis. *Biochemical and Biophysical Research Communication* **52**, 356-364.

Goldfine, H. (1972). Comparative aspects of bacterial lipids. *Advances in Microbial Physiology* **8**, 1-58.

Goodfellow, M. and Board, R.G. (eds.) (1980). "Microbiological Classification and Identification". London: Academic Press.

Goodman, S.R., Marrs, B.L., Narconis, R.J. and Olson, R.E. (1976). Isolation and description of a menaquinone mutant from *Bacillus licheniformis*. *Journal of Bacteriology* **125**, 282-289.

Gordon, R.E. (1977). The genus *Bacillus*. In "CRC Handbook of

Microbiology, 2nd Edition, Volume 1. Bacteria" (eds. A.I. Laskin and H.A. Lechevalier), pp.319-336. Cleveland: CRC Press.

Gordon, R.E., Haynes, W.C. and Pang, C.H.-N. (1973). "The Genus *Bacillus*". United States Department of Agriculture, Washington, D.C.

Guzeva, L.N., Efimova, T.P., Agre, N.S. and Krasilnikov, N.A. (1973). Fatty acids in the mycelia of actinomycetes that form catenate spores. *Mikrobiologiya* **42**, 26-31.

Haest, C.W.M., de Gier, J., Op den Kamp, J.A.F. and van Deenen, L.L.M. (1972). Changes in permeability of *Staphylococcus aureus* and derived liposomes with varying lipid composition. *Biochimica et Biophysica Acta* **255**, 720-733.

Hirschberg, C.B. and Kennedy, E.P. (1972). Mechanism of the enzymatic synthesis of cardiolipin in *Escherichia coli*. *Proceedings of the National Academy of Sciences USA* **69**, 648-651.

Holländer, R., Wolf, G. and Mannheim, W. (1977). Lipoquinones of some bacteria and mycoplasmas, with considerations on their functional significance. *Antonie van Leeuwenhoek* **43**, 177-185.

Houtsmuller, U.M.T. and van Deenen, L.L.M. (1963). Studies on the phospholipids and phospholipase from *Bacillus cereus*. *Proceedings Koninklijke Nederlanse Akademie van Wetenschappen* **66B**, 236-249.

Ichikawa, I. and Matsumoto, M. (1977). Characterization of lipids in a thermophile, *Bacillus stearothermophilus* NYU. *Japanese Journal of Experimental Medicine* **47**, 81-86.

Ito, T. and Ogawa, H. (1959). Chemical studies on the antibiotic esperin. *Bulletin of the Agricultural Chemistry Society of Japan* **23**, 536-547.

Jacobsen, B.K. and Dam, H. (1960). Vitamin K in bacteria. *Biochimica et Biophysica Acta* **40**, 211-216.

Kagawa, Y. and Ariga, T. (1977). Determination of molecular species of phospholipids from thermophilic bacterium PS3 by mass chromatography. *Journal of Biochemistry* **81**, 1161-1165.

Kagawa, Y., Sone, N., Yoshida, M., Hirata, H. and Okamoto, H. (1976). Proton translocating ATPase of a thermophilic bacterium. *Journal of Biochemistry* **80**, 141-151.

Kakinuma, A., Sugino, H., Isono, M., Tamura, G. and Arima, K. (1969). Determination of fatty acid in surfactin and elucidation of the total structure of surfactin. *Agricultural and Biological Chemistry* **33**, 973-976.

Kaneda, T. (1966). Biosynthesis of branched-chain fatty acids. IV. Factors affecting relative abundance of fatty acids produced by *Bacillus subtilis*. *Canadian Journal of Microbiology* **12**, 501-514.

Kaneda, T. (1968). Fatty acids in the genus *Bacillus*. II. Similarity in the fatty acid composition of *Bacillus thuringiensis*, *Bacillus anthracis* and *Bacillus cereus*. *Journal of Bacteriology* **95**, 2210-2216.

Kaneda, T. (1969). Fatty acids in *Bacillus larvae*, *Bacillus lentimorbus* and *Bacillus popilliae*. *Journal of Bacteriology* **98**, 143-146.

Kaneda, T. (1971a). Major occurrence of $cis\Delta^5$ fatty acids in three psychrophilic species of *Bacillus*. *Biochemical and Biophysical Research Communications* **43**, 298-302.

Kaneda, T. (1971b). Incorporation of branched-chain C_6-fatty acid isomers into the related long-chain fatty acids by growing cells of *Bacillus subtilis*. *Biochemistry* **10**, 340-347.

Kaneda, T. (1972a). Positional preference of fatty acids in phospholipids of *Bacillus cereus* and its relation to growth temperature. *Biochimica et Biophysica Acta* **280**, 297-305.

Kaneda, T. (1972b). Positional distribution of fatty acids in phospholipids from *Bacillus subtilis*. *Biochimica et Biophysica Acta* **270**, 32-39.

Kaneda, T. (1977). Fatty acids of the genus *Bacillus*: an example of branched-chain preference. *Bacteriological Reviews* **41**, 391-418.

Kates, M., Kushner, D.J. and James, A.T. (1962). The lipid composition of *Bacillus cereus* as influenced by the presence of alcohols in the culture medium. *Canadian Journal of Biochemistry and Physiology* **40**, 83-94.

Katz, E. and Demain, A.L. (1977). The peptide antibiotics of *Bacillus*: Chemistry, biogenesis and possible functions. *Bacteriological Reviews* **41**, 449-474.

Komura, I., Yamada, K. and Komagata, K. (1975). Taxonomic significance of phospholipid composition in aerobic gram-positive cocci. *Journal of General and Applied Microbiology* **21**, 97-107.

Kostiw, L.L. and Souza, K.A. (1976). Altered phospholipid metabolism in a temperature-sensitive mutant of a thermophilic bacillus. *Archives of Microbiology* **107**, 49-55.

Kröger, A. and Dadàk, V. (1969). On the role of quinones in bacterial electron transport. The respiratory system of *Bacillus megaterium*. *European Journal of Biochemistry* **11**, 328-340.

Kroppenstedt, R.M. and Kutzner, H.J. (1978). Biochemical taxonomy of some problem actinomycetes. *Zentralblatt für Bakteriologie, Parasitenkunde, Infektionskrankheiten und Hygiene. I. Abteilung Supplement* **6**, 125-133.

Lang, D.R. and Lundgren, D.G. (1970). Lipid composition of *Bacillus cereus* during growth and sporulation. *Journal of Bacteriology* **101**, 483-489.

Langworthy, T.A. and Mayberry, W.R. (1976). A 1,2,3,4-tetrahydroxy pentane-substantiated pentacyclic triterpene from *Bacillus acidocaldarius*. *Biochimica et Biophysica Acta* **431**, 570-577.

Langworthy, T.A., Mayberry, W.R. and Smith, P.F. (1976). A sulfonolipid and novel glucosamidyl glycolipids from the extreme thermoacidophile *Bacillus acidocaldarius*. *Biochimica et Biophysica Acta* **431**, 550-569.

Lechevalier, M.P. (1977). Lipids in bacterial taxonomy - a taxonomist's viewpoint. In "CRC Critical Reviews in Microbiology", pp.109-210. Cleveland: CRC Press.

Lee, Y.H. and Oo, K.C. (1973). Metabolism of phospholipids in *Bacillus stearothermophilus*. *Journal of Biochemistry* **74**, 615-621.

Leistner, E., Schmitt, J.H. and Zenk, M.H. (1967). α-Naphthol a precursor of vitamin K_2. *Biochemical and Biophysical Research Communications* **28**, 845-850.

Lester, R.L. and Crane, F.L. (1959). The natural occurrence of coenzyme Q and related compounds. *Journal of Biological Chemistry* **234**, 2169-2175.

Lillich, T.T. and White, D.C. (1971). Phospholipid metabolism in the absence of net phospholipid synthesis in a glycerol-requiring mutant of *Bacillus subtilis*. *Journal of Bacteriology* **107**, 790-797.

Lindgren, V., Holmgren, E. and Rutberg, L. (1977). *Bacillus subtilis* mutant with temperature-sensitive net synthesis of phosphatidyl-ethanolamine. *Journal of Bacteriology* **132**, 473-484.

MacDougall, J.C. and Phizackerley, P.J.R. (1969). Isomers of glucos-aminyl-phosphatidylglycerol in *Bacillus megaterium*. *Biochemical Journal* **114**, 361-367.

Mastroeni, P., Contadini, V. and Teti, D. (1968). Nonesterified fatty acids and phospholipids in vegetable form and spores of *Bacillus megaterium* and *Bacillus subtilis*. *Journal of Bacteriology* **95**, 1961.

Mastroeni, P., Nacci, A., Teti, D. and Teti, M. (1971). Sul significo tassonomico dei fosfolipidi di *Bacillus cereus, Bacillus subtilis* e *Bacillus megaterium*. *Giornale di Batteriologia Virologia ed Immunologia* **64**, 20-30.

Matches, J.R., Walker, H.W. and Ayres, J.C. (1964). Phospholipids in vegetative cells and spores of *Bacillus polymyxa*. *Journal of Bacteriology* **87**, 16-23.

McElhaney, R.N. (1976). The biological significance of alterations in the fatty acid composition of microbial membrane lipids in response to changes in environmental temperature. In "Extreme Environments. Mechanisms of Microbial Adaptation" (ed. M.R. Heinrich), pp.255-281. New York: Academic Press.

McElhaney, R.N. and Souza, K.A. (1976). The relationship between environmental temperature, cell growth and the fluidity and physical state of the membrane lipids in *Bacillus stearothermo-philus*. *Biochimica et Biophysica Acta* **443**, 348-359.

Mindich, L. (1970). Membrane synthesis in *Bacillus subtilis*. I. Isolation and properties of strains bearing mutations in glycerol metabolism. *Journal of Molecular Biology* **49**, 415-432.

Minnikin, D.E. and Abdolrahimzadeh, H. (1974a). Effect of pH on the proportions of polar lipids in chemostat cultures of *Bacillus subtilis*. *Journal of Bacteriology* **120**, 999-1003.

Minnikin, D.E. and Abdolrahimzadeh, H. (1974b). The replacement of phosphatidylethanolamine and acidic phospholipids by an ornithine-amide lipid and a minor phosphorus-free lipid in *Pseudomonas fluorescens* NCMB 129. *FEBS Letters* **43**, 257-260.

Minnikin, D.E. and Goodfellow, M. (1980). Lipid composition in the classification and identification of acid-fast bacteria. In "Microbiological Classification and Identification" (eds. M. Good-fellow and R.G. Board), pp.189-256. London: Academic Press.

Minnikin, D.E. Abdolrahimzadeh, H. and Baddiley, J. (1971a). The interrelation of phosphatidylethanolamine and glycosyl diglycerides in bacterial membranes. *Biochemical Journal* **124**, 447-448.

Minnikin, D.E. Abdolrahimzadeh, H. and Baddiley, J. (1971b). The interrelation of polar lipids in bacterial membranes. *Biochimica et Biophysica Acta* **249**, 651-655.

Minnikin, D.E., Abdolrahimzadeh, H. and Baddiley, J. (1972). Variation of polar lipid composition of *Bacillus subtilis* (Marburg) with different growth conditions. *FEBS Letters* **27**, 16-18.

Minnikin, D.E., Abdolrahimzadeh, H. and Baddiley, J. (1974). The occurrence of phosphatidylethanolamine and glycosyl diglycerides in thermophilic bacilli. *Journal of General Microbiology* **83**, 415-418.

Minnikin, D.E., Abdolrahimzadeh, H. and Wolf, J. (1977). Taxonomic significance of polar lipids in some thermophilic members of

Bacillus. In "Spore Research 1976" (eds. A.N. Barker, J. Wolf, D.J. Ellar and G.W. Gould), pp.879-893. London: Academic Press.

Minnikin, D.E., Collins, M.D. and Goodfellow, M. (1978a). Menaquinone patterns in the classification of nocardioform and related bacteria. *Zentralblatt für Bakteriologie, Parasitenkunde, Infektionskrankheiten und Hygiene. I. Abteilung, Supplement* 6, 85-90.

Minnikin, D.E., Goodfellow, M. and Collins, M.D. (1978b). Lipid composition in the classification and identification of coryneform and related taxa. In "Coryneform Bacteria" (eds. I.J. Bousfield and A.G. Callely), pp.85-160. London: Academic Press.

Mizushima, S., Ishida, M. and Kitahara, K. (1966). Chemical composition of the protoplast membrane of *Bacillus megaterium*. *Journal of Biochemistry* 59, 374-381.

Morman, M.R. and White, D.C. (1970). Phospholipid metabolism during penicillinase production in *Bacillus licheniformis*. *Journal of Bacteriology* 104, 247-253.

Niskanen, A., Kiutamo, T., Mälkki, Y. and Nikkilä, O.E. (1975). Detection of *Bacillus cereus* in foods: gas chromatographic analysis of the bacterial fatty acid composition. *Zentralblatt für Bakteriologie, Parasitenkunde, Infektionskrankheiten und Hygiene. I. Abteilung* 242, 121-124.

Niskanen, A., Kiutamo, T., Räisänen, S. and Raevuori, M. (1978). Determination of fatty acid compositions of *Bacillus cereus* and related bacteria: a rapid gas chromatographic method using a glass capillary column. *Applied and Environmental Microbiology* 35, 453-455.

Ochi, T., Yano, K., Ozaki, M., Higashi, Y., Inoue, K. and Amano, T. (1971). Studies on immunity to megacin A: Susceptibility of phospholipids to phospholipase A activity of megacin A. *Biken Journal* 14, 29-36.

Oo, K.C. and Lee, Y.H. (1972). The phospholipids of a facultatively thermophilic strain of *Bacillus stearothermophilus*. *Journal of Biochemistry* 71, 1081-1084.

Op den Kamp, J.A.F., Bonsen, P.P.M. and van Deenen, L.L.M. (1969a). Structural investigations on glucosaminyl phosphatidylglycerol from *Bacillus megaterium*. *Biochimica et Biophysica Acta* 176, 298-305.

Op den Kamp, J.A.F., Kauerz, M.T. and van Deenen, L.L.M. (1972). Action of phospholipase A_2 and phospholipase C on *Bacillus subtilis* protoplasts. *Journal of Bacteriology* 112, 1090-1098.

Op den Kamp, J.A.F., van Iterson, W. and van Deenen, L.L.M. (1967). Studies on the phospholipids and morphology of protoplasts of *Bacillus megaterium*. *Biochimica et Biophysica Acta* 135, 862-884.

Op den Kamp, J.A.F., Redai, I. and van Deenen, L.L.M. (1969b). Phospholipid composition of *Bacillus subtilis*. *Journal of Bacteriology* 99, 298-303.

Op den Kamp, J.A.F., Verheij, H.M. and van Deenen, L.L.M. (1971). Two isomers of glucosaminylphosphatidylglycerol. Their occurrence in *Bacillus megaterium*, structural analysis and chemical synthesis. *Bioorganic Chemistry* 1, 174-187.

Oshima, M. and Ariga, T. (1975). ω-Cyclohexyl fatty acids in acidophilic thermophilic bacteria. *Journal of Biological Chemistry* 250, 6963-6968.

Oshima, M. and Miyagawa, A. (1974). Comparative studies on the fatty acid composition of moderately and extremely thermophilic bacteria. *Lipids* **9**, 476-480.

Ovchinnikov, Y.A., Ivanov, V.T., Kostetsky, P.V. and Shemyakin, M.M. (1966). On the structure of esperin. *Tetrahedron Letters* 5285-5290.

Paton, J.C., May, B.K. and Elliott, W.H. (1978). Membrane phospholipid assymmetry in *Bacillus amyloliquefaciens*. *Journal of Bacteriology* **135**, 393-401.

Phizackerley, P.J.R., MacDougall, J.C. and Moore, R.A. (1972). 1(O-β-Glucosaminyl)-2,3-diglyceride in *Bacillus megaterium*. *Biochemical Journal* **126**, 499-502.

Rigomier, D. and Lubochinsky, B. (1974). Metabolisme des phospholipides chez des mutants asporogenes de *Bacillus subtilis* au cours de la croissance exponentielle. *Annales de Microbiologie (Institut Pasteur)* **125B**, 295-303.

Rigomier, D., Lacombe, C. and Lubochinsky, B. (1978). Carbiolipin metabolism in growing and sporulating *Bacillus subtilis*. *FEBS Letters* **89**, 131-135. .

Rilfors, L., Wieslander, A. and Ståhl, S. (1978). Lipid and protein composition of membranes of *Bacillus megaterium* variants in the temperature range 5 to 70°C. *Journal of Bacteriology* **135**, 1043-1052.

Rohmer, M., Bouvier, P. and Ourisson, G. (1979). Molecular evolution of biomembranes: Structural equivalents and phylogenetic precursors of sterols. *Proceedings of the National Academy of Sciences USA* **76**, 847-851.

Rothman, J.E. and Kennedy, E.P. (1977a). Asymmetrical distribution of phospholipids in the membrane of *Bacillus megaterium*. *Journal of Molecular Biology* **110**, 603-618.

Rothman, J.E. and Kennedy, E.P. (1977b). Rapid transmembrane movement of newly synthesized phospholipids during membrane assembly. *Proceedings of the National Academy of Sciences USA* **74**, 1821-1825.

Rouser, G., Yamamoto, A. and Kritchevsky, G. (1971). Cellular membranes. *Archives of Internal Medicine* **127**, 1105-1121.

Russell, N.J. (1971). Alteration in fatty acid chain length in *Micrococcus cryophilus* grown at different temperatures. *Biochimica et Biophysica Acta* **231**, 254-256.

Saito, K. and Mukoyama, K. (1971). Diglucosyldiglyceride from *B.cereus*. *Journal of Biochemistry* **69**, 83-90.

Salton, M.R.J. and Schmitt, M.D. (1967). Effects of diphenylamine on carotenoids and menaquinones in bacterial membranes. *Biochimica et Biophysica Acta* **135**, 196-207.

Sato, M., Okuda, S., Izaki, K. and Takahashi, H. (1974). Effect of glycine on phospholipid metabolism in *Bacillus subtilis*. *Journal of General and Applied Microbiology* **20**, 1-9.

Shaw, N. (1974). Lipid composition as a guide to the classification of bacteria. *Advances in Applied Microbiology* **17**, 63-108.

Shaw, N. (1975). Bacterial glycolipids and glycophospholipids. *Advances in Microbial Physiology* **12**, 141-167.

Shen, P.Y., Coles, E., Foote, J.L. and Stenesh, J. (1970). Fatty acid distribution in mesophilic and thermophilic strains of the genus *Bacillus*. *Journal of Bacteriology* **103**, 479-481.

Singh, A., Kalra, M.S. and Raheja, R.K. (1978). Lipid composition of *Bacillus* species. *Indian Journal of Dairy Science* 31, 77-81.

Souza, K.A., Kostiw, L.L. and Tyson, B.J. (1974). Alterations in normal fatty acid composition in a temperature-sensitive mutant of a thermophilic bacillus. *Archives of Microbiology* 97, 89-102.

Thomson, R.H. (1971). "Naturally Occurring Quinones". London: Academic Press.

Threlfall, D.R. and Whistance, G.R. (1971). Biosynthesis of iso-prenoid quinones and chromanols. In "Aspects of Terpenoid Chemistry and Biochemistry" (ed. T.W. Goodwin), pp.357-404. London: Academic Press.

Tishler, M. and Sampson, W.L. (1948). Isolation of vitamin K_2 from cultures of a spore-forming soil bacillus. *Proceedings of the Society for Experimental Biology and Medicine* 68, 136-137.

Uchida, K. and Mogi, K. (1973). Cellular fatty acid spectra of *Sporolactobacillus* and some other *Bacillus-Lactobacillus* inter-mediates as a guide to their taxonomy. *Journal of General and Applied Microbiology* 19, 129-140.

Urakami, C. and Umetani, K. (1968). Compositions of phosphatides from *Bacillus natto* at various growth phases. *Biochimica et Biophysica Acta* 164, 64-71.

Viviani, R., Matteuzzi, D. and Gandolfi, M.G. (1965). Acidi grassi dei lipidi di *Bacillus polymyxa* isolato da rumme. *Archivo Veterinario Italiano* 16, 161-166.

Vivoli, G. and Fabio, U. (1967). Composizione in acidi di stipiti microbici appartenenti alla specie *B.cereus* e *B.thuringiensis*. *Nuovi Annali d Igiene e Microbiologia* 18, 166-173.

Vivoli, G., Olivo, R. and Fabio, U. (1966). Studio gascromatografico sulla composizione in acidi grassi di specie microbiche apparte-nenti al genere *Bacillus*. *Nuovi Annali d Igiene e Microbiologia* 17, 190-210.

Walker, P.D. and Wolf, J. (1971). Taxonomy of *Bacillus stearothermo-philus*. In "Spore Research 1971" (eds. A.N. Barker, G.W. Gould and J. Wolf), pp.247-262. London: Academic Press.

Watanuki, M. and Aida, K. (1972). Significance of quinones in the classification of bacteria. *Journal of General and Applied Micro-biology* 18, 469-472.

Weerkamp, A. and Heinen, W. (1972a). Effect of temperature on the fatty acid composition of the extreme thermophiles, *Bacillus caldo-lyticus* and *Bacillus caldotenax*. *Journal of Bacteriology* 109, 443-446.

Weerkamp, A. and Heinen, W. (1972b). The effect of nutrients and precursors on the fatty acid composition of two thermophilic bacteria. *Archiv für Mikrobiologie* 81, 350-360.

Whiteside, T.L., De Siervo, A.J. and Salton, M.R.J. (1971). Use of antibody to membrane adenosine triphosphatase in the study of bacterial relationships. *Journal of Bacteriology* 105, 957-967.

Wilkinson, S.G. (1968). Glycolipids containing glucose and uronic acids in *Pseudomonas* species. *Biochimica et Biophysica Acta* 152, 227-229.

Wilkinson, S.G. (1972). Composition and structure of the ornithine-containing lipid from *Pseudomonas rubescens*. *Biochimica et Bio-physica Acta* 270, 1-17.

Willecke, K. and Pardee, A.B. (1971). Fatty acid-requiring mutant of *Bacillus subtilis* defective in branched chain α-keto acid dehydrogenase. *Journal of Biological Chemistry* **246**, 5264-5272.

Yamada, Y., Inouye, G., Tahara, Y. and Kondo, K. (1976). The menaquinone system in the classification of aerobic gram-positive cocci in the genera *Micrococcus, Staphylococcus, Planococcus* and *Sporosarcina*. *Journal of General and Applied Microbiology* **22**, 227-236.

Yao, M., Walker, H.W. and Lillard, D.A. (1970). Fatty acids from vegetative cells and spores of *Bacillus stearothermophilus*. *Journal of Bacteriology* **102**, 877-878.

Chapter 5

THE GENUS *BACILLUS:* A NUMERICAL ANALYSIS

F.G. PRIEST*, M. GOODFELLOW** and CAROLE TODD**

*Department of Brewing and Biological Sciences,
Heriot-Watt University, Edinburgh, UK and
**Department of Microbiology, The University,
Newcastle upon Tyne, UK

Introduction

Prior to the mid 1940's, the classification of aerobic,
endospore-forming, rod-shaped bacteria was highly artificial with over 150 species described on the basis of a
few strains and a small number of tests of unknown reproducibility. Gibson [1934, 1935a and b, 1944] clarified
the classification of *B.pasteurii* and *B.subtilis* while
Gibson and Topping [1938] developed the use of spore shape,
size and location within the sporangium as a means of
differentiating between taxa. The major revision in
Bacillus systematics, however, was provided by Smith *et al.*
[1952] who characterized over 1000 aerobic, endospore-
forming strains representing 158 species and assigned them
to 19 species. Revised and supplemented descriptions of
these species have been published recently together with
information on 77 unassigned strains [Gordon *et al.*, 1973].
The morphological and physiological data contained in this
latest monograph together with the histories of the
isolates provide an invaluable framework for taxonomic
studies on the genus *Bacillus*. Unfortunately, the
relationships between the recognized *Bacillus* species are
not immediately apparent from the data which are presented
in tabular form.
    The introduction of numerical taxonomy to bacterial
systematics led to significant improvements in the classification and identification of many bacterial taxa [Good-
fellow, 1977; Jones and Sackin, 1980] but until recently
[Logan and Berkeley, Chapter 6] taxometric techniques had
not been applied in any systematic way to establish the
relationships amongst the aerobic, endospore-forming rods.
This neglect is puzzling, for taxometric techniques
provide one of the quickest and most accurate ways of
determining relationships between taxa and highlighting
presumptive diagnostic characters for detailed test
standardization studies [Wayne *et al.*, 1974, 1976]. In
addition, preliminary analyses of published data on

Bacillus strains [Sneath, 1962] and the examination of
isolates from marine habitats [Bonde, 1975; Boeyé and
Aerts, 1976] have underlined the relevance of the numerical
method to the genus *Bacillus*. In the present study the
results of a numerical analysis of the data cited by Gordon
et al. [1973] are provided in an attempt to highlight phena
and demonstrate relationships between them.

Computation

Physiological data for 560 *Bacillus* strains were coded
for 35 two state characters and analysed using the Clustan
1A package of Wishart [1968] on an IBM 370/168 computer
using the simple matching coefficient [S_{sm}, Sokal and
Michener, 1958] which counts both positive and negative
similarities. Clustering was achieved using the unweighted
average linkage algorithm [Sneath and Sokal, 1973] and
results presented as an abbreviated dendrogram (Fig. 1).

Overall Taxonomic Structure

Over ninety-five per cent of the 560 *Bacillus* strains
were recovered in 26 phena (Fig. 1). All of the phena
coalesced at the 60% similarity level (S-level), a value
much lower than that usually obtained in a S_{sm}, average
linkage analysis of isolates belonging to a single genus.
The low cut-off point probably reflects the small number
of characters used in the analysis and the fact that most
of the tests had originally been selected for their
diagnostic value. Even so, the base S-level can be considered to be very much lower than that used in equivalent
numerical analyses to define phena equated with Gram-
positive genera such as *Actinomyces* [Holmberg and
Hallander, 1973; Holmberg and Nord, 1975], *Corynebacterium* [Jones, 1975], *Mycobacterium* [Tsukamura, 1966;
Meissner *et al.*, 1974; Wayne *et al.*, 1978], *Nocardia*
[Goodfellow, 1971; Tsukamura, 1969] and *Rhodococcus*
[Goodfellow and Alderson, 1977]. The results of the
numerical analysis suggest, therefore, that the genus
Bacillus is heterogeneous, an interpretation that is
supported by a comparison of the numerical phenetic data
with that derived from deoxyribonucleic acid (DNA) base
determination studies (Fig. 1).

Fig. 1. *(opposite)* A simplified dendrogram showing the relationship
between clusters based on the S_{sm} coefficient and unweighted average
linkage clustering (UPGMA).

The reactions of the *B.carotarum* strains were consistent with the
description provided by Gibson and Gordon [1974] but they did not
fully resemble the *B.carotarum* strains described by Smith *et al.*
[1952], and their identification therefore should be considered
tentative.

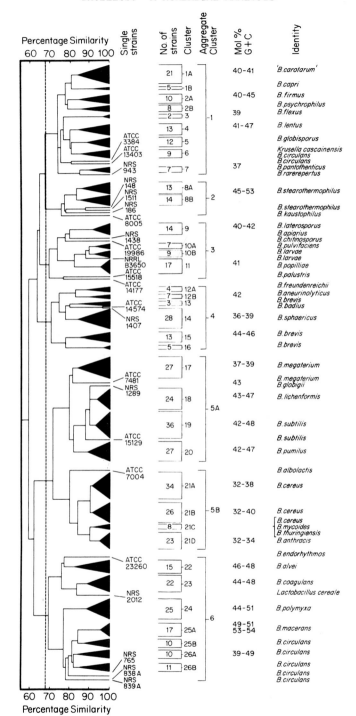

The 26 phena circumscribed in the numerical analysis can be classified into 6 aggregate clusters defined at the 69% S-level. Each of the aggregate clusters accommodates one or more taxospecies containing strains with genomes that fall within an approximate 10% mol guanine (G) plus cytosine (C) range. Since it is generally accepted that the maximum genetic diversity which should be accommodated by a genus corresponds to around 10% mol GC range [De Ley, 1970; Jones and Sneath, 1970; Bradley and Mordarski, 1976; Bradley, 1979; Priest, Chapter 3] then 5 of the 6 aggregate clusters can be equated with the rank of genus. There are two exceptions, aggregate cluster 5 includes strains that vary by approximately 16% mol GC. This taxon can, however, be split into two groups at the 72% S-level, subgroup 5A (36 to 48 mol % GC) includes *B.megaterium* and *B.subtilis* with the closely related *B.licheniformis* and *B.pumilus* whereas group 5B (32 to 40 mol % GC) comprises *B.cereus* and related taxa (Fig. 1). Aggregate cluster 6 varies by some 20% GC. This is partly due to the heterogeneity of *B.circulans* strains (35-51% GC) and indicates the need for detailed study of this group.

It is encouraging that the numerical classification is in general agreement with the one obtained by Logan and Berkeley (Chapter 6), although in the latter phena are defined at a much higher level of similarity and the composition of the aggregate clusters differs in some instances. These differences are to some extent due to the different strains and the different number and type of tests employed and additional studies based on independent taxonomic methods are needed to evaluate and extend the numerical classifications.

Detailed Taxonomic Structure

Aggregate cluster 1 (38 to 48 mol % GC) All but 3 of the 91 strains in this taxon were recovered in one of 6 phena. The largest was divided into 2 subphena, 1A and 1B. The former contains 21 *B.megaterium* strains and shares little similarity to phenon 17 which accommodates most of the remaining strains of *B.megaterium* including the neotype strain NC1B 9376. Phenon 17 can, therefore, be equated with *B.megaterium sensu stricto*. The numerical classification is in strong agreement with the view that *B.megaterium* is heterogeneous [Gordon *et al.*, 1973].

Some of the strains in subphenon 1A have been shown to differ from those of *B.megaterium sensu stricto* in both base composition and homology studies [Hunger and Claus, Chapter 9), by their ability to reduce nitrate to nitrite and their inability to deaminate phenylalanine. In addition, these strains hydrolyze starch weakly and are unable to solubilize amylose azure [Hunger and Claus, Chapter 9]. Other workers using a variety of taxonomic criteria have similarly noted differences between these groups of strains [Rode, 1968; Watanabe and Takesna, 1970; Gibson and Gordon, 1974; Bonde, 1975; Logan and Berkeley, Chapter 6] strongly suggesting that subphenon 1A represents a distinct species. The characteristics of this taxon conform to the description given for *B.carotarum* in the eighth edition of "Bergey's Manual of Determinative Bacteriology" [Gibson and Gordon, 1974].

Subphenon 1B contains 3 strains, 2 of which (NRS 607 and NRS 824) were originally labelled *B.capri*. Further studies are required to determine the integrity of subphenon 1B as strain NRS 607 has 57 to 71% of its DNA in common with that from subphenon 1A strains [Hunger and Claus, Chapter 9].

Eight of the 9 *B.firmus* strains were recovered in subphenon 2A together with 2 isolates of marine origin, *B.epiphytus* NRS 1117 and *B.filicolonicus* NRS 118. The exact relationship of these species to *B.firmus* requires further study as strains of the latter are not uncommon in marine habitats [Denis, 1971; Bonde, 1975; Boeyé and Aerts, 1976; Bonde, Chapter 8]. The *B.firmus* strains are well separated from the *B.lentus* strains of phenon 4, a result in good agreement with preliminary DNA reassociation data [Priest, Chapter 3]. However, additional strains of *B.firmus* and *B.lentus* need to be examined to determine whether they form discrete species or a spectrum-like relationship [Proom and Knight, 1955; Gordon *et al.*, 1977; Logan and Berkeley, Chapter 6]. Nevertheless, the present numerical data are in good agreement with the results of base composition studies which show that the *B.firmus* strains in subphenon 2A fall within the 5% GC range considered to be consistent with a good species [De Ley, 1970; Jones and Sneath, 1970; Bradley and Mordarski, 1976; Bradley, 1980].

Subphenon 2B contains 5 of the 6 strains of *B.psychrophilus*, including the type strain ATCC 23304, *B.firmus* NRS 749 and the type strain of *B.aminovorans* ATCC 7046. Strain NRS 749 has been described as a *B.firmus/B.lentus* intermediate [Gordon *et al.*, 1977].

Phenon 3 accommodates 2 strains of *B.megaterium*, originally labelled *B.agrestis* NRS 602 and *B.flexus* NRS 665, and *B.circulans* ATCC 11032. Hunger and Claus [Chapter 9] have shown that NRS 602 and NRS 665 have few DNA sequences in common with either *B.carotarum* or *B.megaterium* strains, a result which suggests that the strains in phenon 3 may be the nucleus of a distinct species.

Phenon 4 shows a close relationship with phena 5 and 6 which contain strains of *B.globisporus* and *Krusella cascainensis* respectively. In addition to 4 *B.lentus* strains, phenon 4 contains *B.psychrophilus* NRS 1525, *B.psychrosaccharolyticus* NRS 1518 and NRS 1520 and 6 *B.circulans* strains, 4 of which were previously labelled *B.palustris* ATCC 14175, 14175, 14176 and 15519 and *B.amylolyticus* ATCC 995. Phenon 7, the final taxon in

aggregate cluster 1, is clearly separated from the other
phena and contains 7 strains of *B.pantothenticus* and the
single isolate of *B.rarerepertus*. The remaining 2 strains,
B.circulans ATCC 8384 and ATCC 13403 form single member
clusters.

Aggregate cluster 1 cannot readily be equated with any
of the cluster groups defined by Logan and Berkeley
[Chapter 6]. It shares its closest similarity to their
cluster group II, a relatively loosely defined taxon,
which includes phena equated with *B.alvei* and *B.latero-
sporus* as well as those corresponding to *B.carotarum*,
B.firmus and *B.lentus*. Further comparative studies are
required to resolve such apparent anomalies. The present
numerical findings do, however, support the view that
B.globisporus and *B.psychrophilus* are closely related to
B.pantothenticus [Larkin and Stokes, 1967]. In contrast,
other workers have noted a similarity between *B.panto-
thenticus* and *B.circulans* [Proom and Knight, 1950; Gibson
and Gordon, 1974] while Logan and Berkeley [Chapter 6]
found a relationship between this taxon and the species
B.coagulans and *B.stearothermophilus*.

Aggregate cluster 2 (44 to 53% GC) This taxon accommodates 30 strains of *B.stearothermophilus*, all but 3 of
which were recovered in subphena 8A and 8B. The remaining
strains, NRS 86, NRS 148 and NRS 1511 form single member
clusters as does *B.kaustophilus* ATCC 8005 (Fig. 1). These
numerical findings are yet further evidence of the heterogeneity of *B.stearothermophilus* [Baillie and Walker, 1968;
Minnikin *et al.*, 1977; Logan and Berkeley, Chapter 6].
It is difficult to relate the subphena with any of the 3
distinct subdivisions of the taxon recognized on the basis
of biochemical and serological data [Walker and Wolf, 1971;
Wolf and Chowdhury, 1971]. Subphenon 8A, however, includes
B.stearothermophilus Donk and several strains received as
B.thermoliquefaciens.

Aggregate cluster 3 (38 to 47% GC) Forty-eight of the 53
strains in this taxon formed phena 9 to 11. The 14
strains of *B.laterosporus* comprise phenon 9, a homogeneous
taxon sharply separated from all other phena. In contrast,
strains of *B.larvae* and *B.pulvifaciens*, 2 species
considered to be closely related [Gibson and Gordon, 1974],
were recovered as the closely related subphena 10A and 10B.
Comparative studies on additional strains are needed to
determine whether these taxa merit specific status for
typical strains only seem to differ in their temperature
requirements for growth and by the ability of *B.pulvi-
faciens* to grow in nutrient broth after serial transfer
[Gibson and Gordon, 1974]. The cultures of *B.lentimorbus*
and *B.popilliae*, causal agents of milky diseases
of various scarabaeid beetle larvae, were recovered in
phenon 11, a result which questions their separation into

2 taxospecies [Bulla *et al.*, 1978]. Further studies using
more isolates, and sensitive methods such as nucleic acid
reassociation, are required to establish the exact
relationship between *B.lentimorbus* and *B.popilliae* and
between *B.larvae* and *B.pulvifaciens.*

Bacillus apiarius strains NRS 1438 and 1439 are closely
related and need to be compared with similar strains to
ascertain the status of this taxon. Additional work is
also necessary to determine the affinities of *B.chitino-
sporus* NClB 9652 and *B.palustris* ATCC 14177 and 15518, all
of which form single member clusters.

Aggregate cluster 4 (36 to 47% GC) Phena 12 to 16 en-
compass all 67 strains in this taxon. Phenon 12 was
divided into 2 subphena, the first of which contains
B.freudenreichii ATCC 7053 and NRS 671 together with 3
strains of *B.sphaericus* which were originally labelled
B.rotans NRS 633 and ATCC 4978, and *B.loehnisii* NRS 672
Systematic studies using additional strains are necessary
to establish the relationships of these strains to
B.aneurinolyticus and *B.thiaminolyticus*, representatives
of which form subphenon 12B. Logan and Berkeley [Chapter
6] recovered strains of *B.brevis* in a compact phenon which
also contained *B.aneurinolyticus.* In the present study,
however, strains of *B.brevis* were recovered in phena 13,
15 and 16, the second and largest of which includes the
neotype strain, ATCC 8246. Subphenon 13B comprises the
type strain of *B.badius*, ATCC 14574 and an unassigned
strain, NRS 1407.

The recovery of 27 of the 29 *B.sphaericus* strains as a
distinct phenon is in good agreement with the results of
Logan and Berkeley [Chapter 6]. *Bacillus sphaericus* seems
to form a good taxospecies, a conclusion supported by the
few base ratio determinations which show that 3 strains in
this taxon have values between 36 and 39% GC.

It is interesting, that in this analysis the alkali-
forming strains were found in aggregate cluster 4. However, this result may be more apparent than real as it may
be due to alkali formation being overweighted in the small
battery of tests employed. Using a large number of tests,
Logan and Berkeley [Chapter 6] recovered alkali-forming
strains in cluster groups II and III together with a
variety of additional taxa.

Aggregate cluster 5 (32 to 48% GC) At the 67% S-level
this, the largest aggregate cluster, was subdivided into
2 groups, 5A and 5B, which correspond to cluster groups IV
and I of Logan and Berkeley [Chapter 6]. Group 5A (36 to
48% GC) contains 117 strains, all but 4 of which were
recovered in phena 17 to 20. The *B.megaterium* strains of
phenon 17 are well separated from phena 18, 19 and 20, the
latter are closely related and correspond to *B.licheni-
formis, B.subtilis* and *B.pumilus* respectively. The

varieties of *B.subtilis*, namely *aterrimus, natto* and *niger*,
clustered tightly with the parent species in phenon 11 but
B.globigii NRS 1289, *B.megaterium* ATCC 7481 and *B.subtilis*
ATCC 15129 were recovered as single member clusters. The
exact relationship between *B.amyloliquefaciens* and
B.subtilis remains confused. These species have been
separated in DNA reassociation assays [Seki *et al.*, 1975]
and on the basis of pyrolysis gas liquid chromatography
[O'Donnell *et al.*, 1980], but in the numerical analysis of
Logan and Berkeley [Chapter 6] some strains were found to
be intermediate between taxa. In the present analysis
B.amyloliquefaciens ATCC 15841 was recovered in the centre
of the *B.subtilis* phenon.

It can be seen from Fig. 1 that group 5B (32 to 40% GC)
consists of *B.cereus* and its close relatives in phenon 21.
The phenon was subdivided into 4, with the largest subphenon, 21A, accommodating *B.finitimus* ATCC 19269 and NRRL
B-2245 and 4 strains of *B.thuringiensis* in addition to
strains of *B.cereus* and *B.cereus* var. *mycoides*. Thus
B.thuringiensis strains are separated from *B.anthracis*
ATCC 11949 which occurs in subphenon 21D. *Bacillus albolactis* ATCC 7004, which can be distinguished from the
other strains in the group by its ability to form acid
from lactose, was recovered as a single member cluster.

Aggregate taxon 6 (35 to 55% GC) The taxon accommodates
115 strains, all but 5 of which were recovered in phena 22
to 26. The *B.alvei* and *B.coagulans* strains form 2 phena,
22 and 23 respectively, which are homogeneous and
separated both from one another and from the other phena
in the aggregate taxon (Fig. 1). *Bacillus coagulans* has
also been found to be homogeneous on the basis of nutritional [Knight and Proom, 1950; Proom and Knight, 1955;
Gordon *et al.*, 1974] and DNA reassociation [Seki *et al.*,
1978] studies while most strains have base compositions
within the range 44 to 48% GC [Priest, Chapter 3].
Bacillus endorhythmos ATCC 23260 forms a single member
cluster which shares its closest affinity to *B.alvei*
whereas the strain labelled *Lactobacillus cereale* NRS 2012
shows its highest affinity with *B.coagulans* strains.
Lactobacillus cereale may be related to *B.laevolacticus*
and *B.racemilacticus* which have been found to share a high
overall similarity to strains of *B.coagulans* [Logan and
Berkeley, Chapter 6].

It is evident from Fig. 1 that the 25 strains of *B.poly-
myxa* form a homogeneous and discrete phenon that is closely
related to subphenon 25A which contains strains of *B.
macerans*. *B.polymyxa* is widely accepted as a distinct
species and its affinity to *B.macerans* has been established
by several workers [Porter *et al.*, 1937; Sneath, 1962;
Gordon *et al.*, 1973]. Of the 39 *B.circulans* strains, 25
were distributed in subphena 25B, 26A and 26B but strains
NRS 765, NRS 838A were not recovered in any of the defined

phena. The remaining *B.circulans* strains were distributed
in aggregate clusters 1 and 3. The neotype strain, ATCC
4513, was recovered in subphenon 25B showing a close
affinity to *B.macerans.* These findings are in good accord
with other work that has demonstrated the heterogeneity of
B.circulans and its close affinity to *B.macerans* [Knight
and Proom, 1950; Gordon *et al.*, 1973; Gibson and Gordon,
1974] but it is not entirely consistent with DNA analyses.
Although the heterogeneity of *B.circulans* is supported by
the broad range of base composition (35 to 51% GC), the
neotype strain has 36% GC [Marmur and Doty, 1962; Bonde,
1975] considerably lower than that of *B.macerans* strains
(49-54% GC [Normore, 1973; Baptist *et al.*, 1978]). Thus,
despite their phenotypic similarity, the neotype strains of
B.circulans and *B.macerans* are apparently genetically
diverse. A detailed polyphasic approach to the taxonomy
of these bacteria may resolve this anomaly.

Conclusions

The genus *Bacillus* is generally recognized to be more
heterogeneous than most bacterial genera. This is
apparent from the varieties of carbohydrate metabolism,
nutritional requirement for growth and cell wall composition displayed by the individual species [Gibson and
Gordon, 1974]. Furthermore, the range of genome base
composition (32 - 69% GC) indicates considerable heterogeneity as does the very low S-level recorded in this
numerical analysis. This situation has probably developed
from the monothetic description of the genus; namely that
the ability of a rod-shaped cell to form an endospore
aerobically is both sufficient and necessary for member-
ship. Admittedly, sporulation is an important characteristic of these bacteria based, as it is in *B.subtilis*, on
some 42 operons widely dispersed around the chromosome
[Hranueli *et al.*, 1974]. Nevertheless, it would seem that
the endospore and the physiology of its formation have
diverged considerably amongst the Family Bacillaceae.
Although spores from different species have basically the
same structure [Warth, 1978], they differ in topography
[Bradley and Franklin, 1958; Holt and Leadbetter, 1969],
chemical composition [Warth, 1978] and antigenicity [Wolf
and Barker, 1968]. Moreover, it would appear that the
homology between sporulation loci from different species
is negligible since DNA reassociation between strains from
different *Bacillus* species is very low [Seki *et al.*, 1978]
and largely accountable as homology between ribosomal RNA
cistrons [Priest, Chapter 3]. Exceptions are some of the
closely related species, for example *B.subtilis* and *B.
licheniformis* between which DNA homology is relatively
high [Seki *et al.*, 1975].
  Bacillus is therefore partially defined on the basis of
a feature that is in itself heterogeneous. The result is

a genus that is diverse and unwieldy and to accommodate
the problem, considerable merging of species has been
practised [Gibson, 1944; Smith *et al.*, 1952; Gordon *et
al.*, 1973]. There is little doubt that much of this was
very necessary since it eradicated large numbers of poorly
described and often monotypic species. However, several
of the chapters in this book indicate that it was too
extensive and that, to obtain a natural classification of
aerobic spore-formers, splitting is now required. To
accommodate the increased number of species it will be
necessary to define several genera; a proposal that is
clearly indicated by DNA base composition data [Priest,
Chapter 3] and this numerical analysis. For such a classification to be natural, these genera should have polythetic
descriptions and a study designed to provide some of the
necessary data is currently in progress in our laboratories.

Acknowledgements

This work was supported by a grant from the Science
Research Council.

References

Baillie, A. and Walker, P.D. (1968). Enzymes of thermophilic aerobic
spore-forming bacteria. *Journal of Applied Bacteriology* **31**, 114-
119.
Baptist, J.N., Mandel, M. and Gherna, R.L. (1978). Comparative zone
electrophoresis of enzymes in the genus *Bacillus*. *International
Journal of Systematic Bacteriology* **28**, 229-244.
Boeyé, A. and Aerts, M. (1976). Numerical taxonomy of *Bacillus* iso-
lates from North Sea sediments. *International Journal of
Systematic Bacteriology* **26**, 427-441.
Bonde, G.J. (1975). The genus *Bacillus*. An experiment with cluster
analysis. *Danish Medical Bulletin* **22**, 41-61.
Bradley, D.E. and Franklin, J.G. (1958). Electron microscope survey
of the surface configuration of spores of the genus *Bacillus*.
Journal of Bacteriology **76**, 618-630.
Bradley, S.G. (1980). DNA reassociation and base composition. In
"Microbiological Classification and Identification" (eds. M.
Goodfellow and R.G. Board), pp.11-26. London: Academic Press.
Bradley, S.G. and Mordarski, M. (1976). Association of polydeoxyribo-
nucleotides of deoxyribonucleic acids from nocardioform bacteria.
In "The Biology of the Nocardiae" (eds. M. Goodfellow, G.H.
Brownell and J.A. Serrano), pp.310-336. London: Academic Press.
Bulla, Jr., L.A., Costilow, R.N. and Sharpe, E.S. (1978). Biology of
Bacillus popilliae. *Advances in Applied Microbiology* **23**, 1-18.
De Ley, J. (1970). Molecular techniques and applications in bacterial
taxonomy. In "The Actinomycetales" (ed. H. Prauser), pp.317-327.
Jena: Gustav Fischer Verlag.
Denis, F. (1971). Le *Bacillus* du milieu marin: etude de 120 souches.
Comptes Rendus des Séances de la Société de Biologies **165**, 2404-
2408.

Gibson, T. (1934). An investigation of the *B.pasteuri* group: II. Special physiology of the organisms. *Journal of Bacteriology* **28**, 313-322.

Gibson, T. (1935a). The urea-decomposing microflora of soils. I. Description and classification of the organisms. *Zentralblatt für Bakteriologie, Parasitenkunde, Infektionskrankheiten und Hygiene, Abteilung II* **92**, 364-380.

Gibson, T. (1935b). An investigation of the *B.pasteuri* group. III. Systematic relationships of the group. *Journal of Bacteriology* **29**, 491-502.

Gibson, T. (1944). A study of *Bacillus subtilis* and related organisms. *Journal of Dairy Research* **13**, 248-260.

Gibson, T. and Gordon, R.E. (1974). *Bacillus*. In "Bergey's Manual of Determinative Bacteriology. Eighth Edition" (eds. R.E. Buchanan and N.E. Gibbons), pp.529-550. Baltimore: The Williams and Wilkins Co.

Gibson, T. and Topping, L.E. (1938). Further studies of the aerobic spore-forming bacilli. *Proceedings of the Society of Agricultural Bacteriologists*, 43-44.

Goodfellow, M. (1977). Numerical taxonomy of some nocardioform bacteria. *Journal of General Microbiology* **69**, 33-80.

Goodfellow, M. (1971). Numerical taxonomy. In "CRC Handbook of Microbiology, 2nd Edition. Volume I. Bacteria" (eds. A.I. Laskin and H.A. Lechevalier), pp.579-596. Cleveland: CRC Press.

Goodfellow, M. and Alderson, G. (1977). The actinomycete - genus *Rhodococcus*: a home for the '*rhodochrous*' complex. *Journal of General Microbiology* **100**, 99-122.

Gordon, R.E., Haynes, W.C. and Pang, C.H.-N. (1973). "The Genus *Bacillus*". United States Department of Agriculture, Washington, D.C.

Gordon, R.E., Hyde, J.L. and Moore, J.A., Jr. (1977). *Bacillus firmus* - *Bacillus lentus:* a series or one species? *International Journal of Systematic Bacteriology* **58**, 256-262.

Holmberg, K. and Hollander, H.O. (1973). Numerical taxonomy and laboratory identification of *Bacterionema matruchotti, Rothia dentocariosa, Actinomyces naeslundii, Actinomyces viscosus* and some related bacteria. *Journal of General Microbiology* **76**, 43-63.

Holmberg, K. and Nord, C.E. (1975). Numerical taxonomy and laboratory identification of *Actinomyces* and *Arachnia* and some related bacteria. *Journal of General Microbiology* **91**, 17-44.

Holt, S.C. and Leadbetter, E.R. (1969). Comparative ultrastructure of selected aerobic spore-forming bacteria: a freeze-etching study. *Bacteriological Reviews* **33**, 346-378.

Hranueli, D., Piggot, P.J. and Mandelstein, J. (1974). Statistical estimate of the total number of operons specific for *Bacillus subtilis* sporulation. *Journal of Bacteriology* **119**, 684-690.

Jones, D. (1975). A numerical taxonomic study of coryneform and related bacteria. *Journal of General Microbiology* **87**, 52-96.

Jones, D. and Sackin, M.J. (1980). Numerical methods in the classification and identification of bacteria with a special reference to the Enterobacteriaceae. In "Microbiological Classification and Identification" (eds. M. Goodfellow and R.G. Board), pp.73-106. London: Academic Press.

Jones, D. and Sneath, P.H.A. (1970). Genetic transfer and bacterial taxonomy. *Bacteriological Reviews* **34**, 40-81.

Knight, B.C.J.G. and Proom, H. (1950). A comparative survey of the nutrition and physiology of mesophilic species in the genus *Bacillus*. *Journal of General Microbiology* **4**, 508-538.

Larkin, J.M. and Stokes, J.L. (1967). Taxonomy of psychrophilic strains of *Bacillus*. *Journal of Bacteriology* **94**, 889-895.

Marmur, J. and Doty, P. (1962). Determination of the base composition of deoxyribonucleic acid from its thermal denaturation temperature. *Journal of Molecular Biology* **5**, 109-118.

Meissner, G. and Twenty-one Colleagues. (1974). A co-operative numerical analysis of nonscoto- and nonphotochromogenic slowly growing mycobacteria. *Journal of General Microbiology* **83**, 207-235.

Minnikin, D.E., Abdolrahimzadeh, H. and Wolf, J. (1977). Taxonomic significance of polar lipids in some thermophilic members of *Bacillus*. In "Spore Research 1976" Vol. 2 (eds. A.N. Barker, J. Wolf, D.J. Ellar, G.J. Dring and G.W. Gould), pp.879-893. London: Academic Press.

Normore, W.M. (1973). Guanine - plus - cytosine (GC) content of the DNA of bacteria, fungi, algae and protozoa. In "Handbook of Microbiology" (eds. A.I. Laskin and H.A. Lechevalier), Vol. 2, pp.285-740. Cleveland: CRC Press.

O'Donnell, A.G., Norris, J.R., Berkeley, R.C.W., Claus, D., Kanedo, T., Logan, N.A. and Nozaki, R. (1980). Characterization of *Bacillus subtilis*, *Bacillus pumilus*, *Bacillus licheniformis* and *Bacillus amyloliquefaciens* by pyrolysis gas liquid chromatography: characterisation tested using DNA-DNA hybridization; biochemical tests; and API systems. *International Journal of Systematic Bacteriology* **30**, 448-459.

Porter, R., McCleskey, C.S. and Levine, M. (1937). The facultative sporulating bacteria producing gas from lactose. *Journal of Bacteriology* **33**, 163-183.

Proom, H. and Knight, B.C.J.G. (1950). *Bacillus pantothenticus* (n.sp.). *Journal of General Microbiology* **41**, 539-541.

Proom, H. and Knight, B.C.J.G. (1955). The minimum nutritional requirements of some species in the genus *Bacillus*. *Journal of General Microbiology* **13**, 474-480.

Rode, L.J. (1968). Correlation between structure and spore properties in *Bacillus megaterium*. *Journal of Bacteriology* **95**, 1979-1986.

Seki, T., Oshima, T. and Oshima, Y. (1975). Taxonomic study of *Bacillus* by deoxyribonucleic acid - deoxyribonucleic acid hybridization and interspecific transformation. *International Journal of Systematic Bacteriology* **25**, 258-270.

Seki, T., Chung, C.K., Mikami, H. and Oshima, Y. (1978). Deoxyribonucleic acid homology and taxonomy of the genus *Bacillus*. *International Journal of Systematic Bacteriology* **28**, 182-189.

Smith, N.R., Gordon, R.E. and Clark, F.E. (1952). "Aerobic Spore-forming Bacteria". United States Department of Agriculture, Monograph No. 16, Washington, D.C.

Sneath, P.H.A. (1962). The construction of taxonomic groups. In "Microbial Classification" (eds. G.C. Ainsworth and P.H.A. Sneath), pp.289-322. Cambridge: Cambridge University Press.

Sneath, P.H.A. and Sokal, R.R. (1973). "Numerical Taxonomy, The

Principles and Practice of Numerical Classification". San Francisco: W.H. Freeman.

Sokal, R.R. and Michener, C.D. (1958). A statistical method for evaluating systematic relationships. *University of Kansas Science Bulletin* **38**, 1409-1438.

Tsukamura, M. (1966). Adansonian classification of mycobacteria. *Journal of General Microbiology* **45**, 253-273.

Tsukamura, M. (1969). Numerical taxonomy of the genus *Nocardia*. *Journal of General Microbiology* **56**, 265-287.

Walker, P.D. and Wolf, J. (1971). Taxonomy of *Bacillus stearothermophilus*. In "Spore Research 1971" (eds. A.N. Barker, G.W. Gould and J. Wolf), pp.247-262. London: Academic Press.

Warth, A.D. (1978). Molecular structure of the bacterial spore. *Advances in Microbial Physiology* **17**, 1-45.

Watanabe, K. and Takesna, S. (1970). Surface changes of two types of *Bacillus megaterium* spores differing in their response to n-butane. *Journal of General Microbiology* **63**, 375-377.

Wayne, L.G., Andrade, L., Froman, S., Käppler, W., Kubala, E., Meissner, G. and Tsukamura, M. (1978). A co-operative numerical analysis of *Mycobacterium gastri*, *Mycobacterium kansasii* and *Mycobacterium marinum*. *Journal of General Microbiology* **109**, 319-327.

Wayne, L.G. and Eighteen Colleagues. (1976). Highly reproducible techniques for use in systematic bacteriology in the genus *Mycobacterium*: Tests for niacin and catalase and for resistance to isoniazid, thiophene 2-carboxylic acid hydrazide, hydroxylamine, and p-nitrobenzoate. *International Journal of Systematic Bacteriology* **26**, 311-318.

Wayne, L.G. and Nineteen Colleagues. (1974). Highly reproducible techniques for use in systematic bacteriology in the genus *Mycobacterium*: Tests for pigment, urease, resistance to sodium chloride, hydrolysis of Tween 80 and β-galactosidase. *International Journal of Systematic Bacteriology* **24**, 412-419.

Wishart, D. (1968). *A Fortran 11 Program for Numerical Taxonomy*. St. Andrews: University of St. Andrews.

Wolf, J. and Barker, A.N. (1968). The genus *Bacillus*: aids to the identification of its species. In "Identification Methods for Microbiologists, Part B" (eds. B.M. Gibb and B.A. Shapton), pp. 93-109. London: Academic Press.

Wolf, J. and Chowdhury, M.S.U. (1971). Taxonomy of *B.circulans* and *B.stearothermophilus*. In "Spore Research 1971" (eds. A.N. Barker, G.W. Gould and J. Wolf), pp.349-350. London: Academic Press.

Chapter 6

CLASSIFICATION AND IDENTIFICATION OF MEMBERS OF THE GENUS *BACILLUS* USING API TESTS

N.A. LOGAN and R.C.W. BERKELEY

Department of Bacteriology, The Medical School, University Walk, Bristol, UK

Introduction

One of the first attempts to classify aerobic spore-forming bacteria was made by Ford and his associates [Lawrence and Ford, 1916; Laubach *et al.*, 1916] and was stimulated by the state of chaos then prevailing in the systematics of these microorganisms. However, apart from the work of Gibson [1934a, b, 1935a, b, 1937, 1944], who made considerable contributions to the understanding of the *Bacillus pasteurii** and *B.subtilis* groups, little progress was made in the taxonomy of the genus *Bacillus* for 30 years.

In 1946 Smith and his co-workers published a determinative key for the identification of *Bacillus* species [Smith *et al.*, 1946]; an updated version followed 6 years later [Smith *et al.*, 1952]. These publications were based on a study of 1134 strains bearing 158 species names. All but 20 of the strains could be assigned to 19 species [Gordon *et al.*, 1973]. The large numbers of synonyms arose because, in the early literature, species descriptions were often poor, and later authors, when describing new species, could not work from them, or else disregarded them completely [Smith *et al.*, 1952]. In addition, the extent of strain variation was not always fully appreciated [Smith *et al.*, 1952; Norris and Wolf, 1961; Gordon *et al.*, 1973; Wilson and Miles, 1975].

In the classifications of Ford, Gibson and Smith, emphasis was placed on the use of spore size, shape and position in identification. Smith *et al.* [1946, 1952] divided the genus into 3 groups on this basis and, in a revised and supplemented report of Smith's work, Gordon *et al.* [1973] continued to use this division.

---

*References to original descriptions of species can be found in the 8th edition of "Bergey's Manual of Determinative Bacteriology" [Gibson and Gordon, 1974] or in Gordon *et al.* [1973] unless otherwise indicated.

Despite such a scheme, the soundness of which has been
proven by experience and further taxonomic studies [see
Wolf and Barker, 1968], *Bacillus* species have been, and
still are, identified no further than "aerobic spore-
forming rod" in many laboratories, especially medical ones
[Weinstein and Colburn, 1950;   Allen and Wilkinson, 1969;
Ormay and Novotny, 1969;   Bisset and Bartlett, 1978].   In
many instances the name *B.subtilis* has been used as a
synonym for 'aerobic spore-forming rod' [Lawrence and Ford,
1916;   Soule, 1928;   Clark, 1937;   Davenport and Smith,
1952;   Burdon and Wende, 1960;   Goepfert *et al.*, 1972].
   Smith *et al.* [1952], Gordon [1973] and Gordon *et al.*
[1973] were aware that their keys did not enable the identification of atypical strains of the recognized species,
a problem experienced by Bonde [1965, 1973, 1975], and
they emphasized the importance of subjecting unknown
strains to many tests, and of studying known strains at
the same time for comparison.   Notwithstanding such advice,
new species are still described without appropriate
reference strains having been examined [see Gordon *et al.*,
1973;   Gordon, 1975].   The classification and identification of *Bacillus* species is still complicated [Goepfert *et
al.*, 1972;   Green, 1975].   This, and the belief that all
Bacillus species (other than *B.anthracis* which nevertheless may have been missed, in some instances, as a result)
isolated from clinical material, are contaminants [Wein-
stein and Colburn, 1950;   Farrar, 1963;   Allen and Wilkinson, 1969;   Ihde and Armstrong, 1973;   Turnbull *et al.*,
1977;   Bisset and Bartlett, 1978] has meant that the
routine identification of aerobic spore-forming rods has
been greatly neglected.   In this chapter the results of
studies aimed at facilitating this task are described.
The primary objective was to enable identification of those
organisms currently allocated to the genus *Bacillus* using
a series of reproducible, rapid and inexpensive tests.
   Although there is ample evidence that this genus is
heterogeneous [Gibson and Gordon, 1974;   Bradley, 1980;
Priest, Chapter 3] it is not intended to base any reallo-
cation of species solely upon the data described below.

Reproducibility of Classical Tests

   In 1975 the International Committee on Systematic
Bacteriology (ICSB) Sub-Committee on the genus *Bacillus*
prepared a list of methods for performing tests, classi-
cally used in the taxonomy of this group, drawn largely
from the monograph of Gordon *et al.* [1973].   These tests
were examined in an international reproducibility trial in
order to evaluate their usefulness as a set of standard
methods for identifying *Bacillus* strains.
   Five laboratories, expert in the identification of
aerobic spore-forming bacteria, subjected 18 coded representative strains of *Bacillus* to 41 different types of

TABLE 1

Reproducibility of classical Bacillus *tests performed by 5 laboratories on 18 representative* Bacillus *strains*[a]

Test[b]	Number of character states	Overall disagreement[c]
Shape of cells	6	1
Length of cells	4	30
Width of cells	3	5
Spore formation	1	1
Shape of spores	3	12
Position of spores	3	17
Spore diameters ≤0.9μ	1	0
Swelling of sporangium	1	1
Presence of parasporal crystals	1	0
Gram strain	4	5
Presence of shadow forms	1	15
Motility	1	5
Cells occurring singly	1	0
Cells in chains	1	0
Aerobic/anaerobic growth	4	22
Growth at pH 5.7	1	9
Growth in azide dextrose broth	1	0[d]
Resistance to lysozyme	1	8
Growth in nutrient broth	1	0
Maximum and minimum growth temperatures	10	14
Growth in NaCl	4	11
Casein decomposition	1	0
Gelatin liquefaction	1	1
Dihydroxyacetone production	1	6
Indole production	1	0
Nitrate reduction	3	2
Catalase	1	0
Reaction in litmus milk	5	31
Voges-Proskauer (V-P)	1	1
Final pH in V-P broth	9	15
Hippurate hydrolysis	1	5
Acid from carbohydrates	9	16
Gas from carbohydrates	9	4
Starch hydrolysis	1	3
Propionate utilization	1	6
Citrate utilization	1	5
Tyrosine decomposition	1	2
Phenylalanine deamination	1	1
Lecithinase	1	3
Crystalline dextrins	1	0[d]
Vacuoles in vegetative cells	1	0

[a]Reproducibility trial organized by ICSB Sub-Committee on the genus *Bacillus*.
[b]For details of test methods see Gordon *et al.* [1973].
[c]A disagreement is defined as 3 positive and 2 negative scores, or vice versa, for each test on each strain.
[d]Test not performed by all participating laboratories.

test. The results are summarized in Table 1. Since not
all of the participants were able to perform every test,
the figures quoted are minimum estimates of disagreement.

Observations on the shape and position of spores were
inconsistent largely because such determinations are subjective. In addition it appeared that, for measurements of
cell size, calibration was a problem and Sneath and Collins
[1974] noted the same difficulty in a report of a *Pseudomonas* Working Party test reproducibility study. Work in
this laboratory has indicated that eyepiece micrometers,
and photographic methods, for cell measurement, are unsatisfactory and that the use of a carefully calibrated
image-splitting eyepiece [see Quesnel, 1971] gives more
reproducible results.

Sneath and Collins [1974] noted that laboratories
employed a variety of descriptive terms and they considered
that the use of a standard glossary might improve the
reproducibility of observations on cell morphology.
Similar problems were encountered in the *Bacillus* work and
the need for clearer definitions of descriptive terms was
indicated.

The test for presence of shadow forms ("ghost cells")
was found to be completely unreliable and tests for anaerobic growth, growth at pH 5.7, growth in azide dextrose
broth, resistance to lysozyme, growth in different concentrations of NaCl, casein decomposition, gelatin liquefaction, acid from carbohydrates, gas from carbohydrates,
lecithinase and crystalline dextrins, all gave problems in
standardization of media and were, in most cases, not
reproducible as a result. In several instances (azide,
lysozyme and lecithinase) a particular brand of ingredient
was not universally obtainable, and the vessels and
closures specified in some tests were not available in all
laboratories. In the tests for maximum and minimum growth
temperatures, not all ten temperatures could be accurately
maintained by some participants; in the tests for
dihydroxyacetone production, reaction in litmus milk,
final pH in Voges-Proskauer broth, starch hydrolysis and
propionate and citrate utilization, difficulties in
reading and interpreting the reactions were encountered.
It is evident that the preparation and use of special test
media for *Bacillus* is extremely time-consuming, expensive
and inconvenient when use is infrequent, and this may
explain why members of this genus are not always fully
identified in routine laboratories.

In an attempt to overcome the problem of test media
standardization the possibility of using the miniaturized
test materials of the API systems to identify *Bacillus*
species was investigated.

The API System

The quality control measures taken throughout the manufacture of API materials result in highly standardized test media which should considerably improve test reproducibility [Janin, 1976]. In addition, a long shelf life (approximately 18 months), rapidity and convenience in use, applicability to a wide range of organisms, and comparative cheapness [see Smith *et al.*, 1972; Cox *et al.*, 1976; Miller and Lu, 1976; Robertson *et al.*, 1976, Kilian, 1978] make these materials very attractive, especially to the occasional user. Preliminary studies, with 75 strains including strains representative of the genus, indicated that separation of *Bacillus* species could indeed be achieved using the API 20 Enterobacteriaceae strip, the API 50 Enterobacteriaceae gallery and the API ZYM strip in conjunction with four further enzyme test strips, ZYM II, Aminopeptidase (AP)1, AP2 and AP3, which are not commercially available at the time of going to press.

The API 20E strip is a ready-to-use microtube system, developed from Buissière's [1972] modification of the Ivan Hall tube, and contains dehydrated media for performing 23 standard biochemical tests. Twelve of these tests were used and, of the remainder, 9 were duplicated by the API 50E gallery. The latter is similar to the API 50E strip but contains 49 tests, 38 of which are carbohydrate fermentations, and one control.

The API ZYM test strip developed from the Auxotab system described by Buissière *et al.* [1967], comprises a series of semi-quantitative micromethods designed for the detection of enzymatic activities in a wide variety of biological specimens. Each strip consists of 20 cupules containing 19 test substrates and one control. The other 4 enzyme test strips have a similar format and each contains 10 substrates (except the AP2 strip which contains 9 substrates and one control).

The API 20E and API 50E tests and the enzyme test substrates are listed in Table 2.

Reproducibility of API Tests

The reproducibility of the selected API tests when applied to *Bacillus* species was investigated in an international trial in which 6 laboratories applied the 119 API tests to 60 coded, representative strains of *Bacillus*. Three of the participating laboratories were not particularly experienced with the genus *Bacillus* and 3 were not familiar with API systems. The results of the tests were subjected to analyses of variance to give an indication of the relative effects of haphazard and systematic error on reproducibility [see Sneath and Collins, 1974]. The results of API tests were not distributed normally; it was necessary, therefore, to interpret F ratio values with great care. A test was regarded as reproducible if its variation was not significant at the 5% level.

In the API 20E strip the reproducibility of 5 tests was

TABLE 2

Tests in the API system used for this study

API 20E

1.	ONPG	7.	urease
2.	arginine dihydrolase	8.	tryptophan deaminase
3.	lysine decarboxylase	9.	indole
4.	ornithine decarboxylase	10.	Voges-Proskauer
5.	Simmons' citrate	11.	gelatin liquefaction
6.	hydrogen sulphide	12.	nitrate reduction

API 50E

0.	control	25.	aesculin
1.	glycerol	26.	salicin
2.	erythritol	27.	D(+) cellobiose
3.	D(-) arabinose	28.	maltose
4.	L(+) arabinose	29.	lactose
5.	ribose	30.	D(+) melibiose
6.	D(+) xylose	31.	saccharose
7.	L(-) xylose	32.	D(-) trehalose
8.	adonitol	33.	inulin
9.	methyl-D-xyloside	34.	D(+) melezitose
10.	galactose	35.	D(+) raffinose
11.	D(+) glucose	36.	dextrin
12.	D(-) fructose	37.	amylose
13.	D(+) mannose	38.	starch
14.	L(-) sorbose	39.	glycogen
15.	rhamnose	40.	methyl red
16.	dulcitol	41.	DNase
17.	*meso*-inositol	42.	mucate
18.	mannitol	43.	gluconate
19.	sorbitol	44.	lipase
20.	methyl-D-mannoside	45.	tetrathionate reductase
21.	methyl-D-glucoside	46.	pectate
22.	N-acetyl-D-glucosamine	47.	citrate (Christensen)
23.	amygdalin	48.	malonate
24.	arbutin	49.	acetate

ENZYME TESTS
API ZYM

1.	control	11.	phosphatase (acid)
2.	phosphatase (alkaline)	12.	phosphoamidase
3.	esterase (C_4)	13.	α-D-galactosidase
4.	esterase lipase (C_8)	14.	β-D-galactosidase
5.	lipase (C_{14})	15.	β-D-glucuronidase
6.	L-leucine aminopeptidase	16.	α-D-glucosidase
7.	L-valine aminopeptidase	17.	β-D-glucosidase
8.	L-cystine aminopeptidase	18.	N-acetyl-β-D-glucosaminidase
9.	trypsin	19.	α-D-mannosidase
10.	chymotrypsin	20.	α-L-fucosidase

TABLE 2   (Cont'd)

The following are not commercially available at present

ZYM II

1. β-D-xylosidase
2. phosphodiesterase
3. α-D-xylosidase
4. β-D-fucosidase
5. β-L-fucosidase
6. N-acetyl-α-D-glucosaminidase
7. lactosidase
8. sulphatase
9. α-L-arabinosidase
10. β-D-cellobiosidase

AP1 Aminopeptidase substrates

1. L-tyrosyl-β-NA*
2. pyrrolidonyl-β-NA
3. L-phenylalanine-β-NA
4. L-lysine-β-NA
5. L-hydroxyproline-β-NA
6. L-histidine-β-NA
7. glycine-β-NA
8. L-aspartyl-β-NA
9. L-arginyl-β-NA
10. L-alanyl-β-NA

AP2 Aminopeptidase substrates

1. N-γ-L-glutamyl-β-NA
2. N-benzoyl-L-leucyl-β-NA
3. S-benzyl-L-cysteine-β-NA
4. DL-methionyl-β-NA
5. glycylglycine-β-NA-H-Br
6. glycyl-L-phenylalanyl-β-NA
7. glycyl-L-prolyl-β-NA
8. L-leucylglycyl-β-NA
9. L-seryl-L-tyrosyl-β-NA
10. control

AP3 Aminopeptidase substrates

1. N-CBZ-L-arginine-4-
   methoxy-β-NA
2. L-glutamine-β-NA-HCl
3. α-L-glutamyl-β-NA
4. L-isoleucine-β-NA
5. L-ornithine-β-NA
6. L-proline-β-NA
7. L-serine-β-NA
8. L-threonine-β-NA
9. L-tryptophan-β-NA
10. N-CBZ-glycylglycyl-L-
   arginine-β-NA

---

*β-NA = β-naphthylamide.

good. Three tests, gelatin liquefaction, Voges-Proskauer
(V-P) and nitrate reduction, were less consistent;   the
last two require the addition of reagents to reveal the
reactions and the need for greater care and standardization in the addition of reagents, and in the reading of
results, was indicated. Four tests (lysine decarboxylase,
ornithine decarboxylase, H_2S production and tryptophan deamination) had no diagnostic value because all strains
gave the same results. Other authors, working on Enterobacteriaceae, have also reported inconsistent results with
the API 20E gelatin liquefaction [Butler *et al.*, 1975;
Marymont *et al.*, 1978] and V-P tests [Butler *et al.*, 1975;
Marymont *et al.*, 1978; Murray, 1978] and difficulties
with the nitrate reduction test [Washington *et al.*, 1971;
Holmes *et al.*, 1978].

Sixteen tests in the API 50E gallery gave inconsistent results due, it appeared, to disparities in the judgement of end points. Two tests (erythritol and dulcitol) were without diagnostic value because all strains gave the same results.

Of the 58 enzyme tests, 20 were not satisfactorily reproducible and 12 had no diagnostic value. The difficulties encountered with these enzyme tests were probably due to variations in inoculum size and problems in end point assessment. In addition, some of the substrates of the ZYM II strip were photolabile and, unless low lighting conditions were employed, false positive results could be obtained.

Inconsistent results were obtained with 32% of the API tests compared with 61% of the classical tests.

Many of the factors that underlay test inconsistency were highlighted as a result of the trial and attempts to eliminate them were made in a smaller, intra-laboratory study. The 119 tests were each made at 3 different times on 20 coded, representative strains and the results subjected to analyses of variance as before.

Test reproducibility was increased but, in the API 20E strip, the V-P and nitrate reduction tests still showed some inconsistency. All of the tests in the API 50E gallery gave reproducible results, with 3 tests (erythritol, D-arabinose and amylose) now appearing to be of no diagnostic value. Fifty-two of the enzyme tests gave reproducible results and 2 were poorly reproducible. The improvement in results with the enzyme tests was largely due to careful standardization of inoculum. Four had no diagnostic value, but are not named here as later work, with a wider variety of strains, indicated that all of these tests may be useful. Bonde [1978] considered the commercial API ZYM system to be of little use in *Bacillus* taxonomy; he found 2 tests (α- and β-D-glucosidases) to be of diagnostic value but all others were consistently positive or negative for all strains. Westley *et al.* [1967], however, considered that profiles of aminopeptidase activity could be useful for the identification of *Bacillus* species.

In the intra-laboratory trial the haphazard variation was of the same order as between-replicate variation for several tests suggesting that any further improvement of such tests might be difficult [Sneath and Collins, 1974]. It was concluded that more detailed specifications of procedures, and greater attention to directions for performing the tests would reduce discrepancies in results [Logan *et al.*, 1978].

The findings of the reproducibility trials indicated that the API tests were more reproducible than the classical *Bacillus* tests.

Taxonomy Based on API Tests

Further work was undertaken to investigate the usefulness of the API tests in the erection of a taxonomic
scheme as a basis for the identification of those
organisms currently grouped in the genus *Bacillus*. For
such a scheme considerations of vigour and pattern [see
Sneath, 1968] are not so important as they would be if an
attempt to erect a formal taxonomy was being made.

A total of 600 cultures of aerobic spore-forming rods
were collected and studied using these methods; 509 of
the strains bore species names on receipt and 91 were
received unnamed. One hundred and thirty-seven of the
test strains were recent isolates from clinical specimens,
food, beverages, food poisoning outbreaks, anthrax outbreaks, industrial plant and salt marshes. Thirty-four
strains, chosen to represent the recognized species, were
duplicated and carried through separately to provide an
internal check on the study.

The fastidious insect pathogens, *B.larvae*, *B.popilliae*
and *B.lentimorbus*, and species requiring a very low or
high pH for growth (*B.acidocaldarius* and *B.alcalophilus*
respectively) were omitted from this first taxonomic study.

In addition to the 119 API tests, 20 morphological and
physiological tests were selected from the list prepared
by the ICSB Sub-Committee on the genus *Bacillus* on the
basis that they had high separation values and were reproducible and convenient. These tests covered spore and
vegetative cell morphology, catalase, motility, and gas
production from carbohydrates [for detailed list see
Logan, 1980]. The results of the API tests were recorded
in a ranked form (0, negative; to 5, strong positive) and
a range specified, larger than the sample range, so that
operational taxonomic units (OTUs) at the extremes did not
show zero similarity [Gower, 1978; Ware and Hedges, 1978].
Cell measurements were also recorded in a six category
fashion but other tests were scored in a binary (0,
negative; 1, positive) form.

Similarity coefficients were computed using the general
similarity coefficient (S_G) of Gower [1971]. Clusters
were formed by unweighted pair-group average linkage
analysis [Sokal and Michener, 1958] using the CLASP
program (written by G.J.S. Ross, F.B. Lauckner and D.
Hawkins, Rothamsted Experimental Station, Harpenden, Hert-
fordshire, UK) and run on the ICL 4-75 computer at the
University of Bristol.

From a first analysis of 244 representative strains 103
were chosen, from well-defined clusters, to act as
reference OTUs. Random samples of the remaining 497
strains were then run, with the 103 reference OTUs, until
all the data had been analyzed [Sneath, 1964; Sneath and
Sokal, 1973].

Hierarchical clustering techniques generally reproduce

distances between close neighbours faithfully [Sneath and Sokal, 1973] but give poor representation of relationships between major clusters [Rohlf, 1967, 1968, 1970]. However, ordination methods such as principal components analysis and principal coordinate analysis tend to falsify distances between close neighbours [Webb *et al.*, 1967; Rohlf, 1968; Sneath and Sokal, 1973]. For the purposes of publication, Sneath and Sokal [1973] recommended the use of both types of analysis, a phenogram being useful for summarizing the taxonomic relationships, with ordinations being valuable for understanding the taxonomic structure in greater detail. This approach was adopted in this chapter, the cluster analysis serving principally as a basis for breaking the set of strains into groups of manageable size for principal coordinate analysis.

A phenogram, summarizing the results of 4 computer runs with the 103 reference OTUs, is shown in Fig. 1. The 34 pairs of duplicate cultures showed similarities between 94% and 99% to their respective partners, giving an average discrepancy of 2.5% (p 1.2%) which is acceptably small [Sneath and Johnson, 1972].

The effect of using S_G, and of specifying a data range larger than the sample range, was to maximize within-cluster and between-cluster similarities. This has resulted in all the clusters joining above the 82% similarity (S) level. It can be seen that the phenogram falls into 2 aggregate clusters of approximately equal size, each taxon divisible into 3 groups of clusters (Groups I to VI) which are of convenient size for principal coordinate analysis.

The division into 6 groups reflects the activities of the strains on carbohydrates: Group I comprises *B.cereus* and related organisms, which produce acid from a limited number of carbohydrates; Group II contains strains of *B. brevis*, *B.firmus*, *B.lentus* and *B.alvei*, which show ranged activity (with *B.brevis* showing the lowest); Group III, which includes *B.sphaericus* and *B.pasteurii* strains, corresponds with Group 3 of Smith *et al.* [1946, 1952] and its members have a tendency to biochemical inertness [Wolf and Barker, 1968]. In the other part of the phenogram, Group IV consists of the "*B.subtilis* spectrum" [Gibson, 1944; Gordon *et al.*, 1973] and *B.megaterium* strains, which show fermentative activity against a wide range of carbohydrates; Group V comprises the "*B.circulans* complex" [Gibson and Topping, 1938] and strains of *B.polymyxa*, *B. macerans* and *B.macquariensis*, all of which are highly active in carbohydrate breakdown; Group VI contains strains of the thermophilic species *B.stearothermophilus* and *B.coagulans* and the mesophilic species *B.pantothenticus*, all producing acid, without gas, from many carbohydrates.

The clustering pattern shown in Fig. 1 agrees, broadly speaking, with the classifications of Smith *et al.* [1946, 1952] and Gordon *et al.* [1973], though considerable mixing

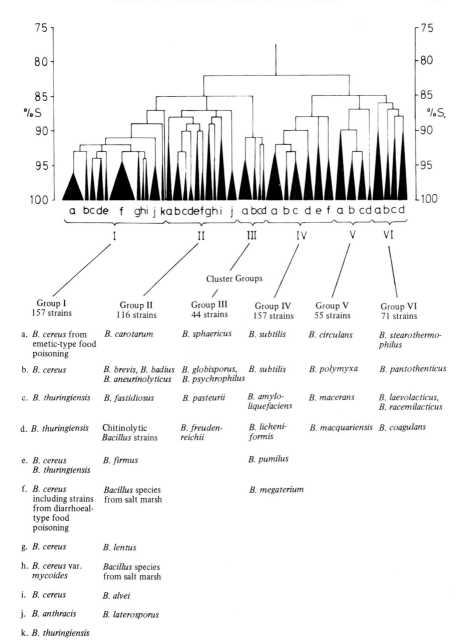

Fig. 1. Phenogram from an average linkage analysis of 600 strains of *Bacillus* based on the results of 139 observations with similarities computed using the general similarity coefficient (S_G) of Gower [1971].

of their morphological Groups 1 (sporangia not definitely swollen by oval to cylindrical spores) and 2 (sporangia definitely swollen by oval spores) has occurred. *Bacillus cereus* and *B.subtilis* strains (both belonging to morphological Group 1) are in separate parts of the phenogram; Cluster Groups I and IV respectively. The separation of these groups of species is in agreement with the findings of Sneath [1962], Somerville and Jones [1972] and Seki *et al.* [1975, 1978]. The analysis of Sneath was based largely on the published results of Smith *et al.* [1952] and was unweighted so that morphological criteria were not emphasized. In this book (Chapter 5) Priest and his colleagues also report an analysis of published data [Gordon *et al.*, 1973]; which is substantially the same as that of Smith *et al.* [1952]. Despite the relatively few tests available to them, the grouping by Priest and his co-workers shows a surprising measure of agreement with that published here.

Members of morphological Group 2 are to be found in cluster Groups II (*B.brevis*, *B.alvei* and *B.laterosporus*), V (*B.circulans*, *B.polymyxa* and *B.macerans*) and VI (*B. stearothermophilus*, *B.coagulans*). The inclusion of *B. pantothenticus* in cluster Group VI is in disagreement with Smith *et al.* [1952], who placed it in morphological Group 3 (sporangia swollen by round spores), but in accordance with Wolf and Barker [1968] who transferred it to Smith's Group 2 on account of its biochemical activity.

Cluster Group II of Fig. 1 is somewhat heterogeneous and is probably the least satisfactory group.

In the consecutive computer runs with reference OTUs (see above) *B.alvei* and *B.lentus* usually clustered together but sometimes linked with the *B.circulans* group and, at other times, with *B.firmus*. Smith *et al.* [1946, 1952], Knight and Proom [1950], Bradley and Franklin [1958], Gordon *et al.* [1973] and Gibson and Gordon [1974] noted the similarity of *B.alvei* to *B.circulans* and Gordon *et al.* [1973, 1977] and Gibson and Gordon [1974] questioned the status of *B.lentus* as a clearly defined species rather than part of the *B.firmus* spectrum.

B.laterosporus strains formed a tight cluster and showed little affinity to any other species.

Smith *et al.* [1952] placed *B.firmus* and *B.lentus* in morphological Group 1 but close to the border with morphological Group 2. Gordon *et al.* [1973], however, considered the retention of *B.firmus* in morphological Group 1 to be unsatisfactory but could not locate it elsewhere, having examined an inadequate number of strains.

Principal Coordinate Analyses

Members of each of the cluster groups I to VI were subjected to principal coordinate analysis [Gower, 1966] to show, more clearly, the relationships existing within these groups. By this method a three-dimensional ordination may

be employed to summarize a multidimensional structure
representing the relationships implied by the entire set
of characters.

Cluster Group I

Sneath and Sokal [1973] pointed out the danger of
establishing taxa simply from the subjective inspection of
ordinations, but it is evident from Fig. 2 that Group I
has split into 3 and that the strains of *B.anthracis* bear
less affinity to those of *B.cereus* than do strains of
B.cereus var. *mycoides*, *B.praussnitzi* and *B.thuringiensis*.
Smith *et al.* [1946, 1952] designated *B.anthracis*, *B.thu-
ringiensis* and *B.mycoides* as varieties of *B.cereus*, and
B.albolactis and *B.fluorescens* as biotypes. Gordon *et al.*
[1973] also adopted this nomenclature because they
"preferred to emphasize the similarities among strains
rather than their differences". The remaining named
strains in Cluster Group I were received as *B.praussnitzi*
which is a synonym of *B.mycoides* [Laubach *et al.*, 1916;
Smith *et al.*, 1952].

The nomenclature of Smith *et al.* [1952] is supported by
the results shown in Fig. 2, with the exception that *B.
anthracis* forms a clearly separated group comprising 20
strains known to have been virulent and 2 avirulent
strains. More avirulent *B.anthracis* strains, however,
must be studied to confirm this [Logan, 1980].

These findings support the work of Burdon [1956] who
considered that it was always possible to differentiate
between avirulent *B.anthracis* strains and *B.cereus* strains,
provided that reference strains were available for comparison, and he explicitly disagreed [Burdon and Wende,
1960] with Smith's designation of *B.anthracis* as a variety
of *B.cereus*. Smith *et al.* [1952] conceived of *B.cereus* as
a stable "parent" species with *B.cereus* var. *anthracis*
becoming *B.cereus* when its virulence was lost. The
species status of *B.anthracis* was also supported by Proom
and Knight [1955], on the basis of their nutritional
studies, and by Sterne and Proom [1957], Ivánovics and
Földes [1958] and Leise *et al.* [1959] as a result of consideration of other physiological characters. In contrast,
Dowdle and Hansen [1961], using the phage-fluorescent antiphage reaction, obtained results indicating that organisms
named *B.anthracis* and *B.cereus* formed a continuous
spectrum. Furthermore, Lammana and Eisler [1960] failed
to distinguish between spores of the two species by agglutination tests.

In DNA reassociation studies, Somerville and Jones
[1972] and Kaneko *et al.* [1978] found a close relationship
between *B.cereus*, *B.anthracis* and *B.thuringiensis* and the
latter group considered that strains in these taxa should
all be classified as *B.anthracis*.

Gordon *et al.* [1973] believed that the status of

B.*anthracis* might, in the future, "still depend on the
judgment of the taxonomist - whether he will emphasize
either the similarities or the differences among strains".
There are historical and medical arguments in favour of
recognizing B.*anthracis* as a valid species and it is
largely for these reasons [Gordon, 1975] that it retains
this status in the 8th edition of "Bergey's Manual of
Determinative Bacteriology" [Gibson and Gordon, 1974].

The evidence presented above, however, seems to indicate
that there are also scientific reasons for its separate
status but only a limited number of strains has, so far,
been examined and a definite statement cannot be made at
this time.

On the basis of DNA reassociation data Seki *et al*. [1978]
considered the insect pathogen B.*thuringiensis* to be a
variant of B.*cereus* and noted a low homology between the
B.*cereus* group and the other species of *Bacillus*; a point
also made by Somerville and Jones [1972] and supported by
the data presented here in the phenogram (Fig. 2).

The third group in Fig. 2 comprises B.*cereus* strains
isolated in connection with emetic-type food poisoning outbreaks usually associated with cooked rice [Goepfert *et al*.,
1972; Gilbert and Taylor, 1976; Gilbert *et al*., Chapter
12] and other strains of serotypes 1, 3, 5 and 8 [Taylor
and Gilbert, 1975] commonly associated with such outbreaks
[Taylor and Gilbert, 1975; Gilbert and Parry, 1977].
Other B.*cereus* strains were isolated in connection with
outbreaks of a diarrhoeal-type food poisoning associated
with a wide variety of foods; these strains lie within the
main B.*cereus* group and cannot be separated from it using
API tests [Logan *et al*., 1979].

It must be emphasized that although these strains were
isolated in connection with food poisoning outbreaks, they
were not necessarily the causal agents.

Gilbert and Taylor [1976] noted that, like B.*anthracis*,
B.*cereus* strains incriminated in emetic-type food poisoning
outbreaks usually failed to ferment salicin whereas other
strains of B.*cereus*, including those from diarrhoeal-type
outbreaks, usually did so. Raevuori *et al*. [1977] found
that strains from emetic-type outbreaks failed to produce
acid from mannose, but that 40% of other B.*cereus* strains
were positive in this test. Similar results were obtained
using the API systems; also strains from the emetic-type
outbreaks, and of serotypes 1, 3, 5 and 8, did not produce
acid from dextrin, starch and glycogen while most other
members of the B.*cereus* group were positive for these
tests. The API system appears, therefore, to be of value
in the presumptive identification of strains able to
produce the emetic toxin or potentially capable of so
doing [Logan *et al*., 1979].

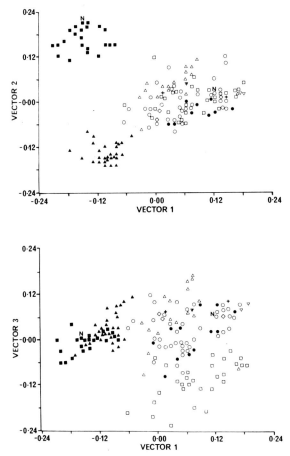

Fig. 2. Plot of the first 3 vectors, accounting for 38% of the total variation, from a principal coordinate analysis of Cluster Group I of Fig. 1. *Bacillus anthracis,* ■ ; *B.cereus,* ○ ; *B.cereus* strains from diarrhoeal-type food poisoning outbreaks, ● ; *B.cereus* strains from emetic-type food poisoning outbreaks, and serotypes 1, 3, 5 and 8, ▲ ; *B.cereus* var. *albolactis,* ◇ ; *B.cereus* var. *fluorescens,* ▽ ; *B. cereus* var. *mycoides,* △ ; *B.praussnitzi,* + ; *B.thuringiensis,* □ ; *B.filicolonicus,* ▼ ; neotype strain, N.

Cluster Group II

Strains of *B.laterosporus* form a group clearly separated from the other members of cluster Group II (Fig. 3) and show little affinity to any other species of *Bacillus*. Seki *et al.* [1978] found low homology between a strain of *B.laterosporus* and other *Bacillus* species in DNA reassociation studies.

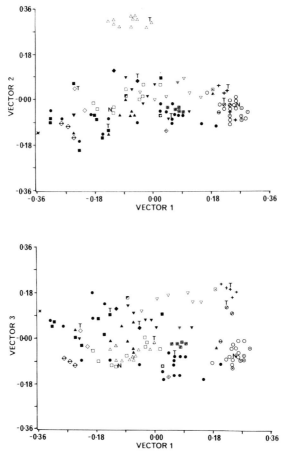

Fig. 3. Plot of the first 3 vectors, accounting for 41% of the total
variation, from a principal coordinate analysis of Cluster Group II of
Fig. 1. *Bacillus firmus,*● ; *B.lentus,*■ ; SM strains,▲ ; *B.
brevis,*○ ; *B.alvei,*□ ; *B.laterosporus,*△ ; *B.carotarum,*▽ ; *B.
insolitus,*◆ ; *B.psychrosaccharolyticus,*◇ ; *B.fastidiosus,* + ;
B.badius,⊘ ; *B.similibadius,*⊖ ; *B.aneurinolyticus,*⊙ ; *B.thi-
aminolyticus,*⬦ ; chitinoclastic *Bacillus* species,⊞ ; *B.macroides,*
⊡ ; *B.cirroflagellosus,*▲ ; *B.medusa,*▼ ; *B.nitritollens,*◈ ;
B.longissimus, × ; *B.epiphytus,*⊡ ; *B.maroccanus,*◪ ; unidentified
or misnamed *Bacillus* species,▼ ; neotype strain, N; type strain, T.

In Sneath's [1962] analysis, based principally on the
data of Smith *et al.* [1952], *B.laterosporus* clustered with
B.alvei and both joined to *B.brevis*. This result is in
accordance with the relationship postulated by Smith *et al.*
[1952] which led to an arrangement similar to that shown in
Fig. 3. The strains of *B.alvei* form a loose group lying

close to the *B.firmus/B.lentus* spectrum but further interpretation is restricted by the small number of strains
studied here.

Strains of *B.brevis* form a tight group which also contains 5 strains of *B.aneurinolyticus*, a species distinguished from *B.brevis*, in the 8th edition of "Bergey's Manual
of Determinative Bacteriology" [Gibson and Gordon, 1974]
solely by its lack of action on casein. Lying between the
B.brevis group, and 5 strains of *B.fastidiosus*, are 2
strains of *B.badius*, 1 of *B.similibadius* [Delaporte, 1972],
and 1 of *B.macroides*. The relationship between *B.aneurino-
lyticus*, *B.badius* and *B.macroides* has been discussed by
Gordon *et al*. [1973] and Gibson and Gordon [1974] who
found few differences between the species but could come to
no conclusions after studying only a small number of
strains. The only strain of *B.cirroflagellosus* studied
lies near the *B.brevis* group.

A spectrum-like arrangement of strains of *B.firmus*,
B.lentus and pigmented isolates (SM strains) from salt
marsh and sea water [Turner and Jervis, 1968] observed by
Gordon *et al*. [1977], supported by the findings of Bonde
[1973, 1975, 1978], is also to be seen in Fig. 3. Most
strains of *B.firmus* show relatively low biochemical
activity and lie near the *B.brevis* group. *B.lentus*
strains, however, have a higher biochemical activity and
lie nearer the other end of the vector 1 axis. The SM
strains show intermediate properties connecting the *B.
firmus* and *B.lentus* groups. This picture is, however,
confused by other strains which appear to be similar to
members of the *B.firmus/B.lentus* spectrum. Thus, chitinoclastic strains from soil form a tight group lying close
to strains of *B.firmus*. Single strains of *B.epiphytus* and
B.nitritollens [Delaporte, 1972] lie near the *B.firmus*
group. Gibson and Gordon [1974] and Bonde [1976] also
noted a similarity between *B.epiphytus* and *B.firmus*.
Single strains of *B.medusa* and *B.maroccanus* lie near the
B.firmus end, and the centre, of the *B.firmus/B.lentus*
spectrum, respectively.

The representation of a species by only one or two
strains is not satisfactory [Gordon *et al*., 1973, 1977]
and for this reason many of the "species" appearing in
Fig. 3 would have been listed in the 8th edition of
"Bergey's Manual of Determinative Bacteriology" as *species
incertae sedis* [Gibson and Gordon, 1974]. Until further
isolates are available for study it is not possible to
establish the taxonomic status of these strains.
Organisms falling into this category include *B.aneurino-
lyticus*, *B.badius*, *B.cirroflagellosus*, *B.epiphytus*, *B.
insolitus*, *B.maroccanus*, *B.macroides*, *B.psychrosaccharo-
lyticus* and *B.thiaminolyticus*. Other such strains, not
listed in the 8th edition of "Bergey's Manual of Determinative Bacteriology", are *B.nitritollens* [Delaporte, 1972]
and *B.longissimus* [Mishustin and Tepper, 1948].

Gordon *et al.* [1973] and Gibson and Gordon [1974] noted a similarity between *B.thiaminolyticus* and *B.alvei*. In Fig. 3 the *B.alvei* group seems to be somewhat heterogeneous with the 3 strains of *B.thiaminolyticus* separated from it. Hayashi and Nakayama [1953], however, reported that *B.thiaminolyticus* was variable in some of the features by which it was distinguished from *B.alvei*. Study of a much larger number of strains of each species is indicated.

Six strains received as *B.carotarum* show differing properties and this may be a reflection of the considerable variation, albeit chiefly morphological, that Gibson [1935a] observed in strains allocated to this species. Smith *et al.* [1946, 1952] accepted *B.carotarum* as a synonym of *B.megaterium*, but other workers [Bonde, 1973, 1975; Gibson and Gordon, 1974; Hunger and Claus, 1978, Chapter 9] noted differences between the two species. The results of API tests do not seem to support a close relationship between these two species.

Cluster Group III

Several of the species in Group III are represented by an inadequate number of strains and great care must be exercised when interpreting the results.

Gibson [1935b] described the *B.pasteurii* group as comprising several distinctive types, namely *B.pasteurii*, *B.loehnisii* and *B.freudenreichii*, completely connected by transitional forms, and believed it to be closely related to *B.sphaericus*. The relationships shown in Figs. 1 and 4 support Gibson's views, though in the small number of strains studied the transitional forms are absent.

Gibson [1935a] noted the similarity of *B.fusiformis* to *B.sphaericus* and Smith *et al.* [1946] designated it as a variety of *B.sphaericus*. After studying further strains Smith *et al.* [1952] considered *B.sphaericus* var. *fusiformis* to be merely a biotype of *B.sphaericus* and the relationship shown in Fig. 4 is in accordance with this.

Gordon *et al.* [1973] and Gibson and Gordon [1974] noted the similarity of *B.globisporus* and *B.psychrophilus* to *B.sphaericus*. It would appear from Fig. 4 that, whilst being closely related to the *B.sphaericus* group, strains of *B.globisporus* and *B.psychrophilus* bear an even greater resemblance to each other, a relationship observed by Gyllenberg and Laine [1971] and supported by studies on vitamin requirements [Adams and Stokes, 1968]. Larkin and Stokes [1967], however, listed 8 criteria (5 of which were used in this study) by which the two psychrophilic species could be differentiated. Gyllenberg and Laine [1970] believed that psychrophilic behaviour was of restricted taxonomic significance in *Bacillus* and, like Larkin and Stokes [1967], considered *B.globisporus* and *B.psychrophilus* to be related to *B.pantothenticus*.

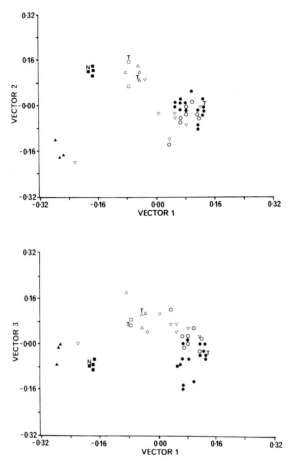

Fig. 4. Plot of the first 3 vectors, accounting for 47% of the total variation, from a principal coordinate analysis of Cluster Group III of Fig. 1. *Bacillus sphaericus,* ● ;   *B.sphaericus* var. *fusiformis,* ○ ;   *B.pasteurii,* ■ ;   *B.freudenreichii,* ▲ ;   *B.globisporus,* △ ; *B.psychrophilus,* □ ;   unidentified or misnamed *Bacillus* species, ▽ ; neotype strain, N;   type strain, T.

Cluster Group IV

It is evident from Fig. 5 that 4 major groups of strains exist in cluster Group IV. *B.megaterium* strains form a tight group lying distant from strains representing the "*B. subtilis* spectrum". This arrangement is in agreement with the analyses of Sneath [1962], Bonde [1973] and Priest *et al*. [Chapter 5] and the DNA-reassociation data obtained by Seki *et al*. [1975, 1978], but other work has indicated a close relationship between *B.megaterium* and *B.cereus*.

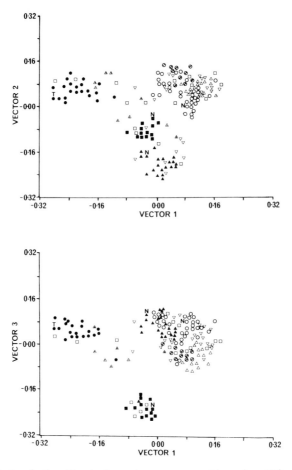

Fig. 5. Plot of the first 3 vectors, accounting for 37% of total variation, from a principal coordinate analysis of Cluster Group IV of Fig. 1. *Bacillus subtilis*, ○ ; *B.licheniformis*, ▲ ; *B.pumilus*, ● ; *B.megaterium*, ■ ; *B.amyloliquefaciens*, ⊘ ; *Bacillus* spp. isolated from clinical specimens and foods, □ ; *Bacillus* species isolated from cocoa, △ ; *Bacillus* species isolated from salt marshes, ▲ ; unidentified or misnamed *Bacillus* species, ▽ ; neotype strain, N; type strain, T.

Smith *et al.* [1946] studied 19 intermediate strains, Knight and Proom [1950] found strains with nutritional requirements intermediate between those of the two species, and Somerville and Jones [1972] and Kaneko *et al.* [1978] found significant homologies in DNA reassociation studies.

The "*B.subtilis*" spectrum falls into 3 closely related groups with *B.subtilis* and *B.amyloliquefaciens* representing one, and *B.licheniformis* and *B.pumilus* the remaining two.

Smith *et al.* [1946] did not consider the characters they studied furnished a basis for recognizing Ford's strain of *B.subtilis* [Lawrence and Ford, 1916] as a separate species. Gibson [1944], however, placed the Ford strain, along with other facultatively anaerobic strains, in the species *B.licheniformis* and Smith *et al.* [1952] revised their opinion in the light of this work. Gibson [1944] and Gordon *et al.* [1973] considered that *B.pumilus* was probably more closely related to *B.subtilis* than was *B.licheniformis* and, in the 8th edition of "Bergey's Manual of Determinative Bacteriology" [Gibson and Gordon, 1974], the difficulties of differentiating the first two species were acknowledged. It is evident from Fig. 5 that, although the boundaries between the 3 groups are somewhat blurred, the strains of *B.licheniformis* and *B.pumilus* show a slightly greater similarity to *B.subtilis* than they do to each other. In Fig. 1 *B.licheniformis* appears to be more closely related to *B.subtilis* than does *B.pumilus*. This relationship is not so evident in Fig. 5, but is supported by DNA reassociation studies and pyrolysis gas-liquid chromatography [O'Donnell *et al.*, 1980].

The work of O'Donnell *et al.* [1980] also argues for the recognition of *B.amyloliquefaciens* as a species separate from *B.subtilis*, in agreement with the findings of Welker and Campbell [1967] and Seki *et al.* [1975, 1978] but this is not obvious in Figs. 1 and 5 where subsequent study of larger numbers of strains has revealed the presence of intermediates obscuring the distinction. Many of these intermediate strains were isolated from cocoa in Africa and Malaysia and from clinical specimens and food in the United Kingdom.

It is of interest that Denis [1971], Bonde [1973, 1975, 1976] and Boeyé and Aerts [1976] also identified many of their isolates from marine environments as *B.firmus* (see cluster Group II above) and as members of the "*B.subtilis* spectrum", especially *B.subtilis* and *B.licheniformis*. Unpigmented strains from salt marshes and clinical and food isolates appear to be intermediate between *B.subtilis*, *B.licheniformis* and *B.pumilus*. Other workers have also reported the existence of such transitional strains [Knight and Proom, 1950; Smith *et al.*, 1952; Burdon, 1956; Hanáková-Bauerová *et al.*, 1965; Gordon *et al.*, 1973; Bonde, 1978] so that *B.subtilis* and related species form a spectrum.

Cluster Group V

Gibson and Topping [1938] found *B.circulans* to be a species complex "exhibiting variation in several directions" and later workers, who investigated biochemical and physiological characters [Smith *et al.*, 1946, 1952; Gray and Hull, 1971; Wolf and Chowdhury, 1971a; Gordon *et al.*, 1973], nutritional requirements [Proom and Knight, 1955],

free amino acid pools [Jayne-Williams and Cheeseman, 1960],
antigenic relationships [Norris and Wolf, 1961; Hill and
Gray, 1967; Wolf and Chowdhury, 1971a, b], spore surface
morphology [Bradley and Franklin, 1958] and DNA homologies
[Seki *et al.*, 1978] have endorsed this description. The
concept of *B.circulans* as a heterogeneous group is further
supported by the plots shown in Fig. 6. Gibson and Gordon
[1974] observed that *B.palustris* had the general character
of *B.circulans*. A single strain, received as *B.palustris*
var. *gelaticus*, lies near to the *B.circulans* group in Fig.
6 as does a strain of *B.longisporus* [Delaporte, 1972].

A close relationship between *B.macerans* and *B.circulans*
has been observed by several authors [Smith *et al.*, 1946,
1952; Knight and Proom, 1950; Wolf and Chowdhury, 1971a;
Gordon *et al.*, 1973; Gibson and Gordon, 1974] and it can
be seen in Fig. 6 that the strains of *B.macerans* form a
diffuse group lying near to *B.circulans*.

The *B.polymyxa* group, also somewhat heterogeneous, is
widely accepted as a clearly demarcated species [Gibson
and Gordon, 1974] but lies close to *B.macerans* - a relationship noted by Porter *et al.* [1937], Smith *et al.* [1946,
1952] and Gordon *et al.* [1973] and observable in the
analyses of Sneath [1962] and Priest *et al.* [Chapter 5].
In addition, Bradley and Franklin [1958] were unable to
separate strains of *B.polymyxa* from *B.macerans* on the
basis of spore surface morphology and Hino and Wilson
[1958] and Ouellette *et al.* [1969] described a strain
which possessed some of the special properties of each
species.

Marshall and Ohye [1966] described *B.macquariensis* as
having a spore surface morphology reminiscent of *B.polymyxa*
and Gibson and Gordon [1974] noted that it shared many of
the properties of *B.circulans*. However, although strains
of *B.macquariensis* join to the *B.circulans* - *B.macerans* -
B.polymyxa series in Fig. 1 they do not show close relatedness to this group in the phenogram or in the two plots
shown in Fig. 6.

Cluster Group VI

Gaughran [1949] and Allen [1950] suggested that the
biochemistries of mesophiles and thermophiles might be
closely related. Allen [1950, 1953] believed that thermophilic variants of mesophilic species occur naturally and
claimed that they could also be derived in the laboratory,
and considered that *B.stearothermophilus* was merely a
thermophilic variant of *B.circulans*. Walker and Wolf
[1971] cite several other reports of stable variants, capable of growth at temperatures up to, and including, 70°C,
arising from species other than *B.coagulans* and *B.stearo-
thermophilus*, the two thermophilic species of the genus
Bacillus recognized by Gordon and Smith [1949] and Smith
et al. [1952].

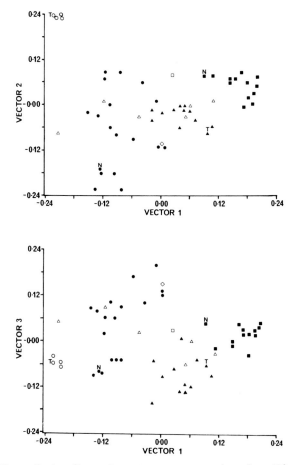

Fig. 6. Plot of the first 3 vectors, accounting for 40% of the total variation, from a principal coordinate analysis of Cluster Group V of Fig. 1. *Bacillus circulans*, ● ; *B.polymyxa*, ■ ; *B.macerans*, ▲ ; *B.macquariensis*, ○ ; *B.palustris* var. *gelaticus*, □ ; *B.longisporus*, ◇ ; unidentified or misnamed *Bacillus* species, △ ; neotype strain, N; type strain, T.

Although growth temperatures were not taken into account in the present study, the two thermophilic species have clustered together in Fig. 1 and some strains of *B.coagulans* lie very close to *B.stearothermophilus* in Fig. 7. This last observation may, however, be explained by the possible misnaming of cultures prior to receipt, indeed several strains received as *B.coagulans* appeared to be typical members of the *B.subtilis* group.

The ability of organisms to grow at 65°C and above is not a satisfactory criterion for separation of species

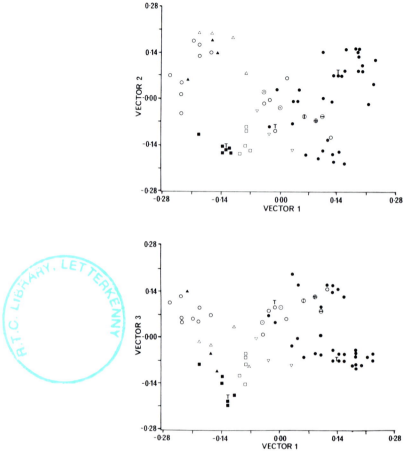

Fig. 7. Plot of the first 3 vectors, accounting for 38% of total variation, from a principal coordinate analysis of Cluster Group VI of Fig. 1. *Bacillus stearothermophilus,* ● ; *B.coagulans,* ○ ; *B.panto-thenticus,* ■ ; *B.racemilacticus,* ▲ ; *B.laevolacticus,* △ ; *B.caldo-lyticus,* ⊖ ; *B.caldotenax,* ⏀ ; *B.caldovelox,* ⊕ ; *B.thermodeni-trificans,* ⊙ ; *Bacillus* species isolated from clinical specimens and foods, □ ; unidentified or misnamed *Bacillus* species, ▽ ; type strain, T.

[Gordon and Smith, 1949; Smith *et al.*, 1952; Gibson and Gordon, 1974] and the emphasis on this character has re-sulted in heterogeneity within the species *B.stearothermo-philus* and the exclusion of organisms that are similar to this species, in most respects, but have temperature maxima between 55 and 65°C [Gibson and Gordon, 1974].

Walker and Wolf [1971] regarded the recognition of only one thermophilic (that is growing at 65°C) species,

B.stearothermophilus, in the 7th edition of "Bergey's
Manual of Determinative Bacteriology" [Smith and Gordon,
1957] as a "dramatic restriction" and described 3 groups,
within this species, on the basis of biochemical tests
[Walker and Wolf, 1961, 1971] and serology [Walker and
Wolf, 1971; Wolf and Chowdhury, 1971b]. This division of
B.stearothermophilus was further supported by studies of
esterase patterns [Baillie and Walker, 1968] and polar
lipids [Minnikin et al., 1977]. The diffuse spread of *B.
stearothermophilus*, apparent in Fig. 7, may represent 2 or
3 groups but more strains must be studied to elucidate this
and to ascertain whether such groups, if they do exist, are
comparable to those described by Walker and Wolf [1971].
Comparison of the groupings of strains common to both
studies, however, suggests that the largest group of
strains (which lies at the top, and the bottom, of the
plots of vector 1 against vectors 2 and 3 respectively),
and which contains the type strains, is equivalent to
Group 3a of Walker and Wolf [1971]. The 2 strains of *B.
thermodenitrificans* (*Denitrobacterium thermophilum* of
Ambrož, 1913) lie close to strains of *B.stearothermophilus*
that were placed in Group 1b of Walker and Wolf [1971];
this is in agreement with the findings of these authors.

Between the two major groups of *B.stearothermophilus*
strains lie single strains of *B.caldolyticus*, *B.caldotenax*
and *B.caldovelox* [Heinen and Heinen, 1972]. This evidence,
such as it is, does not support the recognition of the last
3 as separate species.

Gibson and Gordon [1974] mentioned the existence of
strains of *B.coagulans* that diverged from the typical
strains in one or more respects and which appeared to blur
the distinction between this species and *B.stearothermo-
philus*. Several authors have observed considerable variation, in *B.coagulans*, of spore size and shape [Smith et
al., 1952; Wolf and Barker, 1968; Gordon et al., 1973],
spore surface morphology [Bradley and Franklin, 1958] and
of physiological and biochemical characteristics [Gordon
and Smith, 1949; Wolf and Barker, 1968; Lemille et al.,
1969]. Others, however, considered the species to be homogeneous, on the basis of nutritional studies [Knight and
Proom, 1950; Proom and Knight, 1955; Gordon et al., 1973]
and DNA homologies [Seki et al., 1978].

It is difficult to interpret the findings of the present
study because, as previously noted, a large number of
strains that were received as *B.coagulans* clustered elsewhere in the analyses summarized by Fig. 1. It may, however, be seen in Fig. 7 that the strains representing *B.
coagulans*, other than those showing close affinity to the
B.stearothermophilus group, form a diffuse group to which
strains of *B.laevolacticus* and *B.racemilacticus* show
similarity.

Smith et al. [1952] placed *B.pantothenticus* in morphological group 3 but noted that, "in a natural classifica-

tion, it would seem to be intermediate between (morphological) groups 2 and 3". Wolf and Barker [1968] considered that it should lie wholly within morphological group 2 on account of its biochemical activity. Proom and Knight [1950] considered *B.pantothenticus* to be most similar to *B.circulans* and Gibson and Gordon [1974] noted that these 2 species had many properties in common.

In Figs. 1 and 7, however, the *B.pantothenticus* strains show relatedness to both *B.stearothermophilus* and *B.coagulans*. Several strains of *Bacillus* species, isolated from canned meat, water supply and infant bile, show high similarity to the *B.pantothenticus* group.

Identification

As *Bacillus* species often show large variation between strains, diagnostic keys may be of little use and diagnostic tables, showing percentages of positive reactions would be more appropriate when computer facilities are not available.

When a computer is available, the method of identification using similarity coefficients, described by Ross [1975], may be used. Gower [1968] developed a technique for adding points to a principal coordinate analysis using the similarity coefficients between the new unit and each of the reference units; this is a useful supplement to identification by similarity coefficients and makes it possible to distinguish between intermediate and outlying strains when working with groups. Both facilities are available in the CLASP program [Ross, 1975].

Identification of a new unit as a member of a group enlarges that group, and this is valuable for updating the data base [Morse, 1975]. When more strains have been studied it will be possible to publish diagnostic tables for use with the API tests described in this chapter.

Conclusion

This classification of the genus *Bacillus*, based largely on the tests in the API System, is in general agreement with the taxonomic schemes of Smith *et al.* [1946, 1952] and Gordon *et al.* [1973] and is a tribute to the excellence of their work. Further support is given by the DNA-reassociation studies of Somerville and Jones [1972], Seki *et al.* [1975, 1978], Kaneko *et al.* [1978] and O'Donnell *et al.* [1980]. It may be concluded, therefore, that tests in the API System provide a good basis for taxonomic work with the genus *Bacillus*.

Previous authors have successfully applied miniaturized test systems, such as API, API ZYM and Auxotab, to a wide range of bacterial groups including Actinomycetaceae [Kilian, 1978], anaerobes [Guillermet *et al.*, 1976; Tharagonnet *et al.*, 1977], *Bacillus* [Joubert *et al.*, 1968;

Joly, 1973], coryneforms [Perrier and Gounot, 1975], Enterobacteriaceae and other Gram-negative rods [Washington *et al.*, 1971; Smith *et al.*, 1972; Holmes *et al.*, 1978], Lactobacillaceae [Laban *et al.*, 1978], *Neisseria* species [D'Amato *et al.*, 1978], *Pediococcus* [Dolezil and Kirsop, 1977], *Staphylococcus* [Peny and Buissière, 1970; Brun *et al.*, 1978] and *Streptococcus* species [Humble *et al.*, 1977]. It is not surprising, therefore, that the tests used here, although many of them were designed for identification of Enterobacteriaceae, can be used in the study of *Bacillus*.

The genus *Bacillus*, it is widely agreed, is heterogeneous and while this study has provided sufficient information, taken together with that to be found already in the literature, to enable a division of *Bacillus* into 3 or more different genera, such action would not be timely. Further work, using the API System to study representatives of lesser known areas of the genus, is currently in progress in this laboratory and other groups of workers, using different methods, are engaged upon systematic studies of the group. When all the information from this work is available it should be possible to allocate aerobic spore-forming rods, on a sound basis, to newly described genera.

Division of such newly formed genera into species may be difficult in several instances as it is clear that a spectrum of strains exists in many areas of this taxon. Spectra are apparent in the *B.firmus/B.lentus* group (cluster Group II), the *B.subtilis* group (cluster Group IV), the *B.circulans* complex (cluster Group V) and the thermophilic organisms (cluster Group VI). It is of course probable that further work will reveal the existence of additional examples.

For the present it can be said with some confidence that organisms belonging to the genus, as currently described, can be identified using the API System. It has already been shown that strains of *B.cereus* capable of causing emetic or diarrhoeal types of food poisoning, as indicated by monkey feeding experiments, may be separated using tests in the API system [Logan *et al.*, 1979]. Other work, outlined above (and to be published in greater detail), shows that avirulent as well as virulent strains of *B. anthracis* may be separated from *B.cereus*. It has also been demonstrated that strains of several species encountered in, for example, clinical specimens, foods and beverages may also be identified readily. The role of *Bacillus* species as opportunistic pathogens [see Chapters 1 and 12] is being increasingly recognized and the establishment of a convenient and rapid identification system was an important aspect of the present work for, as Gordon [1976] stressed, the taxonomist should accept his or her responsibility to the routine diagnostic laboratory when potentially pathogenic organisms are involved.

When the studies on a larger number of strains representative of the lesser known types in the genus have been completed, those tests of most diagnostic value for *Bacillus*
will be selected. In this way a rapid and inexpensive
system based on a readily available and carefully standardized test system for identification of *Bacillus* species
can be developed for routine use in hospital and other
laboratories.

Acknowledgement

We wish to thank Professor J.R. Norris, Secretary of the
ICSB Sub-Committee on the Taxonomy of the Genus *Bacillus*,
for making the results of the international reproducibility
trial of classical tests available to us.

References

Adams, J.C. and Stokes, J.L. (1968). Vitamin requirements of psychro-
philic species of *Bacillus*. *Journal of Bacteriology* **95**, 239-240.
Allen, B.T. and Wilkinson, H.A. (1969). A case of meningitis and
generalised Schwartzman reaction caused by *Bacillus sphaericus*.
Johns Hopkins Medical Journal **125**, 8-13.
Allen, M.B. (1950). The dynamic nature of thermophily. *Journal of
General Physiology* **33**, 205-214.
Allen, M.B. (1953). The thermophilic aerobic sporeforming bacteria.
Bacteriological Reviews **17**, 125-173.
Ambrož, A. (1913). *Denitrobacterium thermophilum* spec. *nova*, ein
Beitrag zur Biologie der thermophilen Bakterien. *Zentralblatt für
Bakteriologie* **37**, 3-16.
Baillie, A. and Walker, P.D. (1968). Enzymes of thermophilic aerobic
spore-forming bacteria. *Journal of Applied Bacteriology* **31**, 114-
119.
Bisset, K.A. and Bartlett, R. (1978). The isolation and characters of
L-forms and reversions of *Bacillus licheniformis* var. *endoparasi-
ticus* (Benedek) associated with the erythrocytes of clinically
normal persons. *Journal of Medical Microbiology* **11**, 335-349.
Boeyé, A. and Aerts, M. (1976). Numerical taxonomy of *Bacillus*
isolates from North Sea sediments. *International Journal of
Systematic Bacteriology* **26**, 427-441.
Bonde, G.J. (1965). Classification of *Bacillus* spp. from marine sedi-
ments. *Journal of General Microbiology* **41**, Proceedings xxii.
Bonde, G.J. (1973). The genus *Bacillus*. An examination of 460 meso-
philic strains. International Service System Research Council
Series No. 2. Copenhagen.
Bonde, G.J. (1975). The genus *Bacillus*. An experiment with cluster
analysis. *Danish Medical Bulletin* **22**, 41-61.
Bonde, G.J. (1976). The marine *Bacillus*. *Journal of Applied
Bacteriology* **41**, vi-vii.
Bonde, G.J. (1978). Current status of *Bacillus* classification.
Abstracts of the XII International Congress of Microbiology, p.86.
Munich.
Bradley, S.G. (1980). DNA reassociation and base composition. In

"Microbiological Classification and Identification" (eds. M. Good-fellow and R.G. Board), pp.11-26. London: Academic Press.

Bradley, D.E. and Franklin, J.G. (1958). Electron microscope survey of the surface configuration of spores of the genus *Bacillus*. *Journal of Bacteriology* **76**, 618-630.

Brun, Y., Fleurette, J. and Forey, F. (1978). Micromethod for biochemical identification of coagulase negative staphylococci. *Journal of Clinical Microbiology* **8**, 503-508.

Buissière, J. (1972). Perfectionnement du tube d'Ivan Hall pour l'étude en série de la croissance et de la physiologie des bactéries. *Compte Rendus des Séances de l'Academie des Sciences (Paris)* **274** Série D, 1426-1429.

Buissière, J., Fourcard, A. and Colobert, L. (1967). Usage de substrats synthétiques pour l'étude de l'équipement enzymatique de microorganismes. *Compte Rendus des Séances de l'Academie des Sciences (Paris)* **264** Série D, 415-417.

Buissière, J. and Nardon, P. (1968). Microméthode d'identification des bactéries. I. Intérêt de la quantification des caractères biochimiques. *Annales de l'Institut Pasteur* **115**, 218-231.

Burdon, K.L. (1956). Useful criteria for the identification of *B. anthracis* and related species. *Journal of Bacteriology* **71**, 25-42.

Burdon, K.L. and Wende, R.D. (1960). On the differentiation of anthrax bacilli from *Bacillus cereus*. *Journal of Infectious Diseases* **107**, 224-234.

Butler, D.A., Lobregat, C.M. and Gavan, T.L. (1975). Reproducibility of the Analytab (API 20E) system. *Journal of Clinical Microbiology* **2**, 322-326.

Clark, F.E. (1937). The relationship of *Bacillus siamensis* and similar pathogenic spore-forming bacteria to *Bacillus subtilis*. *Journal of Bacteriology* **33**, 435-443.

Cox, N.A., Mercuri, A.J. and McHan, F. (1976). Miniaturized multitest systems and modifications to expand their use. In "Rapid Methods and Automation in Microbiology" (eds. H.H. Johnston and S.W.B. Newsom), pp.163-168. Learned Information (Europe) Ltd., Oxford.

D'Amato, R.F., Eriquez, L.A., Tomfohrde, K.A. and Singerman, E. (1978). Rapid identification of *Neisseria gonorrheae* and *Neisseria meningitidis* by using enzymatic profiles. *Journal of Clinical Microbiology* **7**, 77-81.

Davenport, R. and Smith, C. (1952). Panophthalmitis due to an organism of the *Bacillus* group. *British Journal of Ophthalmology* **36**, 389-392.

Delaporte, B. (1972). Trois nouvelles espèces de *Bacillus*: *Bacillus similibadius* n.sp., *Bacillus longisporus* n.sp. et *Bacillus nitritollens* n.sp. *Annales de l'Institut Pasteur* **123**, 821-834.

Denis, F. (1971). Le *Bacillus* du milieu marin: étude de 120 souches. *Comptes Rendus des Séances de la Société de Biologie* **165**, 2404-2408.

Dolezil, L. and Kirsop, B.H. (1977). The use of the API *Lactobacillus* system for the characterization of pediococci. *Journal of Applied Bacteriology* **42**, 213-217.

Dowdle, W.R. and Hansen, P.A. (1961). A phage-fluorescent antiphage staining system for *Bacillus anthracis*. *Journal of Infectious Diseases* **108**, 125-135.

Farrar, W.E. (1963). Serious infections due to "non-pathogenic" organisms of the genus *Bacillus*. *American Journal of Medicine* **34**, 134-141.

Gaughran, E.R.L. (1949). Temperature activation of certain respiratory enzymes of thermophilic bacteria. *Journal of General Physiology* **32**, 313-327.

Gibson, T. (1934a). An investigation of the *B.pasteuri* group. I. Description of strains isolated from soils and manures. *Journal of Bacteriology* **28**, 295-311.

Gibson, T. (1934b). An investigation of the *B.pasteuri* group. II. Special physiology of the organisms. *Journal of Bacteriology* **28**, 313-322.

Gibson, T. (1935a). The urea-decomposing microflora of soils. I. Description and classification of the organisms. *Zentralblatt für Bakteriologie, Parasitenkunden, Infektionskrankheiten und Hygiene* Abt. II **92**, 364-380.

Gibson, T. (1935b). An investigation of the *B.pasteuri* group. III. Systematic relationships of the group. *Journal of Bacteriology* **29**, 491-502.

Gibson, T. (1937). The identity of *B.subtilis* and its differentiation from other spore-forming bacteria. *Proceedings. Society of Agricultural Bacteriologists*.

Gibson, T. (1944). A study of *Bacillus subtilis* and related organisms. *Journal of Dairy Research* **13**, 248-260.

Gibson, T. and Gordon, R.E. (1974). *Bacillus*. In "Bergey's Manual of Determinative Bacteriology" (eds. R.E. Buchanan and N.E. Gibbons), pp.529-550. Baltimore: The Williams and Wilkins Co.

Gibson, T. and Topping, L.E. (1938). Further studies of the aerobic spore-forming bacilli. *Proceedings. Society of Agricultural Bacteriologists*, 43-44.

Gilbert, R.J. and Parry, J.M. (1977). Serotypes of *Bacillus cereus* from outbreaks of food poisoning and from routine foods. *Journal of Hygiene (Cambridge)* **78**, 69-74.

Gilbert, R.J. and Taylor, A.J. (1976). *Bacillus cereus* food poisoning. In "Microbiology in Agriculture, Fisheries and Food" (eds. F.A. Skinner and J.G. Carr), pp.197-213. Society for Applied Bacteriology Symposium Series No. 4. London: Academic Press.

Goepfert, J.M., Spira, W.M. and Kim, H.U. (1972). *Bacillus cereus:* food poisoning organism. A review. *Journal of Milk and Food Technology* **35**, 213-227.

Gordon, R.E. (1973). The genus *Bacillus*. In "Handbook of Microbiology" (eds. A.I. Laskin and H.A. Lechevalier), Vol. 1. Organismic Microbiology, pp.71-88. Cleveland: The Chemical Rubber Co. Press.

Gordon, R.E. (1975). Some taxonomic observations on the genus *Bacillus*. In "Biological Regulation of Vectors: The Saprophytic and Aerobic Bacteria and Fungi" (ed. J.D. Briggs), pp.67-82. United States Department of Health, Education and Welfare.

Gordon, R.E. (1976). A taxonomist's obligation. In "The Biology of the Nocardiae" (eds. M. Goodfellow, G.H. Brownell and J.A. Serrano), pp.66-73. London: Academic Press.

Gordon, R.E., Haynes, W.C. and Pang, C.H.-N. (1973). "The Genus

Bacillus." Washington, D.C.:   United States Department of Agriculture.

Gordon, R.E., Hyde, J.L. and Moore, J.A. (1977). *Bacillus firmus-Bacillus lentus:* a series or one species? *International Journal of Systematic Bacteriology* **27**, 256-262.

Gordon, R.E. and Smith, N.R. (1949). Aerobic spore-forming bacteria capable of growth at high temperatures. *Journal of Bacteriology* **58**, 327-341.

Gower, J.C. (1966). Some distance properties of latent root and vector methods used in multivariate analysis. *Biometrika* **53**, 325-338.

Gower, J.C. (1968). Adding a point to vector diagrams in multivariate analysis. *Biometrika* **55**, 582-585.

Gower, J.C. (1971). A general coefficient of similarity and some of its properties. *Biometrics* **27**, 857-874.

Gower, J.C. (1978). Some remarks on proportional similarity. *Journal of General Microbiology* **107**, 387-389.

Gray, T.R.G. and Hull, D.A. (1971). Taxonomy of *Bacillus circulans* with special reference to spore morphology. In "Spore Research 1971" (eds. A.N. Barker, G.W. Gould and J. Wolf), pp.219-226. London and New York: Academic Press.

Green, D.M. (1975). Anthrax Bacillus. In "Medical Microbiology" (eds. R. Cruickshank, J.P. Duguid, B.P. Marmion and R.H.A. Swain), 12th Edition, Vol. 2, pp.449-453. Edinburgh, London and New York: Churchill Livingstone.

Guillermet, F.N., Nardon, P., Dumont, J. and Desbresles, A.M.B. (1976). Biochimie de bacteries anaerobies. *Revue de l'Institut Pasteur de Lyon* **9**, 275-289.

Gyllenberg, H.G. and Laine, J.J. (1970). Psychrophilic behaviour as a taxonomic criterium particularly with respect to the genus *Bacillus. Spisy Přirodovědecke Fakulty Universita V. Brne K.* **47**, 95-98.

Gyllenberg, H.G. and Laine, J.J. (1971). Numerical approach to the taxonomy of psychrophilic bacilli. *Annales Medicinae Experimentalis et Biologiae Fenniae* **49**, 62-66.

Hanáková-Bauerová, E., Kocur, M. and Martinec, T. (1965). Concerning the differentiation of *Bacillus subtilis* and related species. *Journal of Applied Bacteriology* **28**, 384-389.

Hayashi, R. and Nakayama, H. (1953). Studies on the aerobic mesophilic bacteria with distinctly bulged sporangium. I. Special reference to *Bacillus thiaminolyticus. Yamaguchi Medical School Bulletin* **1**, 57-63.

Heinen, U.J. and Heinen, W. (1972). Characteristics and properties of a caldoactive bacterium producing extracellular enzymes and two related strains. *Archiv für Mikrobiologie* **82**, 1-23.

Hill, I.R. and Gray, T.R.G. (1967). Application of the fluorescent-antibody technique to an ecological study of bacteria in soil. *Journal of Bacteriology* **93**, 1888-1896.

Hino, S. and Wilson, P.W. (1958). Nitrogen fixation by a facultative bacillus. *Journal of Bacteriology* **75**, 403-408.

Holmes, B., Willcox, W.R. and Lapage, S.P. (1978). Identification of Enterobacteriaceae by the API 20E system. *Journal of Clinical Pathology* **31**, 22-30.

Humble, M.W., King, A. and Phillips, I. (1977). API ZYM: a simple rapid system for the detection of bacterial enzymes. *Journal of Clinical Pathology* **30**, 275-277.

Hunger, W. and Claus, D. (1978). *Bacillus megaterium*: A species complex. *Abstracts of the XII International Congress of Microbiology*, Munich.

Ihde, D.C. and Armstrong, D. (1973). Clinical spectrum of infection due to *Bacillus* species. *American Journal of Medicine* **55**, 839-845.

Ivánovics, G. and Földes, J. (1958). Problems concerning the phylogenesis of *Bacillus anthracis*. *Acta Microbiologica Hungarica* **5**, 89-109.

Janin, P.R. (1976). Development of a bacteriological identification system: Theory and practice. In "Rapid Methods and Automation in Microbiology" (eds. H.H. Johnston and S.W.B. Newsom), pp.155-162. Oxford: Learned Information (Europe) Ltd.

Jayne-Williams, D.J. and Cheeseman, G.C. (1960). The differentiation of bacterial species by paper chromatography. IX. The genus *Bacillus*: A preliminary investigation. *Journal of Applied Bacteriology* **23**, 250-268.

Joly, B. (1973). Contribution à l'identification des espèces du genre *Bacillus* par l'étude normalisée des caractères enzymatiques. Intérêt en taxonomie. Thèse de doctorat en pharmacie. Clermont-Ferrand.

Joubert, L. and Buissière, J. (1969). Le chimiotype bacterien en epidemiologie. *Bulletin de l'Academie Veterinaire de France* **42**, 81-91.

Joubert, L., Buissière, J. and Chirol, C. (1968). Chimiotypes de *Bacillus anthracis* et enquêtes épidémiologiques. *Bulletin de la Société des Sciences Vétérinaires et de Médecine Comparée de Lyon* **70**, 159-172.

Kaneko, T., Nozaki, R. and Aizawa, K. (1978). Deoxyribonucleic acid relatedness between *Bacillus anthracis, Bacillus cereus* and *Bacillus thuringiensis*. *Microbiology and Immunology* **22**, 639-641.

Kilian, M. (1978). Rapid identification of *Actinomycetaceae* and related taxa. *Journal of Clinical Microbiology* **8**, 127-133.

Knight, B.C.J.G. and Proom, H. (1950). A comparative survey of the nutrition and physiology of mesophilic species in the genus *Bacillus*. *Journal of General Microbiology* **4**, 508-538.

Laban, P., Favre, C., Ramet, F. and Larpent, J.P. (1978). Lactobacilli isolated from French saucisson (taxonomic study). *Zentralblatt für Bakteriologie, Parasitenkunden, Infektionskrankheiten und Hygiene* **Abt I 166**, 105-111.

Lammana, C. and Eisler, D. (1960). Comparative study of the agglutinogens of the endospores of *Bacillus anthracis* and *Bacillus cereus*. *Journal of Bacteriology* **79**, 435-441.

Larkin, J.M. and Stokes, J.L. (1967). Taxonomy of psychrophilic strains of *Bacillus*. *Journal of Bacteriology* **94**, 889-895.

Laubach, C.A., Rice, J.L. and Ford, W.W. (1916). Studies on aerobic spore-bearing non-pathogenic bacteria. Part II. *Journal of Bacteriology* **1**, 493-533.

Lawrence, J.S. and Ford, W.W. (1916). Studies on spore-bearing non-pathogenic bacteria. Part I. *Journal of Bacteriology* **1**, 273-320.

Leise, J.M., Carter, C.H., Friedlander, H. and Freed, S.W. (1959).

Criteria for the identification of *Bacillus anthracis*. *Journal of Bacteriology* **77**, 655-660.

Lemille, F., de Barjac, H. and Bonnefoi, A. (1969). Essai sur la classification biochimique de 97 *Bacillus* de groupe I, appartenant à 9 espèces différentes. *Annales de l'Institut Pasteur* **116**, 808-819.

Logan, N.A. (1980). Studies on the Taxonomy and Identification of Members of the Genus *Bacillus*: Application of Miniaturized, Standardized, Commercially Available Test Systems. Ph.D. Thesis, University of Bristol.

Logan, N.A., Berkeley, R.C.W. and Norris, J.R. (1978). Results of an international reproducibility trial using the API system applied to the genus *Bacillus*. *Journal of Applied Bacteriology* xxviii-xxix.

Logan, N.A. Capel, B.J., Melling, J. and Berkeley, R.C.W. (1979). Distinction between emetic and other strains of *Bacillus cereus* using the API system and numerical methods. *FEMS Microbiology Letters* **5**, 373-375.

Marshall, B.J. and Ohye, D.F. (1966). *Bacillus macquariensis* n.sp., a psychrotrophic bacterium from sub-Antarctic soil. *Journal of General Microbiology* **44**, 41-46.

Marymont, J.H. III, Marymont, J.H. Jr. and Gavan, T.L. (1978). Performance of Enterobacteriaceae identification systems. An analysis of College of American Pathologists survey data. *American Journal of Clinical Pathology* **70**, 539-547.

Miller, R.E. Jr. and Lu, L.-P. (1976). Evaluation of a multitest microtechnique for yeast identification. *American Journal of Medical Technology* **42**, 238-242.

Minnikin, D.E., Abdolrahimzadeh, H. and Wolf, J. (1977). Taxonomic significance of polar lipids in some thermophilic members of *Bacillus*. In "Spore Research 1976" (eds. A.N. Barker, J. Wolf, D.J. Ellar, G.J. Dring and G.W. Gould), Vol. 2, pp.879-893. London: Academic Press.

Mishustin, E.N. and Tepper, E.Z. (1948). Description of a new species of bacterium, *Bac. longissimus*. *Mikrobiologiya* **17**, 413-414.

Morse, L.E. (1975). Recent advances in the theory and practice of biological specimen identification. In "Biological Identification with Computers" (ed. R.J. Pankhurst), pp.11-52. London and New York: Academic Press.

Murray, P.R. (1978). Standardization of the Analytab Enteric (API 20E) system to increase accuracy and reproducibility of the test for biotype characterization of bacteria. *Journal of Clinical Microbiology* **8**, 46-49.

Norris, J.R. and Wolf, J. (1961). A study of the antigens of the aerobic spore-forming bacteria. *Journal of Applied Bacteriology* **24**, 42-56.

O'Donnell, A.G., Norris, J.R., Berkeley, R.C.W., Claus, D., Kaneko, T., Logan, N.A. and Nozaki, R. (1980). Characterization of *Bacillus subtilis, Bacillus pumilus, Bacillus licheniformis* and *Bacillus amyloliquefaciens* by pyrolysis gas-liquid chromatography: Characterization tested using DNA-DNA hybridization; Biochemical tests; and API systems. *International Journal of Systematic Bacteriology* **30**, 448-459.

Ormay, L. and Novotny, T. (1969). The significance of *Bacillus cereus*

food poisoning in Hungary. In "The Microbiology of Dried Foods"
(eds. E.H. Kampelmacher, M. Ingram and D.A.A. Mossel). International Association of Microbiological Societies.

Ouellette, C.A., Burris, R.H. and Wilson, P.W. (1969). Deoxyribonucleic acid base composition of species of *Klebsiella, Azotobacter*
and *Bacillus*. *Antonie van Leeuwenhoek Journal of Microbiology and
Serology* **35**, 275-286.

Peny, J. and Buissière, J. (1970). Microméthode d'identification des
bactéries. II. Identification du genre *Staphylococcus*. *Annales
de l'Institut Pasteur* **118**, 10-18.

Perrier, J. and Gounot, A.M. (1975). Méthode normalisée et minia-
turisée pour l'identification des bactéries corynéformes. *Revue de
l'Institut Pasteur de Lyon* **8**, 133-154.

Porter, R., McCleskey, C.S. and Levine, M. (1937). The facultative
sporulating bacteria producing gas from lactose. *Journal of
Bacteriology* **33**, 163-183.

Proom, H. and Knight, B.C.J.G. (1955). The minimal nutritional
requirements of some species in the genus *Bacillus*. *Journal of
General Microbiology* **13**, 474-480.

Quesnel, L.B. (1971). Microscopy and Micrometry. In "Methods in
Microbiology" (eds. J.R. Norris and D.W. Ribbons), Vol. 5A, pp.1-
103. London: Academic Press.

Raevuori, M., Kiutamo, T. and Niskanen, A. (1977). Comparative
studies of *Bacillus cereus* strains isolated from various foods and
food poisoning outbreaks. *Acta Veterinaria Scandinavica* **18**, 397-
407.

Robertson, E.A., Macks, G.C. and MacLowry, J.D. (1976). Analysis of
cost and accuracy of alternative strategies for Enterobacteriaceae
identification. *Journal of Clinical Microbiology* **3**, 421-424.

Rohlf, F.J. (1967). Correlated characters in numerical taxonomy.
Systematic Zoology **16**, 109-126.

Rohlf, F.J. (1968). Stereograms in numerical taxonomy. *Systematic
Zoology* **17**, 246-255.

Rohlf, F.J. (1970). Adaptive hierarchical clustering schemes.
Systematic Zoology **19**, 58-82.

Ross, G.J.S. (1975). Rapid techniques for automatic identification.
In "Biological Identification with Computers" (ed. R.J. Pankhurst),
pp.93-102. London and New York: Academic Press.

Seki, T., Chung, C.-K., Mikami, H. and Oshima, Y. (1978). Deoxyribonucleic acid homology and taxonomy of the genus *Bacillus*. *International Journal of Systematic Bacteriology* **28**, 182-189.

Seki, T., Oshima, T. and Oshima, Y. (1975). Taxonomic study of
Bacillus by deoxyribonucleic acid-deoxyribonucleic acid hybridiza-
tion and interspecific transformation. *International Journal of
Systematic Bacteriology* **25**, 258-270.

Smith, N.R. and Gordon, R.E. (1957). *Bacillus*. In "Bergey's Manual
of Determinative Bacteriology, 7th Edition" (eds. R.S. Breed,
E.G.D. Murray and N.R. Smith), pp.613-634. London: Baillière,
Tindall and Cox Ltd.

Smith, N.R., Gordon, R.E. and Clark, F.E. (1946). "Aerobic Mesophilic
Sporeforming Bacteria". Washington, D.C.: United States Department of Agriculture Miscellaneous Publication 559.

Smith, N.R., Gordon, R.E. and Clark, F.E. (1952). "Aerobic Spore-forming Bacteria." Washington, D.C.: United States Department of Agriculture, Monograph no.16.

Smith, P.B., Tomfohrde, K.M., Rhoden, D.L. and Balows, A. (1972). API System: a multitube micromethod for identification of *Enterobacteriaceae*. *Applied Microbiology* **24**, 449-452.

Sneath, P.H.A. (1962). The construction of taxonomic groups. In "Microbial Classification" (eds. G.C. Ainsworth and P.H.A. Sneath), pp.289-322. Cambridge: Cambridge University Press.

Sneath, P.H.A. (1964). New approaches to bacterial taxonomy: use of computers. *Annual Review of Microbiology* **18**, 335-346.

Sneath, P.H.A. (1974). Test reproducibility in relation to identification. *International Journal of Systematic Bacteriology* **24**, 508-523.

Sneath, P.H.A. and Collins, V.G. (1974). A study in test reproducibility between laboratories: Report of a Pseudomonas Working Party. *Antonie van Leeuwenhoek Journal of Microbiology and Serology* **40**, 481-527.

Sneath, P.H.A. and Johnson, R. (1972). The influence on numerical taxonomic similarities of errors in microbiological tests. *Journal of General Microbiology* **72**, 377-392.

Sneath, P.H.A. and Sokal, R.R. (1973). "Numerical Taxonomy." San Francisco: W.H. Freeman and Company.

Sokal, R.R. and Michener, C.D. (1958). A statistical method for evaluating systematic relationships. *University of Kansas Science Bulletin* **44**, 467-507.

Somerville, H.J. and Jones, M.L. (1972). DNA competition studies within the *Bacillus cereus* group of bacilli. *Journal of General Microbiology* **73**, 257-265.

Soule, M.H. (1928). Microbic dissociation. *Journal of Infectious Diseases* **42**, 93-148.

Sterne, M. and Proom, H. (1957). Induction of motility and capsulation in *Bacillus anthracis*. *Journal of Bacteriology* **74**, 541-542.

Taylor, A.J. and Gilbert, R.J. (1975). *Bacillus cereus* food poisoning: A provisional serotyping scheme. *Journal of Medical Microbiology* **8**, 543-550.

Tharagonnet, D., Sisson, P.R., Roxby, C.M., Ingham, H.R. and Selkon, J.B. (1977). The API ZYM system in the identification of Gram-negative anaerobes. *Journal of Clinical Pathology* **30**, 505-509.

Turnbull, P.C.B., French, T.A. and Dowsett, E.G. (1977). Severe systemic and pyogenic infections with *Bacillus cereus*. *British Medical Journal* **1**, 1628-1629.

Turner, M. and Jervis, D.I. (1968). The distribution of pigmented *Bacillus* species in saltmarsh and other saline and non-saline soils. *Nova Hedwigia* **16**, 293-297.

Walker, P.D. and Wolf, J. (1961). Some properties of aerobic thermophiles growing at 65°C. *Journal of Applied Bacteriology* **24**, iv-v.

Walker, P.D. and Wolf, J. (1971). Taxonomy of *Bacillus stearothermophilus*. In "Spore Research 1971" (eds. A.N. Barker, G.W. Gould and J. Wolf), pp.247-262. London: Academic Press.

Ware, G.C. and Hedges, A.J. (1978). A case for proportional similarity in numerical taxonomy? *Journal of General Microbiology* **104**, 335-336.

Washington, J.A. II, Yu, P.W. and Martin, W.J. (1971). Evaluation of

accuracy of multitest micromethod system for identification of
Enterobacteriaceae. *Applied Microbiology* **22**, 267-269.

Webb, L.J., Tracey, J.G., Williams, W.T. and Lance, G.N. (1967).
Studies in the numerical analysis of complex rain-forest communities. I. A comparison of methods applicable to site/species data.
Journal of Ecology **55**, 171-191.

Weinstein, L. and Colburn, C.G. (1950). *Bacillus subtilis* meningitis
and bacteremia: report of a case and review of the literature on
'subtilis' infections in man. *Archives of Internal Medicine* **86**,
585-594.

Welker, N.E. and Campbell, L.L. (1967). Unrelatedness of *Bacillus
amyloliquefaciens* and *Bacillus subtilis*. *Journal of Bacteriology*
94, 1124-1130.

Westley, J.W., Anderson, P.J., Close, V.A., Halpern, B. and Lederberg,
E.M. (1967). Aminopeptidase profiles of various bacteria. *Applied
Microbiology* **15**, 822-825.

Wilson, G.S. and Miles, A.A. (1975). Topley and Wilson's Principles
of Bacteriology, Virology and Immunity. 6th Edition. London:
Edward Arnold.

Wolf, J. and Barker, A.N. (1968). The genus *Bacillus*: Aids to the
identification of its species. In "Identification Methods for
Microbiologists, Part B" (eds. B.M. Gibbs and D.A. Shapton), pp.93-
109. London: Academic Press.

Wolf, J. and Chowdhury, M.S.U. (1971a). The *Bacillus circulans* complex:
Biochemical and immunological studies. In "Spore Research 1971"
(eds. A.N. Barker, G.W. Gould and J. Wolf), pp.227-245. London and
New York: Academic Press.

Wolf, J. and Chowdhury, M.S.U. (1971b). Taxonomy of *B.circulans* and
B.stearothermophilus. In "Spore Research 1971" (eds. A.N. Barker,
G.W. Gould and J. Wolf), pp.349-350. London and New York:
Academic Press.

Chapter 7

PYROLYSIS GAS-LIQUID CHROMATOGRAPHIC STUDIES

A.G. D'DONNELL* and J.R. NORRIS*

*Agricultural Research Council, Meat Research Institute,
Langford, Bristol, UK*

Introduction

Although members of the genus *Bacillus* have in common the
formation of spores and certain other general properties
they are diverse in other physiological and biochemical
respects. Furthermore, individual members exhibit a considerable degree of variation. As a result the classification of this genus is unsatisfactory.

The first steps towards achieving a classification were
taken by Ford and his co-workers [1916] but the most
significant advances were made by Smith and his colleagues
[1946, 1952]. These workers, by recognizing the degree
of variation among strains of the same species, were able
to reduce over 150 named species to 19 with 5 subspecific
varieties. Their studies on the morphology of spores and
sporangia enabled them to divide the genus *Bacillus* into
3 major groups of species. A further division of group 1,
the numerically largest group, was made using the diameter
and cytoplasmic appearance of the cells.

With the introduction of chemotaxonomic methods such as
fatty acid analysis of the cell wall [Kaneda, 1967, 1968];
paper chromatography [Jayne-Williams and Cheeseman, 1960];
deoxyribonucleic acid analysis [Welker and Campbell,
1967a; Seki *et al.*, 1975, 1978]; enzyme profiling
[Baptist *et al.*, 1978]; phage typing [Norris, 1961] and
immunological studies [Norris and Wolf, 1957; Norris and
Baillie, 1964; Simon *et al.*, 1977] to *Bacillus* classification, new tools for looking at relationships between
organisms have become available. Surprisingly, considering the diversity of techniques, these new methods have,
to a considerable extent, confirmed rather than repudiated
the existing classification. This report describes the
application of a further technique, pyrolysis gas-liquid
chromatography (p.g.l.c.), to the problems of *Bacillus*
taxonomy.

For present addresses see List of Contributors.

Pyrolysis Gas-liquid Chromatography

Pyrolysis

The pyrolysis process involves heating a sample such as a micro-organism in an inert gas atmosphere, thereby causing thermal fragmentation and subsequent volatilization of the molecules of which it is composed. By heating in an inert gas, commonly nitrogen, molecules are cleaved at their weakest points to yield small, characteristic fragments of the original material [Drucker, 1976]. To maintain the maximum correlation between the molecular structure of the original substance and the pyrolysis products (pyrolysate) it is important to reduce, as far as possible, the level of fragment-fragment interactions which occur during pyrolysis. Such secondary reactions are largely dependent on the following experimental conditions: (a) pyrolyser design [Levy, 1967]; (b) sample size [Janak, 1960; Giacobbo and Simon, 1964]; and (c) the nature, pressure and flow rate of carrier gas [Janak, 1960; Levy, 1966]. Unfortunately, the most important of these parameters, pyrolyser design, is in practice the most difficult to control since most workers are restricted to the use of commercially available equipment.

Pyrolysers Ideally a pyrolyser should have as small a "dead volume" (the space between the pyrolysis zone and the stationary phase) as possible. The pyrolysis zone and the stationary phase must not, however, be so close as to cause volatilization of the stationary phase when a sample is pyrolysed. The overall effect of having a small "dead volume" is to reduce the time the pyrolysate is in the hot zone which in turn reduces the extent of secondary reactions.

In general, pyrolysers are of two main types: continuous mode and pulse mode units. Continuous mode units usually consist of a cylindrical oven held at a constant temperature into which a miniaturized sample boat is introduced. Major disadvantages of continuous mode systems are: lack of control over the pyrolysis temperature and length of time taken to reach a final, equilibrium temperature. Most studies on micro-organisms have involved pulse mode pyrolysers.

Pulse mode units operate by very rapidly raising the temperature of the sample. and are of three basic types: filament pyrolysers, Curie-point pyrolysers and laser pyrolysers. Filament pyrolysers and Curie-point pyrolysers have been employed in the present study to investigate the genus *Bacillus*. The brief outlines given below may be supplemented with more detailed reviews by Quinn [1976] and Gutteridge and Norris [1979].

Filament pyrolysers In this system samples are either

layered directly onto a filament or positioned within a
quartz tube held inside the filament, then introduced to
the column inlet. Pyrolysis is brought about by resistive
heating of the platinum filament. The pyrolysate is then
rapidly flushed onto the cooler column by a flow of inert
carrier gas.

Curie-point pyrolysers This technique involves coating a
clean ferromagnetic wire with the sample to be analysed
and placing it in a high frequency field situated at the
column inlet. The effect of the field on the wire is to
create eddy currents which bring about rapid surface
heating and pyrolysis of the sample. When the Curie-point
(the temperature at which the wire becomes paramagnetic)
is reached, the wire stops heating - this cut-off point
being determined by the chemical composition of the wire
[Dyson and Littlewood, 1968]. Since the wires are dis-
posable, variation due to ageing of the heating element,
as occurs with the filament pyrolyser, is eliminated.
 Curie-point wires are less easily cleaned than filaments and require more care in sample application and in
positioning the wire in the centre of the magnetic field.
Curie-point pyrolysis has not often been applied to micro-
organisms but this possibly reflects the absence of suitable, commercial apparatus.

Gas-liquid Chromatography

 Gas-liquid chromatography (g.l.c.) involves the separation of a mixture of volatiles using a gaseous moving
phase and a liquid stationary phase. In microbial chemotaxonomy this technique has been used primarily to study
the end products of bacterial fermentation. The fermentation end products of aerobic organisms are insufficiently
diverse to be useful in classification. Anaerobic
organisms generally use organic compounds as terminal
electron acceptors reducing them to acids, alcohols and
ketones which are of more taxonomic value [Drucker, 1976].
A further application of g.l.c. in microbiology has been
in the analysis of cell components. However, samples must
be carefully prepared since they are essentially non-
volatile.

Chromatography Two types of g.l.c. are used in the
analysis of microbial specimens. The first and most
extensively used is a low resolution system. Low resolution columns consist of tubing packed with granular porous
support material, such as firebrick or diatomaceous earth,
coated with a liquid stationary phase. In practice this
system comprises two identical columns, one for analysis
and one for balance. The purpose of the latter is to
compensate for increased detector current caused by loss
of phase during a temperature programmed analysis.

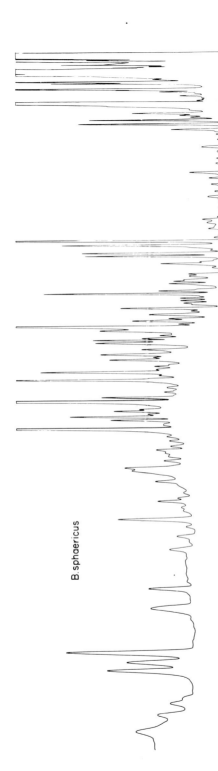

B. sphaericus

Fig. 1 High resolution pyrogram of *B. sphaericus.*

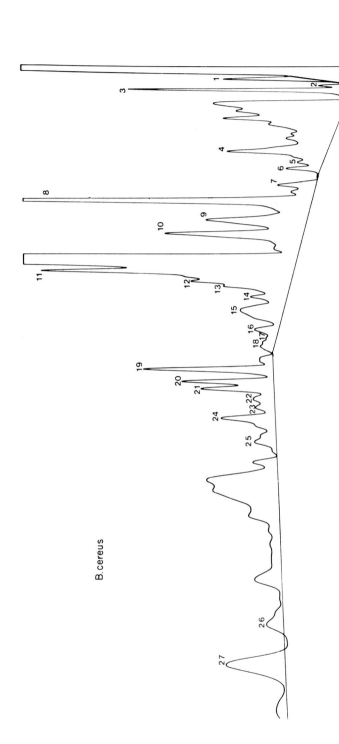

Fig. 2 Low resolution pyrogram of *B.cereus* showing baseline and chosen set of peaks.

The second system, high resolution chromatography, involves the use of very long, small-bore columns known as capillary columns. Capillary columns can be prepared in 2 ways: wall coated open tubular (WCOT) columns which have the inner wall coated directly with the liquid phase, and support coated open tubular (SCOT) in which the inner wall is coated with a porous support layer impregnated with the liquid phase.

The terms "high" and "low" resolution as applied to these systems refer to the ability of the column to sepa- rate the individual components of a pyrolysate and resolve them as individual peaks (see Figs. 1 and 2).

In most of the p.g.l.c. studies in microbiology, Carbo- wax 20M has been the liquid phase used although there is no clear, theoretical justification for this [Gutteridge and Norris, 1979]. Moreover, the need to work at tempera- tures in excess of the tolerance limit of Carbowax 20M (200°C), in order to separate the high boiling point com- pounds found in bacterial pyrolysates, suggests that Carbo- wax 20M is not an ideal material for the liquid phase. Reiner and Kubica [1969] used Carbowax 20M terminated with terephthalic acid (Carbowax 20M-TPA) which has a greater thermostability, probably as high as 250°C. In the studies on the genus *Bacillus* reported here low resolution, Carbo- wax 20M-TPA columns were used.

Sample preparation Since the pyrograms of bacteriological culture media closely resemble those of bacteria when using low resolution systems [Fontanges *et al.*, 1967], it is essential to prepare micro-organisms in such a way as to ensure that contamination by medium is minimal. In this investigation the extracellular products of growing cul- tures were considered valuable for characterizing organisms. Therefore it was necessary to sample directly from an agar plate culture rather than use a washed cell preparation. Many strains of aerobic sporeformers grow into the agar or produce large mucoid colonies and are therefore difficult to remove cleanly from the plate. To overcome this problem, cultures were grown on membrane filters as described by Oxborrow *et al.* [1976].

Microbial Applications

Pyrolysis gas-liquid chromatography was first proposed as a means of differentiating micro-organisms by Oyama [1963] during the development of a system aimed at detec- ting life on Mars. This work was limited by the lack of suitable pyrolysis equipment and it was not until the work of Reiner [1965] that the potential value of p.g.l.c. in microbial differentiation was appreciated. Reiner coupled a pyrolyser previously described by Barbour [1965] to a short 15% Carbowax 20M column and found that on pyrolysis, 18 antigenic types of *Escherichia coli*, one strain of

Shigella, 4 types of Group A streptococci and 10 different mycobacteria, all produced unique pyrograms. The differences between the samples were in the quantities of certain characteristic peaks in the pyrograms which duplicate and replicate analyses showed to be reproducible.

Subsequent improvements in the gas chromatographic technique and the availability of better commercial equipment have since encouraged other workers to use p.g.l.c. for differentiating numerous types of bacteria [MacFie *et al.*, 1978; Reiner *et al.*, 1972; Stack *et al.*, 1978] and fungi [Burns *et al.*, 1976; Vincent and Kulik, 1973].

The usefulness of p.g.l.c. in characterising aerobic sporeformers has been demonstrated by Oxborrow *et al.* [1976, 1977a, 1977b]. These workers have shown that characterization of *Bacillus* species is possible provided chromatographic conditions and major growth parameters remain constant. They also noted that changes in the physiological state of the samples, such as sporulation or pigmentation, resulted in marked changes in the pyrogram. Figures 3 and 4 illustrate the changes observed when cells in different physiological states are pyrolysed. Such alterations to pyrograms have a pronounced effect on the data analysis and steps are therefore taken to ensure that samples examined are of comparable physiological age. In some cases this has involved increasing or decreasing the growth time to accommodate fast or slow growing strains. The need for careful standardization of samples is perhaps the most exacting requirement in applying p.g.l.c. to *Bacillus* characterization. Nevertheless, careful monitoring of samples and application of discriminatory analysis techniques make p.g.l.c. a powerful system for tackling some of the "problem areas" of *Bacillus* taxonomy.

Data Analysis

The complex nature of the bacterial pyrogram is such that development of p.g.l.c. was limited for many years by a lack of suitable data processing techniques. The application of multivariate statistics has provided a data handling system capable of coping with this complexity.

The peaks on a pyrogram can be considered as separate pieces of information about the original sample, rather like the results of a series of biochemical tests, and a pyrogram therefore represents a "fingerprint" or profile of the original micro-organism. Discrimination between micro-organisms obtained using qualitative differences in pyrograms (the presence or absence of a particular peak) has been reported in the literature [Garner and Chi Yuan, 1964; Vincent and Kulik, 1970, 1973; Kulik and Vincent, 1973; Sekhon and Carmichael, 1972, 1973; Taylor, 1976; Cone and Lechowich, 1970; Emswiler and Kotula, 1978]. With the exception of the work by Cone and Lechowich [1970] and Emswiler and Kotula [1978] all these studies were on

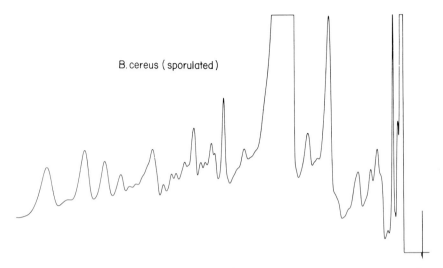

Fig. 3 Pyrogram of *B.cereus* grown for 5 d on nutrient agar at 30°C.

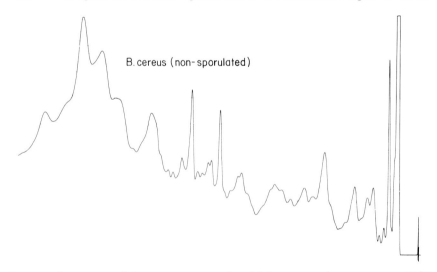

Fig. 4 Pyrogram of *B.cereus* grown for 14 h on nutrient agar at 30°C.

fungi, where significant chemical differences may be
expected [Gutteridge and Norris, 1979]. The authors
analyses have been concerned with quantitative differences
between pyrograms as reflected in their peak heights as
they are unconvinced of the existence of qualitative
differences between bacterial pyrograms obtained using low
resolution chromatography.

Figures 5 and 6 represent typical p.g.l.c. fingerprints
from a low resolution system. The only difference between

these pyrograms is in the amount of material pyrolysed.
To eliminate the difference due to sample size when comparing peak heights a standardization procedure has been
employed in which each peak involved in the analysis is
divided by the sum of all the peak heights used, and the
resultant quotient multiplied by 1,000. In this way each
of the selected peak heights is represented as a fraction
of the sum of chosen peaks (Fig. 7) and pyrograms of
different sample sizes can be compared.

It is necessary to select a set of peaks (Fig. 2, Nos.
1-27) because of changes in the long-term reproducibility
of the analytical system. In practice most of the changes
are a result of column degradation. Quinn [1974] and
Needleman and Stuchbery [1977] attributed this degradation
to condensation of high boiling point tar residues on the
first section of the column, which interferes with the
initial adsorption of the pyrolysate. A function of this
degradation is to cause "tailing" of the larger peaks
[Needleman and Stuchbery, 1977] and a deterioration in the
resolution of pyrolysate components [Quinn, 1974; Needleman and Stuchbery, 1977]. Column degradation can also be
caused by a gradual loss of stationary phase. Meuzelaar
[1974] has emphasized the need for a stationary phase with
similar separation properties to Carbowax 20M but with
greater thermal stability.

The criteria applied in selecting peaks are as follows:
(a)   Choose the maximum number of peaks possible.
(b)   Choose the same peaks from each pyrogram used in the
      study.
(c)   Ensure that all the chosen peaks are present and can
      be measured on every pyrogram.

Peak heights, the variables in the statistical analyses,
are measured relative to a baseline (Fig. 2). The shape
of the baseline used is dependent upon the shape of the
pyrogram and is set so that all of the chosen peaks can be
easily measured. Although setting this baseline is an
arbitrary procedure, once established for a particular
study it is set to all of the pyrograms used in that study
in the same way. The justification for setting a baseline
rather than measuring from a straight line drawn from 0 is
that the area excluded by drawing a baseline across complexes of peaks is probably not linear with respect to
sample size. Figures 8 and 9 represent average linkage
cluster analyses performed on the same pyrograms but
measured in different ways. The change in clusters represented in Fig. 9 may be due to an increased redundancy in
the data or an increased level of variation between
samples caused by altering the method of measurement.

Most p.g.l.c. studies of micro-organisms have used a
similarity coefficient to define relationships between
pyrograms. Meuzelaar et al. [1975] and Stack et al.
[1977], having first standardized to total peak height,
calculated similarities by comparing each peak of the

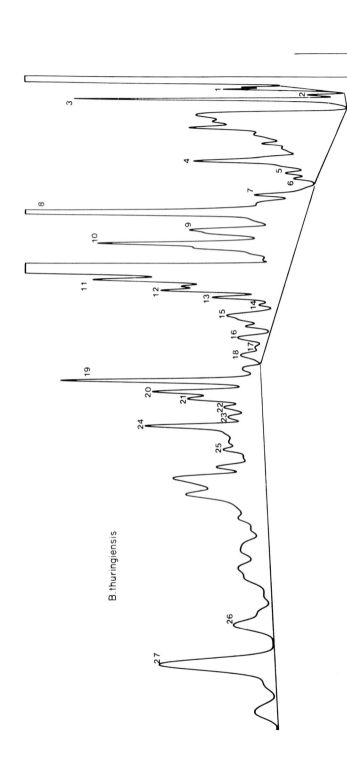

B.thuringiensis

Fig. 5 *B.thuringiensis* grown for 5 d on nutrient agar at 30°C.

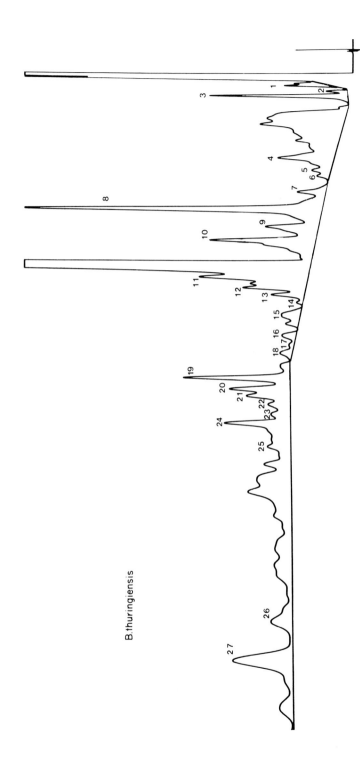

Fig. 6 Conditions identical with those used in Fig. 5 but sample size smaller.

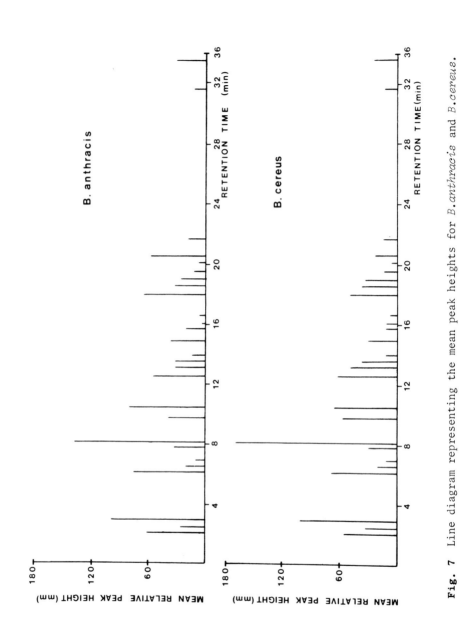

Fig. 7 Line diagram representing the mean peak heights for *B. anthracis* and *B. cereus*.

sample with the corresponding peak of a standard and divi-
ding the smaller value by the larger in each case. Using
this method, absolute identity is scored 1 and similarities
are represented by a range 0 to 0.99. By taking the mean
of all the individual peak height comparisons, an overall
similarity between two organisms is obtained. Although
this technique is capable of providing valuable information, the nature of the bacterial pyrogram makes it
amenable to multivariate statistical analysis and this has
been used in the present analyses.

The aim of multivariate statistics as applied to
p.g.l.c. is to determine whether a number of pyrograms,
represented by a series of peak height measurements, fall
into groups which have some bacteriological significance
[MacFie and Gutteridge, 1978]. Several different types of
multivariate analysis were used.

Principal Components Analysis

Consider a situation in which samples A to J are each
represented by a pyrogram consisting of two peaks. A pictorial representation of the differences between samples
can be achieved by plotting a 2-dimensional graph of peak
1 against peak 2 (Fig. 10). As the number of peaks (variables) defining the positions of the points is increased,
the number of dimensions necessary to display these positions also increases. Principal components analysis is
used to obtain co-ordinates of the pyrograms that are the
weighted sums of peak heights. These are displayed relative to new orthogonal directions of maximum variation.
In this way principal components analysis, by plotting the
points relative to the first 2 derived axes, produces a
2-dimensional representation of the scatter which most
closely resembles the multidimensional configuration
[MacFie and Gutteridge, 1978]. The consideration of
further principal components, which account for the maximum possible variation not accounted for by previously
selected components, may give further information on the
scatter of samples but in fact 3 or 4 principal components
are usually enough to represent well over 90% of the total
variation between samples. Each principal component is
related to the original dimensions (peak heights) by a set
of coefficients known as loadings. Multiplying each standardized peak height by the corresponding loading and cal-
culating the linear sum gives the co-ordinates of a pyrogram relative to that principal component [Gutteridge *et
al.*, 1979]. The usefulness of this technique in detecting
aberrant analyses has been discussed previously [MacFie *et
al.*, 1978; MacFie and Gutteridge, 1978 and Gutteridge *et
al.*, 1979]. A further use of principal components analysis
is in providing a pictorial representation of major
clustering tendencies (Fig. 11).

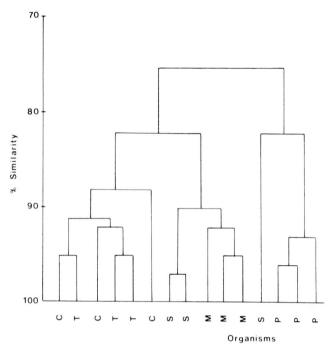

Fig. 8 Average linkage cluster analysis on data from spored cultures. Baseline set across complexes of peaks. C = *B.cereus*, T = *B.thuringiensis*, S = *B.subtilis*, P = *B.pumilus* and M = *B.megaterium*.

Cluster Analysis

In cluster analysis the distances between strains are transformed to similarities and clustering is achieved using one of several clustering functions. In the studies on the genus *Bacillus* reported here, cluster analysis has provided an initial look at the data and both average linkage and furthest neighbour (complete linkage) clustering techniques have been applied. In average linkage cluster analysis, points join a cluster or clusters themselves combine at the average level of similarity between any 2 elements of the new cluster whereas complete linkage cluster analysis allows units to form clusters or clusters to combine at the lowest level of similarity between any 2 elements of the new cluster. These techniques are limited in their application to p.g.l.c. data because in many cases incomplete discrimination is obtained. This is believed to be due to internal variation within groups and the high proportion of redundant peaks [MacFie *et al.*, 1978] (Fig. 12).

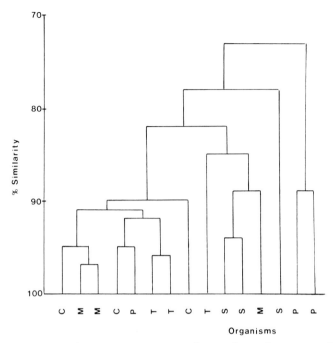

Fig. 9 Average linkage cluster analysis on data from spored cultures. Baseline set straight across. C = *B.cereus*, T = *B.thuringiensis*, S = *B.subtilis*, P = *B.pumilus* and M = *B.megaterium*.

Canonical Variates Analysis

The directions of maximum variation among samples (principal components) are not necessarily the directions that show the greatest variation between groups of pyrograms. Canonical variates analysis, by redefining the distances between groups of points in terms of Mahalanobis D^2 - a generalized concept which takes into account the scatter of samples around the mean - finds successive orthogonal directions that show the maximum ratio of between group to within group variation. The result of this analysis is to produce a 2-dimensional plot representing the positions of samples relative to the directions of maximum variation between groups. The coefficients of the weighted sums used to form the canonical variate scores are all scaled so that the points display a variance of one. Thus, it is possible to draw circular confidence intervals around the population means on the canonical variate plots. This provides a strategy for determining the significance of the observed differences and may provide a method for identifying new isolates [MacFie *et al.*, 1978].

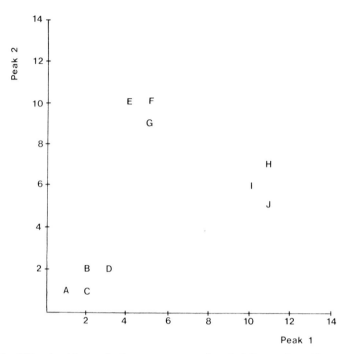

Fig. 10 Illustration of the concepts of multidimensionality using only two variables (peak heights). Peak 1 *versus* Peak 2 for A (1.1), B (2.2), C (2.1), D (3.2), E (4.10), F (5.10), G (5.9), H (11.7), I (10.6), J (11.5).

When using canonical variates analysis it is necessary to have prior knowledge of the taxonomic structure to be applied. By allocating strains to groups then requiring the analysis to find any actual differences in the pyrograms of the group members, canonical variates provides an ideal method of determining whether or not groups of organisms are distinct or similar. In addition, the ability of the analysis to weight a different set of variables in such a way as to impose a particular grouping structure enables this approach to mimic taxonomies derived by other methods.

Stepwise Discriminant Function Analysis

A high proportion of redundant variables (peaks which do not contribute to the discrimination) exist in p.g.l.c. studies. Stepwise discriminant analysis is used to find the subset of variables which allow maximum separation of groups. Peaks are entered into the classification function one at a time. This process is halted when the addition of new variables fails to improve the classification.

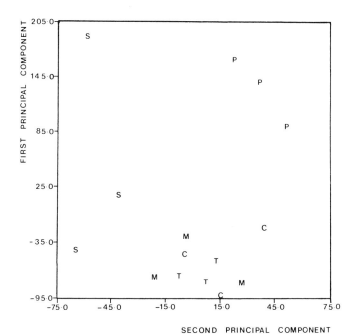

Fig. 11 Principal components analysis on *B.subtilis* (S), *B.pumilus* (P), *B.cereus* (C), *B.thuringiensis* (T) and *B.megaterium* (M). Plot represents 88.4% of the variation between samples.

At each step a one-way analysis of variance, F̲ statistic, is used to determine the next variable. The first variable (Step 1) is that in which the means differ most. Subsequent F̲ statistics are calculated taking into account the variables already used. When a variable is included, the analysis recomputes the classification function. After all the variables have been entered the program lists the distance (Mahalanobis D^2) of each strain from the centre of each of the *a priori* groups. Strains are then allocated to the group to which they are closest. This information provides a useful method of determining the stability of a particular *a priori* structure applied to the canonical variates analysis. The discriminant analysis procedure can be considered successful if a large percentage of the strains are jack-knifed (reallocated) correctly. Under these conditions it is reasonable to assume that group differences do exist and that the peaks selected reflect these differences. Furthermore this program (BMDP7M) [Dixon, 1975] by printing the distances of the individual strains from the group means allows conclusions to be drawn regarding the heterogeneity of the groups found using a particular subset of peaks.

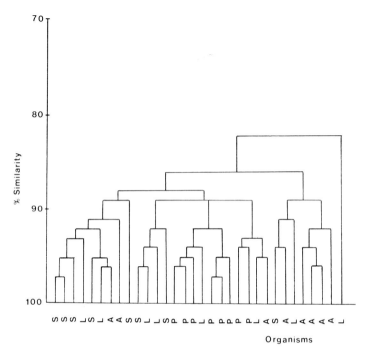

Fig. 12 Average linkage cluster analysis on 32 strains representing 4 groups, S = *B.subtilis*, P = *B.pumilus*, L = *B.licheniformis* and A = *B.amyloliquefaciens*.

Groups Studied Using P.G.L.C.

"Subtilis-*spectrum*" *and* B.amyloliquefaciens

Although a rapid means of obtaining a microbial profile (approx. 1 h from pure culture), p.g.l.c. is limited by its inability to handle large numbers of samples. As a result, establishing a data base involves many months of work and it is for this reason that present work has been concentrated on those areas of *Bacillus* taxonomy which present clearly defined problems.

The first study was to investigate the relationship between the "subtilis-spectrum" (*B.subtilis, B.pumilus* and *B.licheniformis*) and *B.amyloliquefaciens*. The "subtilis-spectrum" represents a well defined group of species which on biochemical analysis share many common properties and show relatively few distinguishing characteristics [Gordon *et al.*, 1973]. *B.amyloliquefaciens* has been described as a species distinct from *B.subtilis* by Welker and Campbell [1967a, 1967b], Seki *et al.* [1975, 1978] and Baptist *et al.* [1978]. This separation has been based primarily on differing DNA-DNA homology, α-amylase production, marker

transduction and enzyme profiles. Gordon *et al.* [1973],
on examination of the data of Smith *et al.* [1952], failed
to substantiate this separation and therefore proposed
that *B.subtilis* and *B.amyloliquefaciens* remain 1 species.

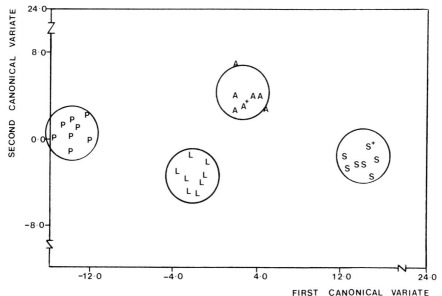

Fig. 13 Plot of the strain means of the 32 organisms used relative to
the first 2 canonical variate axes. S = *B.subtilis*, P = *B.pumilus*,
L = *B.licheniformis* and A = *B.amyloliquefaciens*. Points marked +
represent the position of more than one strain.

Figure 13 shows the discrimination obtained by applying
canonical variates analysis to pyrograms of non-spored
cultures. The circles represent a 95% confidence interval
set to the population mean and are used to show the degree
of separation of groups. Since there are 4 groups only 3
canonical variate axes are required to display 100% of the
variation between groups. In this case plotting the third
canonical variate adds only 2% to the variation since 98%
is displayed by the first two.
Since canonical variates analysis separates groups of
points according to an *a priori* grouping it was necessary
to test the validity of the applied structure. This was
done by analysing the same strains using DNA-DNA reassociation, biochemical tests performed in the classical way
and API systems (API Laboratory Products Ltd., Invincible
Road, Farnborough, Hants. GU14 8QH, UK) [O'Donnell *et al.*,
1980]. Table 1 summarizes the results of the DNA-DNA
pairing studies. In every case but one (a strain of
B.subtilis showing greater homology to *B.amyloliquefaciens*)
the initial identity was confirmed and homologies suggested

TABLE 1

Summary table of DNA-DNA hybridization data on
B.subtilis, B.pumilus, B.licheniformis *and* B.amyloliquefaciens

Organisms	DNA-DNA hybridization
B.subtilis	7/8 correctly identified
B.pumilus	8/8 correctly identified
B.licheniformis	8/8 correctly identified
B.amyloliquefaciens	8/8 correctly identified

the existence of *B.subtilis* and *B.amyloliquefaciens* as
separate groups. Figure 14 shows a complete linkage
cluster analysis performed on the API data. These data
were derived from the API 2OE, 5OE, APIZYM, ZYM II, AP1,
AP2 and AP3 test strips and comprised 119 tests [see Logan
and Berkeley, Chapter 6]. Similarities were calculated
using the co-efficient of Gower [1971] and clustering
achieved by complete linkage using the GENSTAT package.
The separation into groups is consistent with that
obtained using p.g.l.c. and DNA-DNA pairing although 2
B.subtilis strains form a small separate cluster which
joins the main cluster at 65% similarity. On examination
of the strains using the methods of Gordon *et al.* [1973]
it was possible to separate *B.pumilus* and *B.licheniformis*
from each other and from *B.subtilis* and *B.amylolique-
faciens* but impossible to separate further *B.subtilis* from
B.amyloliquefaciens.

A plot such as Fig. 13 is sufficient for the purposes
of discrimination between groups but to allocate unknowns
into groups (identification) using canonical variate axes
the stability of these axes must first be verified.
MacFie *et al.* [1978] have listed among the requirements
for such a system, the need to reduce variables that do
not contribute to the discrimination and a means of testing how representative the axes are of the types of micro-
organism within a particular sampling habitat. Stepwise
discriminant function analysis and jack-knifing satisfy
both these parameters.

Figure 15 shows the effect of including peaks, re-
arranged in order of contribution to the plot, on both the
per cent correctly classified and per cent correctly jack-
knifed. Optimum jack-knifed classification, after including 14 peaks, is 84%. This means that of the 32 strains
analysed and allocated to groups, 5 do not reallocate
correctly. The inclusion of additional peaks to the analysis lowers rather than improves the jack-knifed values
but has no effect on the classification values. A feature
of the program used (BMDP7M) is that it provides a summary
table showing the misclassified strains and their distances
from the centre of the various groups (Table 2).

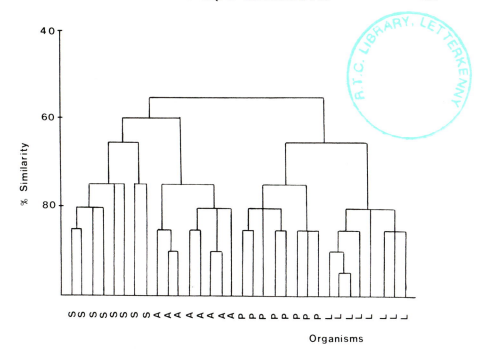

Fig. 14 Complete linkage cluster analysis on the API data. S = *B. subtilis*, P = *B.pumilus*, L = *B.licheniformis* and A = *B.amyloliquefaciens*.

The strain of *B.subtilis* (No. 3) identified by DNA-DNA hybridization as *B.amyloliquefaciens* has, on jack-knifing, been reallocated to the *B.amyloliquefaciens* group. The close relationship between the strains used in this study is reflected in the distances shown in Table 2. For example, *B.subtilis* 6 has been reallocated to the *B.licheniformis* group since its distance from the mean of group 3 (*B.licheniformis*) is 48.1, whereas it is 50.1 units from group 1 (*B.subtilis*). Furthermore, certain strains such as *B.subtilis* 8 and *B.amyloliquefaciens* 32, which on jack-knifing reallocate to their proposed groups, show a very close relationship to different groups. Nevertheless, the overall correctly jack-knifed value of 84% supports the validity of performing canonical variates analysis on the 4 groups and indicates that group differences do exist.

Comparison of Sporulated and Non-sporulated Cultures

For analysis of the "subtilis-spectrum", non-spored cultures were used since the differences between vegetative cells of the various strains were thought to be greater than those between spored cells and thus would provide more information for characterization. Nevertheless, in the

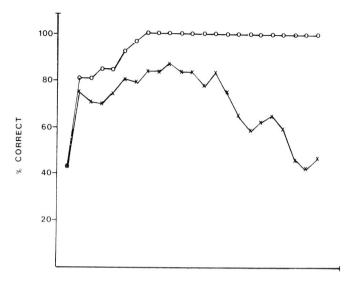

STEP Nº: 1 2 3 4 5 6 7 8 9 10 11 12 13 14 15 16 17 18 19 20 21 22 23

PEAK Nº: 23 17 8 6 15 9 5 13 10 12 1 18 22 14 7 11 3 2 16 20 19 21 4

Fig. 15 Plot of per cent correctly jack-knifed (X) and per cent correctly classified (0) against peak number for the *B.subtilis*, *B. pumilus*, *B.licheniformis* and *B.amyloliquefaciens* study.

case of organisms such as *B.thuringiensis*, the sporangia of which contain a large protein crystal in addition to a typical endospore, the pyrolysis products of spored cultures may be of more interest than products of vegetative cells.   To test whether or not discrimination was possible using sporulated cultures a small study, involving 3 strains each of *B.subtilis*, *B.pumilus*, *B.cereus*, *B.thuringiensis* and *B.megaterium*, was undertaken.   Each culture was grown on nutrient agar + 2 g litre^{-1} glucose and incubated at 30°C for 14 hours and 5 days to obtain both non-sporulated and sporulated cultures.   The baseline and set of peaks used for the analyses of spored cultures were different from those used to analyse the non-spored cultures. This was due to changes in the pyrograms associated with sporulation and storage polymer production (see Figs. 3 and 4).   Figures 16 and 17 show the result of applying canonical variates to each set of data.   Figure 16 shows that in this study non-spored cultures of *B.subtilis* and *B.pumilus* do not separate from each other, a result apparently contradictory to results shown previously (Fig. 13).   In canonical variates analysis the inclusion of strains which are very different can cause distortion of the plot in such a way as to force together closely related groups.   A situation in which the first canonical variate accounts for 60% of the variation leaves 40% to be divided amongst the remaining

TABLE 2

*Output from program BMDP7M showing distances of strains
1-32 from each of the 4 group means representing* B.subtilis,
B.pumilus, B.licheniformis *and* B.amyloliquefaciens.

Organism	Strain No.	Jack-knifed Mahalanobis D^2				Jack-knifed group
		1	2	3	4	
B.subtilis (1)	1	28.3	141.4	67.8	63.5	1
	2	45.9	130.8	54.0	46.3	1
	3	64.9	143.8	76.1	48.9	4**
	4	34.5	137.8	83.5	81.4	1
	5	16.2	158.9	58.9	48.7	1
	6	50.1	131.0	48.1	54.7	3**
	7	9.9	175.5	83.6	24.6	1
	8	81.8	162.4	127.5	81.9	1
B.pumilus (2)	9	182.7	18.9	69.2	138.0	2
	10	283.8	30.8	115.0	188.1	2
	11	121.4	14.7	32.2	122.9	2
	12	162.3	18.1	62.9	143.4	2
	13	101.8	27.8	38.8	85.5	2
	14	139.4	42.8	101.8	131.3	2
	15	148.0	84.7	62.4	179.9	3**
	16	102.5	31.8	41.8	82.6	2
B.licheniformis (3)	17	104.3	81.4	36.5	140.2	3
	18	58.5	67.4	21.7	79.0	3
	19	80.0	26.3	14.0	83.1	3
	20	47.5	93.2	58.8	70.0	1**
	21	82.3	46.3	42.5	52.5	3
	22	90.0	39.0	23.4	98.7	3
	23	36.1	74.7	20.0	45.1	3
	24	46.8	74.4	17.5	70.9	3
B.amyloliquefaciens (4)	25	93.4	92.5	84.8	79.0	4
	26	49.2	188.1	94.8	60.6	1**
	27	102.4	209.7	168.5	41.6	4
	28	99.1	257.6	151.6	71.3	4
	29	45.9	127.0	97.0	20.0	4
	30	35.4	83.5	47.4	15.4	4
	31	21.7	92.7	47.9	13.8	4
	32	39.4	140.9	62.7	36.5	4

**Represents a strain showing greater affinity to a group other than
that to which it was originally assigned.

variates. The inclusion of a further group showing very
different characteristics to those already present would
result in a greater proportion of the variation being expressed in the first canonical variate. Consequently the
variation available for distribution among remaining variates would be reduced and groups previously separated may
now overlap.

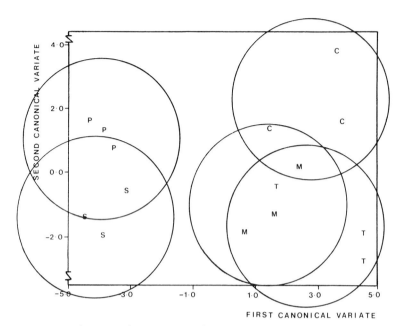

Fig. 16 Canonical variates analysis on 5 groups of non-spored cells.
S = *B.subtilis*, P = *B.pumilus*, C = *B.cereus*, T = *B.thuringiensis* and
M = *B.megaterium*.

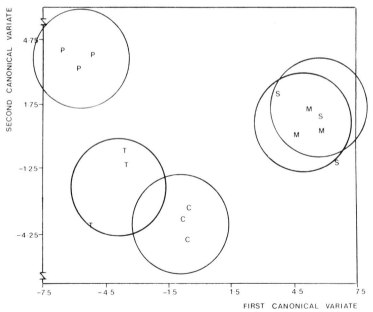

Fig. 17 Canonical variates analysis on 5 groups of sporulated cultures. S = *B.subtilis*, P = *B.pumilus*, C = *B.cereus*, T = *B.thuringien-
sis* and M = *B.megaterium*.

Figure 17 indicates that spored cells are in fact sufficiently diverse to allow characterization of species. With
regard to the overlap between *B.subtilis* and *B.megaterium*
it is conceivable that sporulated cultures of these species
are very similar, but the most reasonable explanation is an
inadequate definition of the groups, since only 3 strains
were used for this purpose.
Comparison of the dendrogram obtained using data obtained from spored cultures (Fig. 8) with that from non-
spored cultures (Fig. 18) indicates that sporulation has a
stabilizing effect on the samples which makes it easier to
obtain standardized preparations. Furthermore, spored
cultures of those organisms which in their vegetative state
produced storage polymers (see Figs. 3 and 4) were easier
to analyse since baseline rise and time taken to clean the
column prior to subsequent analyses were greatly reduced.
As a consequence of these studies pyrolysis of spored
samples has been employed whenever possible.

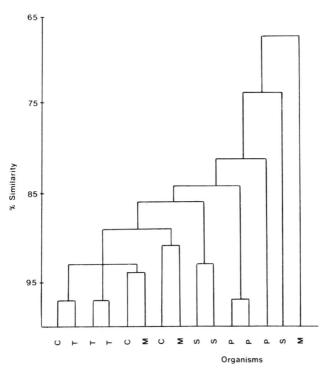

Fig. 18 Average linkage cluster analysis on non-spored data. Baseline drawn across complexes of peaks. C = *B.cereus*, T = *B.thuringien-
sis*, S = *B.subtilis*, P = *B.pumilus* and M = *B.megaterium*.

Relationships within the B.cereus *Group*

The description of *B.anthracis*, *B.thuringiensis* and
B.mycoides as varieties of *B.cereus* by Smith *et al.* [1946,
1952] has caused a great deal of controversy among microbiologists. Particularly concerned were those interested
in the causative agents of disease in man and insects. As
a result the literature on relationships between these
groups is extensive [see Gordon *et al.*, 1973, pp.23-34].
The primary reason for giving these strains varietal rather
than species status is the instability of distinguishing
characteristics. In addition chemotaxonomic studies such
as DNA-DNA reassociation [Somerville and Jones, 1972;
Seki *et al.*, 1978; Kaneko *et al.*, 1978], spore precipiti-
nogens [Norris, 1961] and fatty acid analysis [Vivoli and
Fabio, 1967] have shown that these organisms share common
chemical components. Such methods look at the chemistry
of specific compounds whereas p.g.l.c. enables the overall
composition to be examined. Consequently the pyrogram may
reflect similarity or dissimilarity not evident using other
techniques.

Eighteen strains of *B.thuringiensis*, 9 *B.mycoides*, 10
B.cereus non-food poisoning strains, 7 *B.anthracis* and 10
B.cereus food poisoning strains (5 vomiting, 5 diarrhoeal
[see Gilbert *et al.*, Chapter 12]) were grown on nutrient agar
until well sporulated. Curie-point pyrolysis was carried
out at 610°C and data treated as before. Cultures of *B.
anthracis* were applied to the Curie-point wires at The
Microbiological Research Establishment, Porton Down, and
autoclaved prior to dispatch to the Meat Research
Institute. As controls to detect possible changes due to
autoclaving, several *B.cereus* strains were similarly prepared. Due to gas chromatographic difficulties it was
necessary to divide the data. Two separate canonical
variate analyses were performed. The groups used as input
were as follows:
1. *B.thuringiensis*; *B.mycoides*; *B.cereus* (non-food)
    (Fig.19)
2. *B.anthracis* (autoclaved); *B.cereus* (non-food);
    B.cereus (autoclaved); *B.cereus* (food) (Fig. 21).

Figure 19 represents 100% of the between group variation and shows that low resolution pyrograms of *B.thurin-
giensis*, *B.cereus* and *B.mycoides* are consistently different
and allow separation into 3 distinct groups. Comparison of
this plot with Fig. 17 stresses the importance of defining
groups in space with large numbers of strains. The positions of the groups along the first canonical variate
suggests that *B.cereus* and *B.mycoides* are more closely related to one another than to *B.thuringiensis*.

Figure 20 summarizes the results of applying stepwise
discriminant analysis, with jack-knifing, to these data.
The order of peak inclusion is given along the bottom. The
percentage of samples correctly allocated, before jack-

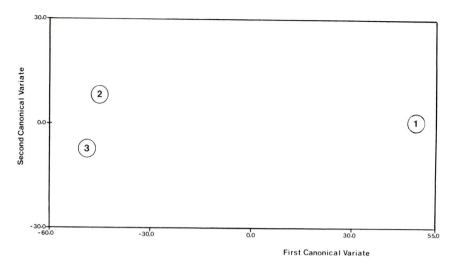

Fig. 19 Canonical variates analysis on strains representing *B.thurin-giensis* (1), *B.mycoides* (2) and *B.cereus* (3). Circles define 95% confidence regions for each species.

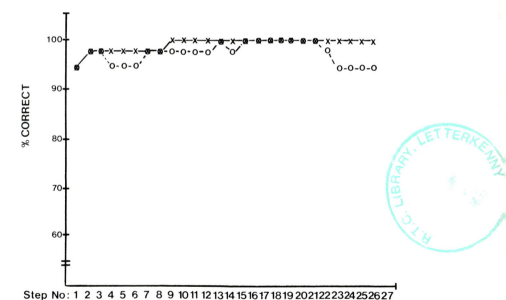

Fig. 20 Plot of per cent correctly jack-knifed (O) and per cent correctly classified (X) against peak number for the *B.thuringiensis*, *B.cereus*, *B.mycoides* study. N^+ = Peak 8 not entered, because no extra discrimination achieved.

knifing, reaches 100% after only 2 peaks (10, 19) have been
included. At this stage only strain 26 is not correctly
reallocated after jack-knifing. All strains are correctly
allocated (before jack-knifing) after 9 peaks have been
included, and with 13 peaks included all strains are also
correctly reallocated after jack-knifing. As more peaks
are included the number of strains correctly reallocated
fluctuates and drops to 35 out of 37. This plot succinctly
expresses the practical problems in selecting the most
stable subset for discrimination. Simple acceptance of the
13 peak solution ignores the fact that the only difference
between this and the 2 peak solution is that sample 26 is
correctly reallocated. Table 3 lists the Mahalanobis
distances (squared) of each strain from each species mean.
This table shows the stability of the 3 groups and confirms
the relationships between them.

Figure 21 shows that there is a sense in which pyrograms
representing *B.cereus*, *B.cereus* (food poisoning strains)
and *B.anthracis* are sufficiently different to allow separation into 3 groups. However, the separation of *B.cereus*
and *B.cereus* (autoclaved) suggests that the separation of
B.anthracis from *B.cereus* is in part due to autoclaving
samples prior to analysis. Possible explanations for
differences between autoclaved and non-autoclaved cells are
(1) changes in the chemical nature of the sample or (2)
changes in the temperature time characteristics of the

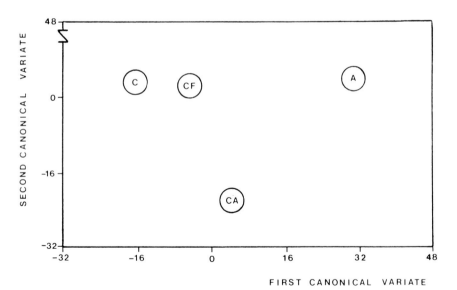

Fig. 21 Canonical variates analysis on strains representing
B.anthracis (autoclaved) (A); *B.cereus* (non-food) (C); *B.cereus*
(autoclaved) (CA); *B.cereus* (food poisoning) (CF).

TABLE 3

Output from program BMDP7M, showing distances, calculated on a subset of 2 peaks, of strains from the group means of B.thuringiensis *(1),* B.mycoides *(2) and* B.cereus *(3)*

Organism	Strain No.	Jack-knifed Mahalanobis D^2			Jack-knifed group
		1	2	3	
B. thuringiensis (1)	1	19.1	755.0	725.9	1
	2	0.3	712.3	667.5	1
	3	0.1	718.5	673.4	1
	4	0.7	762.0	720.7	1
	5	0.3	745.0	702.2	1
	6	0.1	727.7	680.6	1
	7	1.2	775.6	736.1	1
	8	0.1	726.2	681.5	1
	9	0.0	723.5	677.9	1
	10	0.1	733.2	688.8	1
	11	0.1	752.0	706.5	1
	12	0.3	713.6	667.4	1
	13	1.1	792.0	749.9	1
	14	0.1	744.2	698.6	1
	15	0.1	744.2	698.6	1
	16	0.2	752.8	708.6	1
	17	1.0	761.5	707.0	1
	18	0.5	771.9	721.1	1
B.mycoides (2)	19	1225.5	15.1	23.1	2
	20	743.5	0.0	11.7	2
	21	976.2	7.7	31.9	2
	22	714.7	12.8	45.4	2
	23	705.9	3.3	21.4	2
	24	710.2	1.0	15.1	2
	25	704.7	4.0	4.4	2
	26	747.8	29.3	1.2	3**
	27	757.0	1.4	6.8	2
B.cereus (3)	28	721.8	11.1	0.4	3
	29	756.5	10.9	1.1	3
	30	720.1	5.2	1.5	3
	31	715.5	8.2	0.4	3
	32	754.0	7.1	1.5	3
	33	662.3	15.6	0.8	3
	34	660.3	22.3	3.8	3
	35	662.7	16.1	0.9	3
	36	685.0	14.0	0.2	3
	37	669.7	10.3	0.3	3

**Represents a strain assigned to the wrong group by selecting minimum Jack-knifed Mahalanobis D^2.

Curie-point wires. Nevertheless, the plot also shows that
there are sufficient differences between autoclaved samples
of *B.cereus* and *B.anthracis* to enable their easy separation
from each other. Examination of Fig. 22 suggests that this
classification of samples can be achieved using only 3
peaks (12, 2, 18). Table 4 shows that with 3 peaks only 1
sample (96.9% correctly jack-knifed), a *B.cereus* strain reallocated to the *B.cereus* food poisoning group is in-
correctly classified. Examination of the distances of the
10 food poisoning strains from the mean of the *B.cereus*
group suggests that there may be 2 "groups" of *B.cereus*
food poisoning strains. The first of these "groups", that
with a mean distance from *B.cereus* of 4.7 units, was shown
to contain 4 out of 5 strains giving diarrhoeal symptoms
whilst the other "group", mean distance of 17.6 units,
comprised 4 out of 5 strains responsible for vomiting
symptoms. As a result of this observation a 3-group
canonical variates analysis was performed and the output
shown in Fig. 23.

As suspected from Table 4, the food poisoning organisms
form 2 distinct groups. Examination of the output from
stepwise discriminant function analysis and jack-knifing
(Fig. 24; Table 5) indicates the stability of the groups
(95% correctly jack-knifed) and the subset of peaks

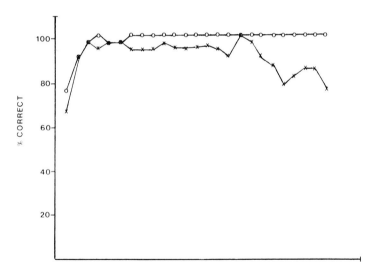

STEP NO 1 2 3 4 5 6 7 8 9 10 11 12 13 14 15 16 17 18 19 20 21 22 23 24 25

PEAK NO 12 2 18 25 15 9 10 6 7 5 4 26 16 22 24 27 1 17 21 13 14 3 23 19 11

Fig. 22 Plot of per cent correctly jack-knifed (X) and per cent
correctly classified (O) against peak number for the *B.cereus*,
B.anthracis (autoclaved), *B.cereus* (autoclaved) and *B.cereus* (food
poisoning) study.

TABLE 4

*Output from program BMDP7M showing distances of strains
1-32 from group means representing* B.anthracis *(autoclaved),*
B.cereus *(food poisoning),* B.cereus *(autoclaved) and* B.cereus

Organism	Strain No.	Jack-knifed Mahalanobis D^2				Jack-knifed group
		1	2	3	4	
B.anthracis (1)	1	9.9	58.9	120.7	25.2	1
	2	9.0	64.4	55.0	43.5	1
	3	4.1	31.9	46.1	14.8	1
	4	3.9	35.2	63.3	11.7	1
	5	2.4	39.4	49.5	20.3	1
	6	4.0	70.4	101.1	32.9	1
	7	0.8	47.2	68.3	20.2	1
B.cereus	8	72.6	6.0	32.3	24.5	2
(food poisoning) (2)	9	32.4	2.0	35.7	2.7	2
	10	44.0	1.7	25.9	11.3	2
	11	82.6	5.7	42.2	24.0	2
	12	49.4	6.5	24.1	12.9	2
	13	38.4	2.1	40.7	5.7	2
	14	61.2	1.8	33.9	15.4	2
	15	41.1	0.6	33.6	5.5	2
	16	42.5	2.4	35.2	6.0	2
	17	28.4	6.2	25.3	3.8	4**
B.cereus	18	75.9	92.7	19.2	101.1	3
(autoclaved) (3)	19	47.6	21.6	8.3	34.5	3
	20	77.9	25.6	4.0	46.5	3
	21	52.2	18.6	7.1	32.0	3
	22	111.5	47.4	5.9	79.6	3
B.cereus (4)	23	11.0	15.4	50.6	3.5	4
	24	13.4	11.9	47.1	2.8	4
	25	18.0	10.8	45.4	1.2	4
	26	22.5	14.9	72.2	4.9	4
	27	21.1	9.3	48.1	0.7	4
	28	27.2	3.9	42.2	1.1	4
	29	33.9	5.8	43.8	3.2	4
	30	18.0	10.8	45.4	1.2	4
	31	24.4	6.8	45.3	0.7	4
	32	25.3	6.0	48.9	0.5	4

**Represents a strain showing greater affinity to a group other than
that to which it was originally assigned.

necessary to achieve this discrimination (12, 5, 17, 19,
1, 18, 27, 9, 25, 11, 14, 8, 15). However, some of the
Mahalanobis D^2 values are very large and this suggests a
high level of variation within groups characterized by
this subset of peaks. It may well be that there is an
alternative grouping structure for these pyrograms.

TABLE 5

*Output from BMDP7M showing distances of strains 1-20 from
group means representing* B.cereus *(emetic),*
B.cereus *(diarrhoeal) and* B.cereus *(non-food)*

Organism	Strain No.	Jack-knifed Mahalanobis D^2			Jack-knifed group
		1	2	3	
B.cereus (emetic) (1)	1	35.9	283.5	136.6	1
	3	127.5	294.3	377.7	1
	5	114.4	326.3	116.6	1
	6	114.1	543.2	183.0	1
	7	73.3	638.7	569.2	1
B.cereus (diarrhoeal)	2	310.8	32.0	364.2	2
(2)	4	323.9	152.1	315.5	2
	8	331.8	18.6	352.0	2
	9	509.9	23.9	530.0	2
	10	296.1	215.0	318.3	2
B.cereus (non-food)	11	272.1	362.3	139.6	3
(3)	12	149.0	369.6	61.7	3
	13	138.8	329.6	23.4	3
	14	142.2	313.8	37.0	3
	15	106.4	314.1	30.1	3
	16	251.0	651.4	39.1	3
	17	268.2	365.9	148.9	3
	18	101.6	1140.8	111.1	1**
	19	268.0	976.7	228.0	3
	20	439.2	548.9	233.9	3

**Represents a strain showing greater affinity to a group other than
that to which it was originally assigned.

Discussion

Pyrolysis gas-liquid chromatography is a technique
developed and exploited by the analytical chemists. Only
recently has it found microbiological applications. Its
extensive use in other fields for analysing biological
materials indicates its future importance in microbiology.
There are, however, certain problems such as column degradation, which make the immediate future of g.l.c., as an
analysis system for microbial pyrolysates, suspect. Consequently several workers are now using the mass spectro-
meter to analyse bacterial pyrolysates [Meuzelaar and
Kistemaker, 1973; Meuzelaar *et al.*, 1976]. Mass spectrometry is particularly promising because of its low
analysis time (approximately 2 mins) and absence of deterioration problems. Nevertheless, column technology is
advancing rapidly and many of the problems associated with
using g.l.c. as an analysis system may possibly be overcome.

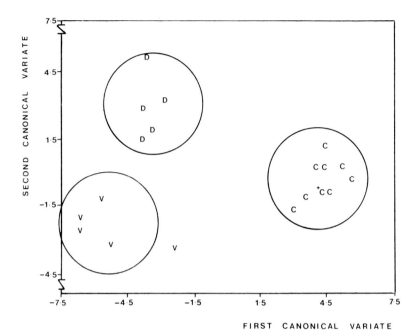

Fig. 23 Canonical variates analysis on 10 strains of *B.cereus* (C), 5 strains of *B.cereus* (emetic) (V) and 5 strains of *B.cereus* (diarrhoeal) (D).

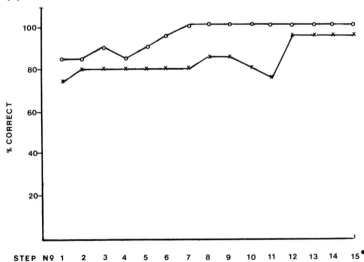

Fig. 24 Plot of per cent correctly jack-knifed (X) and per cent correctly classified (O) against peak number for the *B.cereus* (emetic), *B.cereus* (diarrhoeal) and *B.cereus* (non-food) study. *Analysis halted at step 15 since the inclusion of the remaining peaks failed to improve the classification.

Classical methods of bacterial classification group organisms according to their morphology and behaviour whereas pyrolysis techniques provide chemical profiles of cellular composition. Therefore traditional taxonomies and those derived from p.g.l.c. data might be considerably different. Examination of the natural classification of pyrograms as determined by cluster analysis suggests that this is not the case. Furthermore, the possibility of weighting a different set of peaks in the pyrogram to produce a particular classification (canonical variates analysis) and the ability to test the taxonomic structure obtained (jack-knifing) make p.g.l.c. a powerful tool for testing and improving classifications derived by other methods.

The power of p.g.l.c. and canonical variates analysis for testing relationships derived by other methods is suitably demonstrated by Fig. 13. This plot shows complete separation of *B.subtilis* and *B.amyloliquefaciens* and confirms the results obtained using DNA-DNA reassociation. The pyrolysis process therefore provides structural evidence supporting the species status of *B.amyloliquefaciens*. It is not always the case, however, that the pyrolysis results and DNA-DNA pairing evidence coincide. From the literature available on the status of *B.cereus, B.anthracis, B.mycoides* and *B.thuringiensis* [Somerville and Jones, 1972; Seki *et al.*, 1978; Kaneko *et al.*, 1978], as determined using DNA-DNA reassociation studies, there is strong evidence for proposing that these organisms be considered one species, whereas the techniques described in this chapter support their status as distinct groups. Since it is at present very difficult to relate a particular peak on a pyrogram to its parent molecule in the cell, the influence of cell components, such as the presence or absence of toxins, on the resultant pyrograms is difficult to determine. It may well be that the separation of *B.anthracis* and *B.thuringiensis* from *B.cereus* is due to the presence of particular toxins in the first 2 organisms. Further evidence that toxins may play an important role in determining the quantitative relationships between organisms as reflected in their peak heights is shown in Figs. 21 and 23. Figure 21 shows that it is possible to separate food poisoning *B.cereus* from non-food poisoning *B.cereus* whereas Fig. 23 shows that it is possible, by changing the canonical variates input to reveal differences between the emetic strains of *B.cereus* and the diarrhoeal strains. These differences might be due to a different cell composition or may, in fact, represent differences between the diarrhoeal and emetic toxins responsible for food poisoning.

Underlying all considerations for the use of p.g.l.c. or p.m.s. (pyrolysis mass spectrometry) in microbial classification is the concept that a truly objective analysis demands the application of statistical techniques.

Visual comparison of pyrograms is not an acceptable basis for data analysis. Since canonical variates analysis requires a pre-existing taxonomy these techniques cannot be used to classify organisms. They do, however, provide taxonomists with a valuable method of testing existing classifications. For arbitrary classifications - those in which the *a priori* groups are composed of strains showing little or no relationship to each other - it is possible that, when there are a large number of variates, canonical variates analysis will be able to discriminate between the groups. In such cases, experience has shown that the application of jack-knifing to these groups reveals their instability by yielding very few values that are correctly jack-knifed. For these reasons the techniques described in this chapter should be considered as confirmatory rather than exploratory.

The work reported here has shown that the application of multivariate statistical methods to p.g.l.c. data can result in a classification which corresponds to that obtained using other methods. As a result it is possible to use this system to identify unknowns. When used in this way pyrolysis techniques (p.m.s.) have the potential of identifying unknowns in under 5 mins.

The genus *Bacillus* is currently undergoing intensive investigation using a diversity of techniques. As a result ideas on the relationships between organisms are changing. The range of G + C% found in the genus *Bacillus* (32 to 67) [Gibson and Gordon, 1974; Priest, this volume] suggests that the aerobic sporeformers could be divided into more than one genus. In addition, investigations of novel isolates will probably make it necessary to change the existing classification. Pyrolysis gas-liquid chromatography and discriminatory analysis have a role to play in confirming and expanding upon new developments in *Bacillus* taxonomy.

Acknowledgements

We are grateful to Dr. H.J.H. MacFie for his assistance with the statistical analyses. A.G. O'Donnell thanks the Agricultural Research Council for receipt of a studentship during which this work was undertaken.

References

Baptist, J.N., Mandel, M. and Gherna, R.L. (1978). Comparative zone electrophoresis of enzymes in the genus *Bacillus*. *International Journal of Systematic Bacteriology* **28**, 229-244.

Barbour, W.M. (1965). Refinement of pyrolysis techniques for gas chromatography. *Journal of Gas Chromatography* **3**, 228-231.

Burns, D.T., Stretton, R.J. and Jayatilake, S.D.A.K. (1976). Pyrolysis gas chromatography as an aid to the identification of *Penicillium* species. *Journal of Chromatography* **116**, 107-115.

Cone, R.D. and Lechowich, R.V. (1970). Differentiation of *Clostridium botulinum* types A, B and E by pyrolysis gas-liquid chromatography. *Applied Microbiology* **19**, 138-145.

de Barjac, H. and Bonnefoi, A. (1968). A classification of strains of *Bacillus thuringiensis* Berliner with a key to their differentiation. *Journal of Invertebrate Pathology* **11**, 335-347.

Dixon, W.J. (1975). Biomedical Computer Programs. University of California Press, Los Angeles.

Drucker, D.B. (1976). Gas-liquid chromatographic chemotaxonomy. In "Methods in Microbiology", Vol. 9 (ed. J.R. Norris), pp.51-125. London and New York: Academic Press.

Dyson, N. and Littlewood, A.B. (1968). Seventh international symposium on gas chromatography and its exploitation. *Journal of Gas Chromatography* **6**, 449-454.

Emswiler, B.S. and Kotula, A.W. (1978). Differentiation of serotypes by pyrolysis gas-liquid chromatography. *Applied and Environmental Microbiology* **35**, 97-104.

Fontanges, R., Blandenet, G. and Queignec, R. (1967). Difficultés d'application de la chromatographie en phase gazeuse à l'identification des bactéries. *Annales de l'Institut Pasteur* **112**, 10-23.

Ford, W.W., Lawrence, J.S. and Laubach, C.A. (1916). Studies on aerobic spore-bearing, non-pathogenic bacteria. *Journal of Bacteriology* **1**, 273-316.

Garner, W. and Chi Yuan, F. (1964). Taxonomy of micro-organisms by gas chromatography. *Proceedings of 148th Meeting of American Chemical Society*, Chicago, Illinois.

Giacobbo, H. and Simon, W. (1964). Methodik zur Pyrolyse und anschließenden gaschromatographischen Analyse von Probemengen unter einem Mikrogramm. *Pharmaceutica Acta Helvetiae* **39**, 162-167.

Gibson, T. and Gordon, R.E. (1974). *Bacillus*. In "Bergey's Manual of Determinative Bacteriology, 8th Edition" (eds. R.E. Buchanan and N.E. Gibbons), pp.529-550. Baltimore: The Williams and Wilkins Co.

Gordon, R.E., Haynes, W.C. and Pang, C.H.-N. (1973). "The Genus *Bacillus*". Washington, D.C.: United States Department of Agriculture.

Gower, J.C. (1971). A general coefficient of similarity and some of its properties. *Biometrics* **27**, 857-872.

Gutteridge, C.S., MacFie, H.J.H. and Norris, J.R..(1979). Use of principal components analysis for displaying variation between pyrograms of micro-organisms. *Journal of Analytical and Applied Pyrolysis* **1**, 67-76.

Gutteridge, C.S. and Norris, J.R. (1979). A review: the application of pyrolysis techniques to the identification of microorganisms. *Journal of Applied Bacteriology* **47**, 5-43.

Janak, J. (1960). Identification of the structure of non-volatile organic substances by gas chromatography of pyrolytic products. *Nature* **185**, 684-686.

Jayne-Williams, D.J. and Cheeseman, G.C. (1960). The differentiation of bacterial species by paper chromatography. IX. The genus *Bacillus*: a preliminary investigation. *Journal of Applied Bacteriology* **23**, 250-268.

Kaneda, T. (1967). Fatty acids in the genus *Bacillus*. I. Iso- and anteiso-fatty acids as characteristic constituents of lipids in 10 species. *Journal of Bacteriology* **93**, 894-903.

Kaneda, T. (1968). Fatty acids in the genus *Bacillus*. II. Similarity in the fatty acid compositions of *Bacillus thuringiensis*, *Bacillus anthracis* and *Bacillus cereus*. *Journal of Bacteriology* **95**, 2210-2216.

Kaneko, T., Nozaki, R. and Aizawa, K. (1978). Deoxyribonucleic acid relatedness between *Bacillus anthracis*, *Bacillus cereus* and *Bacillus thuringiensis*. *Microbiology and Immunology* **22**, 639-641.

Kulik, M.M. and Vincent, P.G. (1973). Pyrolysis gas-liquid chromatography of fungi: observations on variability among nine *Penicillium* species of the section *asymmetrica*, subsection *fasiculata*. *Mycopathologica et Mycologia Applicata* **51**, 1-18.

Levy, R.L. (1966). Pyrolysis gas chromatography: a review of the technique. *Chromatographic Reviews* **8**, 48-89.

Levy, R.L. (1967). Trends and advances in design of pyrolysis units for gas chromatography. *Journal of Gas Chromatography* **5**, 107-113.

MacFie, H.J.H. and Gutteridge, C.S. (1978). Analysis of pyrolysis gas-liquid chromatography data using multivariate statistical techniques. *Journal of Applied Bacteriology* **45**, iv-v.

MacFie, H.J.H., Gutteridge, C.S. and Norris, J.R. (1978). Use of canonical variates analysis in differentiation of bacteria by pyrolysis gas-liquid chromatography. *Journal of General Microbiology* **104**, 67-74.

Meuzelaar, H.L.C. (1974). "Identification of bacteria by pyrolysis gas chromatography and pyrolysis mass spectrometry." Ph.D. Thesis, University of Amsterdam.

Meuzelaar, H.L.C. and Kistemaker, P.G. (1973). A technique for fast and reproducible fingerprinting of bacteria by pyrolysis mass spectrometry. *Analytical Chemistry* **45**, 587-590.

Meuzelaar, H.L.C., Kistemaker, P.G. and Tom, A. (1975). Rapid and automated identification of micro-organisms by Curie-point pyrolysis techniques. I. Differentiation of bacterial strains by fully automated Curie-point pyrolysis gas-liquid chromatography. In "New Approaches to the Identification of Micro-organisms" (eds. C.G. Hedén and T. Illéni), pp.165-178. London: Wiley and Sons.

Meuzelaar, H.L.C., Kistemaker, P.G., Eshuis, W. and Engel, H.W.B. (1976). Progress in automated and computerised characterisation of micro-organisms by pyrolysis mass spectrometry. In "Proceedings of the 2nd International Symposium on Rapid Methods and Automation in Microbiology", pp.225-230. Oxford: Learned Information Ltd.

Needleman, M. and Stuchbery, P. (1977). The identification of micro-organisms by pyrolysis gas-liquid chromatography. In "Analytical Pyrolysis" (eds. C.E.R. Jones and C.A. Cramers), pp.77-88. Amsterdam: Elsevier Scientific Publishing Co.

Norris, J.R. (1961). Bacteriophages of *Bacillus cereus* and of crystal forming insect pathogens related to *B.cereus*. *Journal of General Microbiology* **26**, 167-173.

Norris, J.R. and Wolf, J. (1957). Some serological properties of aerobic spore-forming bacilli. *Journal of Applied Bacteriology* **20**, xi.

Norris, J.R. and Baillie, A. (1964). Immunological specificities of spore and vegetative cell catalases of *Bacillus cereus*. *Journal of Bacteriology* **88**, 264-265.

O'Donnell, A.G., Norris, J.R., Berkeley, R.C.W., Claus, D., Kaneko, T., Logan, N.A. and Nozaki, R. (1980). Characterisation of *Bacillus subtilis*, *Bacillus pumilus*, *Bacillus licheniformis* and *Bacillus amyloliquefaciens* by pyrolysis gas-liquid chromatography and by deoxyribonucleic acid (DNA)-DNA hybridization, biochemical tests and API systems. *International Journal of Systematic Bacteriology* **30**, 448-459.

Oxborrow, G.S., Fields, N.D. and Puleo, J.R. (1976). Preparation of pure microbiological samples for pyrolysis gas-liquid chromatography studies. *Applied and Environmental Microbiology* **32**, 306-309.

Oxborrow, G.S., Fields, N.D. and Puleo, J.R. (1977a). Pyrolysis gas-liquid chromatography studies of the genus *Bacillus*. Effect of growth time on pyrochromatogram reproducibility. In "Analytical Pyrolysis" (eds. C.E.R. Jones and C.A. Cramers), pp.69-76. Amsterdam: Elsevier Scientific Publishing Co.

Oxborrow, G.S., Fields, N.D. and Puleo, J.R. (1977b). Pyrolysis gas-liquid chromatography of the genus *Bacillus*: effect of growth media on pyrochromatogram reproducibility. *Applied and Environmental Microbiology* **33**, 865-870.

Oyama, V.I. (1963). Use of gas chromatography for the detection of life on Mars. *Nature* (London) **200**, 1058-1059.

Quinn, P.A. (1974). Development of high resolution pyrolysis gas chromatography for the identification of micro-organisms. *Journal of Chromatographic Science* **12**, 796-806.

Quinn, P.A. (1976). Identification of microorganisms by pyrolysis: the state of the art. In "Proceedings of the 2nd International Symposium on Rapid Methods and Automation in Microbiology", pp.178-186. Oxford: Learned Information Ltd.

Reiner, E. (1965). Identification of bacterial strains by pyrolysis gas-liquid chromatography. *Nature* (London) **206**, 1272-1274.

Reiner, E. and Kubica, G.P. (1969). Predictive value of pyrolysis gas-liquid chromatography in the differentiation of mycobacteria. *American Review of Respiratory Diseases* **99**, 42-49.

Reiner, E., Hicks, J.J., Ball, M.M. and Martin, W.J. (1972). Rapid characterisation of *Salmonella* organisms by means of pyrolysis gas-liquid chromatography. *Analytical Chemistry* **44**, 1058-1061.

Sekhon, A.S. and Carmichael, J.W. (1972). Pyrolysis gas-liquid chromatography of some dermatophytes. *Canadian Journal of Microbiology* **18**, 1593-1601.

Sekhon, A.S. and Carmichael, J.W. (1973). Column variation affecting a pyrolysis gas-liquid chromatographic study of strain variation in two species of *Nannizzia*. *Canadian Journal of Microbiology* **19**, 409-411.

Seki, T., Oshima, T. and Oshima, Y. (1975). Taxonomic study of *Bacillus* by deoxyribonucleic acid-deoxyribonucleic acid hybridization and interspecific transformation. *International Journal of Systematic Bacteriology* **25**, 258-270.

Seki, T., Chung, C.-K., Mikami, H. and Oshima, Y. (1978). Deoxyribonucleic acid homology and taxonomy of the genus *Bacillus*. *International Journal of Systematic Bacteriology* **28**, 182-189.

Simon, M.I., Emerson, S.U., Shaper, J.H., Bernard, P.D. and Glazer, A.N. (1977). Classification of *Bacillus subtilis* flagellins. *Journal of Bacteriology* **130**, 200-204.

Smith, N.R., Gordon, R.E. and Clark, F.E. (1946). Aerobic mesophilic
    sporeforming bacteria. Miscellaneous Publication No. 559, United
    States Department of Agriculture, Washington, D.C.
Smith, N.R., Gordon, R.E. and Clark, F.E. (1952). Aerobic spore-
    forming bacteria. Monograph No. 16, United States Department of
    Agriculture, Washington, D.C.
Somerville, H.J. and Jones, M.L. (1972). DNA competition studies
    within the *Bacillus cereus* group of bacilli. *Journal of General* ·
    Microbiology **73**, 257-265.
Stack, M.V., Donoghue, H.D., Tyler, J.E. and Marshall, M. (1977).
    Comparison of oral streptococci by pyrolysis gas-liquid chromato-
    graphy. In "Analytical Pyrolysis" (eds. C.E.R. Jones and C.A.
    Cramers), pp.57-68. Amsterdam: Elsevier Scientific Publishing Co.
Stack, M.V., Donoghue, H.D. and Tyler, J.E. (1978). Discrimination
    between oral streptococci by pyrolysis gas-liquid chromatography.
    Applied and Environmental Microbiology **35**, 45-50.
Taylor, J.J. (1976). *Ex vivo* determination of potentially virulent
    Sporothrix schenckii. *Mycopathologia* **58**, 107-114.
Vincent, P.G. and Kulik, M.M. (1970). Pyrolysis gas-liquid chromato-
    graphy of fungi: differentiation of species and strains of several
    members of the *Aspergillus flavus* group. *Applied Microbiology* **20**,
    957-963.
Vincent, P.G. and Kulik, M.M. (1973). Pyrolysis gas-liquid chromato-
    graphy of fungi: numerical characterisation of species variation
    among members of the *Aspergillus* group. *Mycopathologica et
    Mycologia Applicata* **51**, 251-265.
Vivoli, G. and Fabio, U. (1967). Composizione in acid: de stipiti
    microbici appartenenti alla species *B.cereus* et *B.thuringiensis*.
    Nuovi Annali d'Igiene e Microbiologia **18**, 166-173.
Welker, N.E. and Campbell, L.L. (1967a). Unrelatedness of *Bacillus
    amyloliquefaciens* and *Bacillus subtilis*. *Journal of Bacteriology*
    94, 1124-1130.
Welker, N.E. and Campbell, L.L. (1967b). Comparison of the α-amylase
    of *Bacillus subtilis* and *Bacillus amyloliquefaciens*. *Journal of
    Bacteriology* **94**, 1131-1135.

Chapter 8

BACILLUS FROM MARINE HABITATS: ALLOCATION TO
PHENA ESTABLISHED BY NUMERICAL TECHNIQUES

G.J. BONDE

Institute of Hygiene, Århus University, Denmark

Introduction

Systematic studies of bacteria from marine habitats date
back to the 1880s and by the second decade of this century
it had become obvious that such organisms often exhibited
distinctive properties such as pigmentation, a marked
plasticity of form, and an obligate requirement for certain
ions. The results of these early studies were reported by
ZoBell [1946]. The early 1950s heralded a very productive
period, bacteria from different environments were enume-
rated, strains from marine sources classified [Brisou,
1955;   Kriss, 1963;   Anderson, 1962] and the importance of
pressure, low temperatures, and certain ions on the
development of bacterial populations studied [MacLeod,
1965;   Stanley and Morita, 1968;   Morita, 1976]. A topic
of growing importance is the role of bacteria in the metabolism and self purification in the sea, in particular
their contribution to clearing pollutants such as oil and
their role in the turnover of dumped wastes [Bonde, 1968;
Schwarz and Colwell, 1975;   Jannasch and Wirsen, 1977].
    The bacteria most frequently isolated from marine
habitats are classified in *Aeromonas, Micrococcus, Myco-
plana, Pseudomonas, Vibrio,* and as coryneform bacteria
[Sieburth, 1964;   Wood, 1967]. *Bacillus* species are, however, also frequently observed and their presence noted
even in the earlier studies [Russel, 1891;   Newton, 1924;
ZoBell and Johnson, 1949;   ZoBell and Morita, 1959;
Bartholomew and Paik, 1966;   ZoBell, 1968]. Much of the
work on *Bacillus* strains from marine habitats was focused
on estimates of total numbers, on the frequency of *Bacillus*
species among the bacterial community or on their specific
properties [Bonde, 1965, 1967, 1971;   Johnson *et al.*, 1968;
Baumann *et al.*, 1972;   Schwarz *et al.*, 1974;   Geslin,
1975]. In addition, the resistance of such isolates to
heavy metals and their role in the breakdown of metal (Hg,
Mn) compounds in sediments has been described by several
workers [Sreenivasan, 1956;   Wood, 1967;   Trimble and
Ehrlich, 1970;   Nelson and Colwell, 1975].
    The search for specific marine *Bacillus* species was the

subject of many investigations; ZoBell and Upham [1944]
listed 5 new species, Kriss [1963] 7 new and 5 old ones,
Dias *et al.* [1968] one species, Denis [1971a, b] 9 new and
11 old species, and Delaporte [1967] one new species.
Other workers [Wood, 1967; Delabre *et al.*, 1973; Bonde,
1975, 1976a; Geslin, 1975; Boeyé and Aerts, 1976] were
reluctant to propose new species but described unidentified strains often on the basis of numerical data. In
some instances no attempt was made to identify the spore-
formers, studies being limited to total counts [Rüger,
1973, 1975].

Generally, high populations of aerobic spore-forming
rods are found in habitats such as the soil, salt lakes
[Geslin, 1975], and fresh water [Schubert, 1975], but
their frequency in marine sediments may sometimes approach
or exceed that found in soils and dust and numbers up to
millions per gram dry weight of sediments have been found
and, in the deeper layers of sediments may account for 100
per cent of the micro-organisms [Rüger, 1973, 1975].
Species such as *B.subtilis* and *B.licheniformis* seem to be
universally distributed, others appear to be of soil origin
[Bonde, 1965, 1976a]. In tropical areas the frequency of
Bacillus species relative to the rest of the bacterial
flora seems to be high (see Table 4).

Marine Samples and Sampling Methods

Samples studied include sea water from various depths
below the surface and sediments collected at the water-
sediment interface. All of the sea water samples were
collected in Danish home waters, whereas the sediment
samples were obtained from widely separate parts of oceans.
The sediment samples can be considered to be more representative of the local conditions and to reflect the microbial
activities of the marine habitats more than the water
samples [Bonde, 1967, 1968, 1971].

The composition of water and the character of the sediments differ considerably in the coastal, estuarine and
deep sea areas [Anderson, 1962; Wood, 1967]. The estuarine areas are affected by terrestrial as well as by
oceanic factors which determine their physico-chemical and
biological characters, and great variations in salinity,
temperature, pH, and oxidation-reduction potential are
found. Estuarine waters are much affected by sunlight and
by the admixture of fresh water from rivers and by the
seepage of ground water. Pollution, often with pathogens,
is not uncommon and terrestrial and fresh water organisms
may be dominant. In estuaries there may exist boundary
layers between oceanic and fresh water, thermoclines and
haloclines, varying in position with changes in wind and
current (in the öre Sund they occur between 10-15 m).
The samples collected from Danish home waters represented
a transition from estuarine to oceanic conditions, the

salinity varying from the brackish waters of fjords and
coastal samples (salinity 8-10 o/oo) to the oceanic salinity of the Belts, Skagerak and Kattegat (10-30 o/oo).
   In the oceans, zones of upwelling and downwelling can
exist due to currents and to the profile of the sea bottom.
These zones greatly influence the nutrient status and composition of the microbial community. Generally, however,
the majority of the oceanic areas are characterized by
stable environmental conditions. Thus, salinity varies
only between 35 and 37 o/oo and the relative proportion of
the major constituents is usually constant [Sverdrup *et
al.*, 1949; Wood, 1967]. The mean depth of ocean waters
is 3,795 m, and the pressure is a linear function of depth
increasing by 0.1 atmosphere/m. Light penetrates sea
water to a depth of about 80 m and below 200 m there is
continual darkness. The temperature falls to 5°C at
depths of 200 to 1000 m; below this the temperature is
between 5°C and -1.5°C; 90% of marine areas are colder
than 5°C.

Position of the sampling stations

   The sediment samples represent estuarine as well as
oceanic conditions:
1. The fjord of Bandholm, depth 0 to 7 m, salinity 8 o/oo,
    situated on the island of Lolland (latitude 11°30' E,
    longitude 54°50' N).
2. The Öre Sund separating Sweden and Denmark (Sjaelland)
    (latitude 55° to 56°N, longitude 11° to 12°E; depths
    0 to 30 m).
3. The Lysegrund area to the north of Sjælland (latitude
    56°10 to 17' N, longitude 11°37' to 12°18' E; depths
    5 to 30 m).
4. The Lillebælt, separating Jylland and Fyn (latitude
    55° to 56° N, longitude 9°40' E; depths 2 to 26 m).
5. The Limfjord traversing the northern part of Jutland
    (latitude 57°N, longitude 10° to 10°30' E; depths 4
    to 9 m).
6. The coastal waters off Gothenburg (latitude 57°40' N,
    longitude 11° to 12°E; depths 20 to 30 m).
7. The Irish sea (off Anglesey), (latitude 52°40' N,
    longitude 5° to 6°W; depths 30 to 50 m) (sampling by
    Professor Fogg).
8. Samples taken off Peru by Professor De Sp. Smith near
    the equatorial zone (latitude 0° to 14°S, longitude
    15° to 81°W; depth 15 to 3,000 m). A few of these
    samples were collected from the coastal areas of the
    Galapagos Islands in quite shallow water (15 to 30 m),
    but most were taken "in the extremely productive zone
    where the deep water upwells, often greenish-black
    because of the heavy deposit of chlorophyll and sul-
    fides" [Smith, 1968].
9. Samples from the Sargasso Sea, taken at depths of

6,750 m from the R/V *Dana* (latitude 26°06' N, longi-
tude 57°32' W).
All of the sea water samples were taken in Danish home
waters; at midstream and coastal stations of the Sound,
Lillebælt, and in the Gulf of Aarhus. Samples of the
contents of fish guts and gills were examined from fish
caught in Danish home waters.

Sampling methods

   Water samples were collected in sterile glass-stoppered
bottles at the coastal stations, and midstream samples at
various depths in sterile, evacuated 300 ml ampoules, the
necks of which were broken at the depth of sampling.
   Samples of sediments were taken with a gravity corer,
which allowed segments to be isolated from varying depths.
The sediments from the Danish and Swedish areas were
collected by a "chandelier-sampler" [Willemoes, 1964] which
is a ball-stoppered gravity core-sampler. The samples from
the Sargasso Sea were taken using a special bottom scraper
which is dragged along the bottom, and closed by a flap.
A few samples in the Öre Sund area and off the coast of
Peru were taken by grab samplers. Sediments were transferred to sterile, screw-capped glass vials and kept at 0°C
during transportation.
   Sampling of water does not pose great contamination
problems but sampling of sediments can be difficult as
cleaning and sterilization between 2 consecutive samplings
during a cruise can only be done by chemical means. Most
samplers, with the exception of the gravity corers, will
not completely prevent pollution from the surrounding
water. Because of this problem only the interior parts of
solid sediment samples can be considered to be representative.

Characteristics of samples

   The samples showed a variation in consistency from
those which were stony, sandy, clayey and poor in organic
material to those which were muddy, black or greenish-
black with a rotten odour. The organic content varied
between 0.30 and 24% and in some of the coastal samples
pollution by sewage was demonstrated. Detailed descriptions of the stations and the samples can be found in
Smith [1968] and in the Reports of the Swedo-Danish
Committee on the Pollution of the Sound [Bonde, 1967,
1971].
   Suspensions in water of the contents of the guts and
gills, after disintegration in a mortar, from mackerel,
brisling, haddock, whiting, flounder, and plaice were also
examined.

Bacteriological Examination of Samples

Estimation of pollution   The degree of pollution was estimated by counting the anaerobic sulphite reducing bacteria
at 48°C (*Clostridium perfringens*) using the method of
Bonde [1962] and by counting coliforms and '*Escherichia
coli*' (thermostable, indole positive coliforms in MacConkey's broth) using the Most Probable Number method.   In
some sediment samples *E.coli* and the coliforms were demonstrated by a presence-absence test by incubating 1 g of
sediment in 50 ml of MacConkey's broth followed by the
identification of colonies streaked onto a solid MacConkey
agar medium [Bonde, 1962, 1967, 1971].

Numbers of heterotrophs   Medium number V of Anderson [1962]
was the primary medium used for counting the heterotrophs.
This contains aged sea water 750 ml, tap water 250 ml,
peptone (0.25%, w/v), yeast extract (0.25%, w/v), FePO₄
(0.01%, w/v), agar (New Zealand) (1.5%, w/v) and is steri-
lized by steaming for 1 h on 3 consecutive days;   it was
compared with both plain nutrient agar and King's agar B
[Bonde, 1962].   In fresh water and some brackish water
samples the latter gave the highest counts, but in 80% of
the samples studied Anderson's sea water agar gave the
higher count.   Water and sediment samples were diluted in
phosphate buffer in a decimal series, and 1 ml aliquots of
the 1:10, 1:100, and 1:1000 dilutions were inoculated onto
Anderson's sea water agar and incubated in duplicate.
From the sediment samples, 1 g was taken from the central
parts of the sample using a sterile spoon and the dilutions
shaken mechanically for 1 h before incubation.   Streak
plates were also prepared from the undiluted clayey and
sandy samples that were expected to contain few micro-
organisms.

Number of aerobic spore-forming rods   Part of all inocula
were pasteurized at 80°C for 5 mins.   The counts of
pasteurized samples often gave counts of the same order of
magnitude as those from non-pasteurized samples on nutrient
agar suggesting that *Bacillus* species were predominant in
these samples.

Isolation and identification of randomly selected bacteria
After counting, four colonies were picked at random for
identification, and pure cultures were ensured on sea water
agar.   The organisms were identified according to the
modified scheme of Shewan *et al.* [1960] [Bonde, 1967].

Bacterial Content of Samples

In some of the coastal samples a high degree of pollution was evident from the counts of *Cl.perfringens* and the
demonstration of *E.coli* and coliforms.   Smith [1968] also

isolated 12 different clostridial species, with *Cl.per-fringens* dominant from the "Peru Current Samples". Samples giving counts of 1000 sulphite reducers per g dry weight or more, and of more than 1000 *E.coli* per 100 ml were considered to be polluted.

The counts of heterotrophs, as well as the species representation were influenced by the plating medium used and the temperature and time of incubation. The number of *Bacillus* strains also depends on the character of the sample and on the site from which it was taken. Selected results are given in Table 1 which gives the counts of *Cl. perfringens* and *Bacillus* spp., and the total count of heterotrophs in some of the sediment samples; those from the Galapagos Islands (Samanco Bay) are polluted coastal samples.

TABLE 1

Counts of Clostridium perfringens, Bacillus *species and the total number of heterotrophs in one gram dry weight of sediment of the "Peruvian" samples*

| Sample | Depth m | Counts per gram dry weight | | |
		Cl.perfringens	*Bacillus* spp.	Total number of heterotrophic bacteria
Samanco Bay	15	606	48,000	1,700,000
598	129	12	23,000	1,300,000
586	537	104	17,000	16,000,000
588	1970	16	600	16,000,000
604	3185	0	50	36,000,000
Sargasso Sea	6750	0	10	90

It can be calculated that the counts of *Bacillus* colonies as a proportion of the total number of heterotrophs varies between 0.004% and 11%, the proportion being higher in coastal areas and in clayey-sandy deep sea sediments than in areas rich in organic matter. These results are in good agreement with the observations of Rüger [1973, 1975]. However, in the coastal regions the *Bacillus* spp. may be of soil and fresh water origin whereas in the sediment they may exist as spores.

Comparable counts from various segments of cores from the Irish Sea samples are given in Table 2. It is evident that while the total number of bacteria decreases markedly with depth, due to a decrease in organic matter and other nutrients, the relative number of *Bacillus* shows a dramatic rise.

The number of *E.coli* and the total count of heterotrophic bacteria in 3 groups of sea water samples are shown in Table 3.

TABLE 2

Counts of Clostridium perfringens, Bacillus *species
and the total number of heterotrophs in one gram dry weight
of sediment from the Irish Sea, taken at various depths
below the water sediment interface*

Depth (cm)	Counts per gram dry weight		
	*Cl.*perfringens	*Bacillus* spp.	Total number of heterotrophic bacteria
0	80	1,700,000	∞
20	4	1,300,000	3,600,000
40	1	800,000	4,200,000
50	0	20,000	1,300,000
60	0	20,000	40,000
80	0	24	20
100	0	20	20

TABLE 3

Counts of Escherichia coli, Bacillus *species
and heterotrophic bacteria in sea water*

	Counts per ml		
	E.coli	*Bacillus* spp.	Heterotrophic bacteria
Coastal, polluted	2400	20-70	40,000
Coastal, non-polluted	23-79	20-70	193- 1,200
Sea samples	0-2	20-70	70-10,000

These results indicate that the *Bacillus* strains are passive inhabitants of the marine environment, are possibly of soil origin and are surviving as spores since their numbers are high where evidence of growth nutrient levels are low. Corrected for the content of organic matter, the number of aerobic spore-forming rods are still higher in sediments than in water, contrary to the numbers of *Cl. perfringens* [Bonde, 1967].

Frequency of Bacillus *strains relative to other bacteria and their distribution in different areas*

Because of the random selection of colonies from the non-pasteurized samples it is possible to get an estimate of the relative frequency of different groups of bacteria in the areas examined.

Frequency in water samples The overall frequency of

TABLE 4

Frequency of Bacillus *strains relative to other bacteria in coastal (estuarine), high sea and terrestrial samples*

Bacillus spp. as percentage of total bacterial population	Area	Author
0	Coastal	Huddleston [1955]
0	Sea	Sieburth [1964]
0.1	Coastal	Murchelano and Brown [1970]
1.5	Sea	Anderson [1962]
4	Sea	Kriss [1963]
5.5	Sea	Simidu and Oisi [1952]
7	Sea	Wood [1967]
8	Coastal	Denis [1971a, b]
9	Coastal	Bonde [1967]
16	Coastal	Wood [1953]
20	Coastal	Gianelli *et al.* [1970]
40.2	Sea	Venkataraman and Sreenivasan [1954]
4 - 2.5	Fresh water	[Geslin, 1975; Schubert, 1975]
30-54	Soil	[Geslin, 1975]
50-80	Tropical soil	[Geslin, 1975]
65	Air, dust	[Geslin, 1975]

Bacillus strains in the sea water samples was 9%, a figure which is compared with those of earlier workers in Table 4. It is evident that the great majority of bacteria in sea water are not *Bacillus* spp. but, as stated earlier, are Gram negative rods, including coliforms and vibrios (50 to 96%), followed in frequency by Gram positive cocci (4 to 31%) though Gram positive asporogeneous rods (coryneform bacteria) may be dominant in some areas [Sieburth, 1964]. The frequency of *Bacillus* strains in sea water increases with temperature and depth but decreases with distance from the coast. Bonde [1967] found *Bacillus* strains with increasing frequency in and about the thermocline; pigmentation (which is a frequent character of bacteria, including *Bacillus* spp., from marine habitats) of isolates, on the other hand, was more frequently encountered in those from the surface [Bonde, 1967].

Frequency in sediments The frequency of *Bacillus* strains in sediments was generally greater than those in water samples [Bonde, 1967, 1971]. In the samples examined, these strains accounted for between 14 and 80% of the total number of heterotrophic bacteria. The frequency of the different types of bacteria in the sampling areas (Table 5) is considerably influenced by the distance of the sampling sites from the coast and by the level of pollution.

The vertical distribution in sediments (Tables 1 and 2) was quite different from that in water, as *Bacillus* spp. were very much more frequent in samples taken below the thermocline, which occurs at 10-15 m in the Öre Sund, and in those from greater depths (Table 6).

TABLE 5

Frequencies (as percentages) of identified bacteria in sediments from the various sampling areas

	Band-holm	Öre Sund	Lille-baelt	Lim-fjord	Gothen-burg	Peru	Sar-gasso
Achromobacter–							
Alcaligenes	0	0	3	0	0	3.5	0
Acinetobacter	8	0	10	7	0	2.7	8
Aeromonas–Vibrio	0	18.4	6	2.4	3.7	0	0
Bacillus	46	14.4	46	80	33.3	30.5	46
Coryneform							
bacteria	8	0	4	3.5	0	11.1	8
Cytophaga–							
Flavobacterium	0	1.2	3	2.3	0	19.4	0
E.coli and							
coliforms	16	28	0	1.2	25.9	0	0
Micrococcus	0	5	12	2.3	3.7	2.7	0
Pseudomonas	31	33	16	0	33.3	25.7	0

TABLE 6

Frequency distribution of main groups of bacteria in samples taken at varying depths from Öre Sound

Depth (m)	Gram negative rods	*Micrococcus* spp.	*Bacillus* spp.	
0- 5	14(10)	0	0	14
5-10	109(90.1)	3(2.5)	9(7.4)	121
10-15	167(86.1)	10(5.1)	17(8.8)	194
15-20	98(74.2)	12(9.1)	22(16.7)	132
20-25	53(71.6)	2(2.7)	19(25.7)	74
25-30	21(61.8)	2(5.9)	11(32.3)	34
30+	4(44.4)	0	5(55.6)	9
				578

(Percentages in brackets)

No *Bacillus* strains were isolated from the samples of fish; the gut contents and gills were dominated by Gram negative rods, including coliforms that ferment lactose

only at 20°C [cf. Bonde, 1967]. These results are in
agreement with the literature on fish in colder areas.

Summary of isolates   The water and sediment samples can
finally be allotted to 5 groups. The number of *Bacillus*
strains isolated from each group is shown below:

I   :   Coastal, polluted samples   58 strains
II :   Coastal, non-polluted samples   71 strains
III: Mid-stream samples from home waters,
        non-polluted   33 strains
IV :   Sediments from the Irish Sea,
        non-polluted   37 strains
V   :   Oceanic samples (Peruvian and
        Sargasso Sea)   72 strains
                                                       ‾‾‾‾‾‾‾‾‾‾
                                                       271 strains

     Also included in this study were a few *Bacillus* strains
collected by Delaporte from tadpoles and sea-urchins.

Classification of the *Bacillus* Isolates

     A total of 271 isolates were added to the author's
collection, listed in the WHO/Unesco Directory [Martin and
Skerman, 1972], which now comprises 737 strains including
194 named *Bacillus* strains from other culture collections.

     In bacteriology, classification based upon phylogenetic
principles has never been successful [Mayr, 1965]. The
free transfer of genetic material in microbial populations,
the short generation time, mutation, and selection have
given rise to a broad spectrum of related groups, which
are not easily separated. Natural (phylogenetic) classifications have, therefore, been replaced by artificial
ones based upon tradition and conventional criteria. In
recent years, however, accumulated knowledge of more and
more characters, of which one could not easily be given
priority over the other, and the advent of electronic computers, have favoured the development of numerical phenetic
classification.

Classification by conventional taxonomy

     In the classification of the genus *Bacillus* several
investigators emphasize the importance of cell-morphology
in making primary divisions [see Gordon *et al.*, 1973;
Lemille *et al.*, 1969; Denis, 1971a; Thibault, 1971]
whereas others apply biochemical tests, in particular the
formation of acetoin and the breakdown of starch [see
Gibson and Topping, 1938; Geslin, 1975; Gordon *et al.*,
1973].

     More than one third of the 271 isolates from marine
habitats were difficult to classify by published schemes
compared with about one quarter of the strains from other
sources in the author's collection.

Tests, attributes and coding for classification

The 271 isolates were examined for the 77 attributes listed in Table 7. In addition, 114 strains were analysed for DNA-base composition according to the method of Marmur and Doty [1962] and for specific enzymatic activities using the API-ZYM system. Production of antibiotics was detected as described by Bonde [1975].

All of the strains grew well on the stock culture agar (Bacto, B 54) which was used as the maintenance medium. However, a few strains from the marine habitats lost the ability to grow on this medium and were maintained on sea water agar. Strains were also kept in the lyophilized state.

All strains were initially incubated on 3 solid media: nutrient agar, glucose nutrient agar, and soyabean agar, or on agar supplemented with sea water agar where appropriate. The temperature requirements and tolerance of the test strains to various concentrations of NaCl were then examined (Table 7). None of the strains grew significantly better on sea water agar though pigmentation was generally more pronounced on this medium. All strains grew well at 30°C (attributes 18 and 22-25, Table 7).

After growth for 24 to 30 h on glucose agar, moist preparations were examined for motility using phase contrast microscopy. Smears were stained according to Gram, by safranine, for the detection vacuoles, and for fat globules according to Hartman [Gordon *et al.*, 1973]. The diameter and length of cells were measured using a micrometer eyepiece (attributes 1-13, 14, 15-18, Table 7). As in a previous paper [Bonde, 1975] the rods were graded into those with a diameter above or below 0.9 μm and length above or below 7 μm (attributes 5, 6, Table 7) and they were divided into Gram positive or Gram variable, and Gram negative (attribute 10, Table 7). The shape, position, and dimension of the endospores were recorded from both unstained and Gram stained preparations.

The simple media were prepared according to Smith *et al.* [1952] and Bonde [1975] and these included fluid media and the basic mineral medium for sugar fermentation: (attributes 19-21, 22-25 and 34-46, Table 7). The following tests, listed in Table 7, were also performed after Smith *et al.* [1952]: (30) growth in Gibson's glucose; (33) hydrolysis of starch; (50) growth on glucose nitrate medium; (51) growth on glucose asparagine medium; (52) reduction of nitrate in Gibson's medium (with inner inverted vial); (53) reduction of methylene blue; (54) lecithinase test; (64) hydrolysis of gelatin and (66) hydrolysis of casein.

As described earlier [Bonde, 1975], colonies were classified into 6 types: (17), 1) flat, glistening, moist and veily, 2) flat, membranous, dull, often wrinkled, 3) raised butyrous, 4) raised membranous and adhering,

TABLE 7

Tests used for the characterization of isolates

List of Attributes and Codes

I   Cell morphology
     1. Vacuoles present 0; absent 1
     2. Reserve fat present 0; absent 1
     3. Spore oval 0; round 1
     4. Spore non-swelling 0; swelling 1
     5. Diameter above 0.9μm 0; below 1
     6. Length below 7μm 0; above 1
     7. Ends pointed 0; round or square 1
     8. Rod straight 0; curved 1
     9. "Ghosts" present 0; absent 1
     10. Gram stain + and variable 0; - and variable 1
     11. Spores central 0; eccentric, terminal 1
     12. Sporulation on glucose agar 0; no sporulation on glucose
         agar 1
     13. Motile 0; non-motile 1

II   Arrangement of cells
     14. Chains 0; pallisades (or snapping) 1; heaps 2

III Characteristics of colonies
     15. Pigment yellow 0; dark 1; red 2; no pigment 3
     16. Potato growth wrinkles, no pigment 0; smooth, no pigment
         1; wrinkled pigment 2; pigment smooth 3
     17. Colonies flat glistening 0; flat dull 1; raised
         butyrous 2; raised membranous 3; rhizoid 4; rotating 5
     18. Growth alike on all media 0; ex-agar bigger 1; glucose
         agar bigger 2; soya agar bigger 3; no growth on glucose
         agar or soya agar 4

IV   Growth in fluid media
     19. Clearing with sediment 0; uniform turbidity 1
     20. Coarse flaky or granular sediment 0; fine sediment 1
     21. Surface: ring and/or pellicle present 0; no ring or
         pellicle 1

V   Physiology
     22. Temperature, no growth 60°C 0; growth 60°C 1
     23. Growth 20°C > 44°C 0; 20°C = 44°C 1; 44°C > 20°C 2
     24. No growth 5°C 0; growth 5°C 1
     25. NaCl. Growth 10% and 5% 0; not 1% 1; neither 5% nor
         10% 2
     26. pH. Growth Sabouraud's medium 0; no growth 1; pH > 8 2
     27. Anaerobiosis. Growth anaerobic 0; no growth anaerobic 1

TABLE 7 (cont'd)

VI Biochemistry
28. Fermentation of Hugh and Leifson's medium 0; no change 1; oxidation 2; alkaline 3
29. Acid from proteose peptone glucose 0; intermediate 1; alkaline 2
30. Growth Gibson's glucose anaerobically 0; aerobically only 1
31. No gas from glucose 0; gas from glucose 1
32. VP + 0; VP - 1
33. Hydrolysis of starch + 0; no hydrolysis - 1
34. Fermentation glucose in mineral medium + 0; - 1
35. Fermentation maltose in mineral medium + 0; - 1
36. Fermentation glycerol in mineral medium + 0; - 1
37. Fermentation arabinose in mineral medium + 0; - 1
38. Fermentation xylose in mineral medium + 0; - 1
39. Fermentation rhamnose in mineral medium + 0; - 1
40. Fermentation ribose in mineral medium + 0; - 1
41. Fermentation laevulose in mineral medium + 0; - 1
42. Fermentation galactose in mineral medium + 0; - 1
43. Fermentation mannose in mineral medium + 0; - 1
44. Fermentation sucrose in mineral medium + 0; - 1
45. Fermentation lactose in mineral medium + 0; - 1
46. Fermentation starch in mineral medium + 0; - 1
47. ONPG + 0; - 1

VII Sources of carbon
48. Citrate growth + 0; - 1
49. Propionate growth + 0; - 1
50. Glucose nitrate + 0; - 1
51. Glucose asparagine growth + 0; - 1

VIII Enzymatic activities
52. Nitrate reduced to NO_2 0; to N_2 1; no reduction 2
53. Reduction of methylene blue 0; no reduction 1
54. Lecithinase + 0; - 1
55. DNAse + 0; - 1
56. Urease + 0; - 1
57. Hemolysis + 0; - 1
58. Tyrosinase + 0; - 1
59. Pectinase + 0; - 1
60. Ornithine decarboxylase + 0; - 1
61. Lysine decarboxylase + 0; - 1
62. Arginine decarboxylase + 0; - 1
63. No phenylalanine deamination 0; - 1

IX Protein decomposition
64. Hydrolysis of gelatin + 0; - 1
65. Liquefaction of iron gelatin + 0; - 1
66. Hydrolysis of casein + 0; - 1
67. Litmus milk, no change 0; alkaline digestion 1; acid coagulation 2

TABLE 7 (cont'd)

X   Splitting of fats
      68. Tribuyrin +0; - 1
      69. Tween 80 + 0; - 1

XI   Miscellaneous
      70. Resistance to lysozyme 0; no resistance 1
      71. Splitting of hippurate +0; - 1
      72. Formation of dihydroxyacetone +0; - 1
      73. No formation of indole 0; formation 1
      74. Kovacs oxidase late or - 0; + 1
      75. Catalase +0; - 1
      76. No growth in azide dextrose 0; growth 1
      77. No formation of crystalline dextrin 0; formation 1

5) rhizoidal and 6) rotating. In fluid media a distinction was made between: (19) clearing with sediment or uniform turbidity; (20) coarse, flaky, granular or fine sediment; (21) ring and/or pellicle present or no ring or pellicle; (29) acidity in glucose peptone medium was measured electrometrically and graded as acid (below pH 6.2), intermediate and alkaline (above pH 7.7) (32). Formation of acetoin was tested by Barritt's modification. Anaerobic growth (27) was tested in a Whitley SS anaerobic jar (L.I.P. Ltd., England) on the solid medium most favourable for growth.

The tests and media were performed according to Gordon et al. [1973]. These included: (26) growth in Sabouraud's medium at pH 5.7; (48) utilization of citrate medium and (49) propionate medium; (58) the tyrosinase test; (63) the phenylalanine deaminase test; (70) lysozyme resistance; (71) splitting of hippurate; (72) formation of dihydroxyacetone, and (75) the catalase test. Growth in azide dextrose broth and the formation of crystalline dextrins were not systematically examined (76, 77). Tests carried out according to Bonde [1962] were: (28) growth in Hugh and Leifson's medium; (52) reduction of nitrate in semisolid medium V; (65) liquefaction of iron gelatine; (67) litmus milk; (73) formation of indole in peptone broth; (74) Kovac's oxidase test, and hemolysis (57). Formation of gas (31) was estimated by inserting inverted inner tubes in the media for tests (29) and (30) (see Table 7).

The following tests, useful for the classification of bacteria in the family Enterobacteriaceae [Edwards and Ewing, 1972], were also applied: (47) the ONPG-test; (55) the DNAse test; (60) ornithine decarboxylase [Falkow modification]; (61) lysine decarboxylase, and (62) arginine dihydrolase. The urease test (56) was performed according to Jessen's method [Bonde, 1975], splitting of fats (68, 69) was detected in Hugo and Beveridge's tributyrin medium [Bonde, 1975] and in Tween 80 medium [Sierra, 1957] while (59) pectinase activity was examined by the method of Hankin *et al.* [1971].

Coding and computation of similarity

The results of the 75 tests (Table 7) were coded as 'disordered multistate' or 'alternatives', giving a total of 173 character states. Because of the importance attached to negative attributes in bacterial classification a simple matching coefficient including double negatives was chosen, the dissimilarity coefficient,

$$\text{Dissimilarity coefficient} = \frac{+-, \ -+}{++, \ --, \ +-, \ -+,} = 0 \text{ to } 1$$

resulting in a square, symmetric matrix of M dissimilarities, δ_{ij}, between N objects, i, j = 1, 2, 3, ..., N (N, e.g. = 271 strains): as $\delta_{ij} = \delta_{ji}$ and $\delta_{ii} = 0$, the number of dissimilarity coefficients between 271 pairs of objects is

$$\binom{271}{2} = 36,585.$$

Kruskal's non-metric multidimensional scaling

Results of previous experiments with 64 characters and 460 *Bacillus* strains including 226 of marine origin, and the single linkage principle [Bonde, 1975] did not give convincing limits to taxa, and the establishment of groups was greatly influenced by personal judgements.

Because of the mixed nature of the characters, and to minimize the tangling between clusters, the non-metric scaling of Kruskal [1964a, b] was considered to be the method of choice; this method is also useful in the analysis of infra-specific variation which may be a problem in this genus [Jardine and Sibson, 1970] as well as being important in the study of isolates from marine habitats.

The problem of multidimensional scaling is to find N points, whose inter-point distances match in some sense the dissimilarities of the N objects. The most modest demand put upon such a configuration is that pairs with small δ_{ij}'s be represented by points with a small distance and large δ_{ij}'s by points with larger distances.

It is always possible to arrange the $\{\delta_{ij}$'s$\}$ according to magnitude

$$\delta_1 < \delta_2 \ \cdots\cdots\cdots < \delta_M \tag{1}$$

In a configuration of N points x_1, x_2, ... x_N in † dimensional space, the geometric distances between x_i and x_j = d_v can also be arranged according to magnitude, as were the δ_{ij}'s

$$d_1, d_2 \ \cdots\cdots\cdots, d_M, \tag{2}$$

d_i having the same rank in the table of the distances as δ_i, and so on. If the d-sequence (2) monotonically increases exactly as the δ-sequence (1) the configuration of points is said to give a perfect fit.

The sequence of distances of d's (1), however, will generally not increase monotonically and therefore a sequence of distances

$$\hat{d}_\nu, \nu = 1, \ldots \ldots M,$$

is chosen, such that these $\hat{d}_\nu$'s increase monotonically and only one sequence of $\hat{d}_\nu$'s minimizes the sum of the differences.

$$S^* = \sum_{\nu=1}^{M} (d_\nu - \hat{d}_\nu)^2, \tag{3}$$

the $\hat{d}_\nu$'s being determined from the d_ν's according to Bartholomew [1959]. The S* is normalized, to secure independence of scale, by division by the sum of d_ν^2's whence

$$S = \sqrt{\frac{\sum_{\nu=1}^{M} (d_\nu - \hat{d}_\nu)^2}{\sum_\nu^{M} d_\nu^2}}, \quad 0 \leq S < 1.$$

S is the measure of goodness of fit introduced by Kruskal [1964a, b], also called the measure of stress; S = 0 signifies $d_i = \hat{d}_i$ or perfect fit.

Kruskal believed that stress, as a function of dimension, would show an abrupt inflection, but this is not the case with the present data [cf. Fig. 5].

Comments on application of Kruskal's method

Technically the first problem is the determination of the configuration which minimizes the stress function S. Kruskal applied the method of steepest descent but this does have some drawbacks, for example, when to stop iteration. Secondly, the Kruskal technique was only applied with a few elements. The present programme version was written in PL/1 and as the expenditure in time and space is largely proportional to the number of dissimilarity coefficients individual scalings were preferentially restricted to about 100 strains at a time, although a scaling of the first 341 strains has been performed (Fig. 1). A more detailed account of the mathematical treatment and of the initial results has been published [Bonde, 1976b].

The programme lists: 1) the data matrix, 2) a matrix of all dissimilarities of any one strain to any other strain, 3) the distribution of dissimilarities, 4) the course of the iterative procedure, 5), 6), 7) the configuration of the elements in 4, 3, and 2 dimensions, the

corresponding total stress, and the contribution of individual strains to stress. Graphs of dimensions 2 and 3 were
also plotted automatically (Figs. 1 and 2) and models in
space were constructed (Figs. 3 and 4).

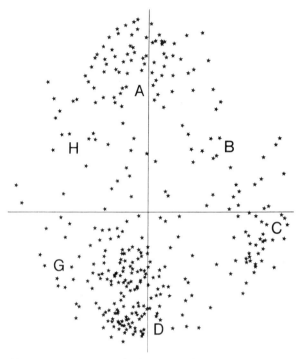

Fig. 1 Two-dimensional plot of the multidimensional scaling of 341
strains.   Indication of cluster formation off A, C, and D (compare
with Fig. 2).

Results of Kruskal scaling

Figure 1 displays a 2-dimensional plot of scaling of the
first 341 strains of the collection and shows 3 condensa-
tions of strains (off the letters A, C, and D). It would,
however, be very difficult to draw any conclusion as to a
division into groups from this plot. A stepwise procedure
was therefore chosen.
By these serial experiments a distinction between
strains that form clusters and intermediates became obvious. The collection was then supplemented with strains
supposed to form clusters with the intermediates. After
20 serial scalings a collection of 231 strains was finally
divided into 10 clusters with relatively few intermediates
(Figs. 2 and 3). Strains of marine origin did not form
separate groups or subgroups. In Fig. 2 the strains from

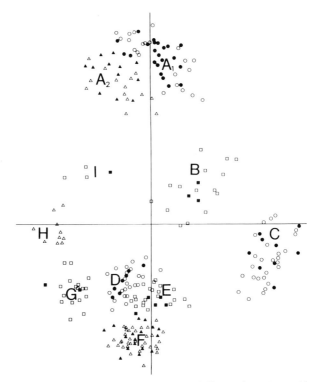

Fig. 2 Two-dimensional plot of the multidimensional scaling of 231
selected strains.   Black symbols are strains from marine sources

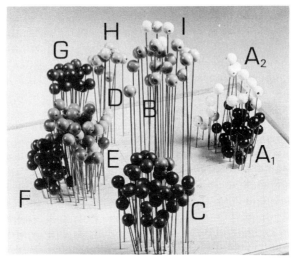

Fig. 3 Three-dimensional model of the 231 strains considered in Fig. 2.

marine sources are indicated as black symbols, strains with other origins as open symbols. Figure 3 shows a model in space of the final grouping.

The groups also include type strains and their distribution indicated that group A_1, holds *B.sphaericus* and *B. brevis*; A_2, *B.firmus* and *B.lentus*; C, *B.cereus*; B, *B. megaterium*; D, *B.pumilus*; E, *B.subtilis*; F, *B.licheniformis*; G, *B.polymyxa*; H, *B.circulans* and I, *B.badius*.

Further examination of each of these groups was executed in 3 ways:

1. Rescaling of each of the groups. Where the groups are inhomogeneous, a division into subgroups will appear. This was the case with group A which split into groups A_1 and A_2, and group D which separated into groups D, E, and F (Fig. 4). On the other hand, rescaling of a homogeneous group will not cause further splitting but will result in a "zero configuration" of randomly scattered points.

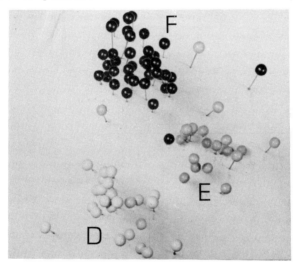

Fig. 4   Three-dimensional model of rescaling of the group D strains of Fig. 1.

2. Application of the stress measure. The stress depends upon: the number and kind of strains and tests, and the number of dimensions. Application of the stress as a measure implies a knowledge of the variance of the stress function, which was considered to be quite small and homogeneous by Stenson and Knoll [1969], so that deviations from the average stress can be taken as evidence that a strain does not belong to a group or is a "bad" member. The total stress gives evidence of the success or failure in generating classes.

3. Within and between group similarity. For each strain

the output lists the number of characters by which it differs from the other strains in the collection, and an average of character-differences within and between groups is computed.

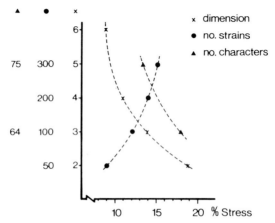

Fig. 5 Variation of stress with dimension, number of strains, and nature and number of characters.

The number of tests varying between members of the same group is 10 to 15, the average number of tests differing between neighbouring groups is 20 - 32, and between more distant groups it is 35 to 50. The minimum difference between 2 strains is 1 character and the maximum 50.

However, the tests differing within groups are of the category proved in the following section to be less useful, and only 2 to 3 significant tests might be sufficient to separate groups.

Position of isolates from marine sources in 10 groups

Of the 231 strains in the 10 well defined groups, 97 or 42% are of marine origin. It appears from Fig. 2 that the 97 isolates (the black symbols) are scattered at random within the groups. Of these isolates 27% belong to group F, 24% to group A_1, 12% to group A_2, 12% to group D, and 11% to group E. The distribution of the strains from marine habitats within the eco-groups of sediments and water (coastal, oceanic, polluted, non-polluted) is given in Table 8.

This distribution suggests an accumulation of certain groups of strains in specific areas. Sea water, irrespective of pollution, is dominated by group F (*B. licheniformis*) strains. In non-polluted sediments from Danish non-polluted areas A_1 strains (in particular those that resemble *B. sphaericus*) are frequent as are A_2 strains in the Irish sea (*B. firmus*, and *firmus*-like strains of marine origin "M").

TABLE 8

Distribution of the isolates from marine habitats
from clusters A₁-I within the different groups of samples

	A₁			A₂	B	C	D	E	F	G	H	I	
	sphaericus	*brevis*	*firmus*	"M"	*megaterium*	*cereus*	*pumilus*	*subtilis*	*licheniformis*	*polymyxa*	*circulans*	*badius*	Number of strains
1. Coastal, polluted	0	1	0	0	0	1	4	0	0	0	0	0	6
2. Coastal, non-polluted	0	1	0	0	2	3	4	0	0	1	0	1	12
3. Sea sediment	10	3	0	0	0	4	0	0	1	0	0	0	18
4. Oceanic sediment	0	0	0	0	1	2	0	3	0	0	0	0	6
5. Irish sea	1	5	6	6	0	0	0	0	0	0	0	0	18
6. Water, polluted	0	0	0	0	0	1	0	4	13	0	0	0	18
7. Water, non-polluted	0	2	0	0	0	0	4	1	12	0	0	0	19
	11	12	6	6	3	11	12	8	26	1	0	1	97

Canonical variates

Identification of *Bacillus* strains by 75 characters is
not feasible and in any case would be a waste of time, as
some characters might be useless. However, numerical
methods are at hand which can utilize the same data matrix,
which is applied in clustering, for a grading of characters.

The method of choice, when it is intended 1) to gain
insight into the way in which clusters differ, 2) to grade
the tests according to significance, 3) to reduce the number of tests, and 4) to give possibilitie's for allocation
of new strains, is an analysis of canonical variates, when
more than 2 groups are at hand. Canonical, in this context, stands for "authoritative" or "ideal" and refers to
the establishment of new coordinates (often called "principal coordinates") in a system maximizing distances between groups. The principle of computation is to maximize
the ratio of the sum of squares between groups to the sum
of squares within groups (F_c). It is therefore attempted
to replace the M-dimensional estimate of strain j in group
i by test ν by one single estimate, Y_{ij}.

$$Y_{ij} = \Sigma_\nu c_\nu X_{ij\nu},$$

i.e. the weighted sum of the individual estimates $X_{ij\nu}$, weighted by the coefficients c_ν, which are chosen in such a manner as to maximize F_c

$$F_c = \frac{\Sigma_i \Sigma_j (Y_i. - Y..)^2}{\Sigma_i \Sigma_j (Y_{ij} - Y_i.)^2}$$

For k-groups, k-1 discriminators (eigenvalues) may be defined, and the contribution to each of these by the co-efficients c_ν (eigenvectors) of individual characters is computed by the programme. The first canonical variate gives the best possible discrimination from one linear function, the second is the next best function which is uncorrelated with the first - and so on for k-1 variates. If most of the variation between groups were given by the first 2 canonical variates a plot of these alone would give a satisfactory division. With the present data, however, a joint application of the first versus the fourth or fifth (Fig. 6), and the third versus the fourth gives a good separation of all clusters (Fig. 7). The first 4 eigen-values contain 90% of information in the data.

DNA-base composition

Preliminary reports of the DNA-base composition of *Bacillus* species are given by Jackson and Bonde [1971] and Bonde [1975] but are considered in detail by Priest (Chapter 3). Considerable overlapping is found within the genus but the ranges and means may give certain guidelines which in the present study may be helpful in the subdivision of group A_1 (*B.sphaericus* - *B.brevis* subgroups) and help in the allocating of intermediate strains into defined groups. Table 9 gives the group and species names, number of the strains examined, and the range and means of their guanine (G) plus cytosine (C) ratios.

Enzymatic tests, the API-ZYM system

Most of these tests were found to be positive or negative for all of the *Bacillus* strains. The negative tests were: cystine-aminopeptidase, trypsin, α-mannosidase and α-fucosidase while the C_4 and C_8 esterases and phospho-amidase were all positive. A few positive tests were randomly distributed in the groups: C_{14}-lipase, leucine-aminopeptidase, valine-aminopeptidase, chymotrypsin, acid phosphatase, α-galactosidase and β-glucuronidase. Besides the β-galactosidase (ONPG) test (no. 47), the only tests of diagnostic value were the α- and β-glucosidases, which support the distinction between subgroups A_1 (-) and A_2 (+).

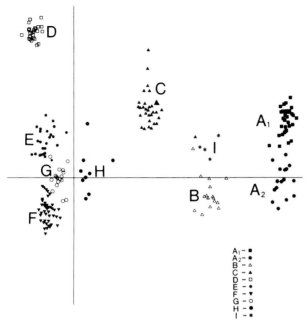

Fig. 6 Plot of canonical variates 1 versus 4. Most groups are separated.

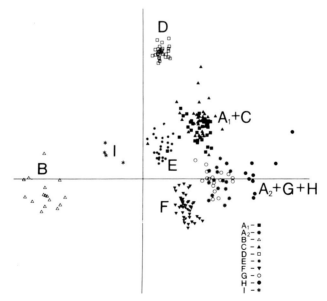

Fig. 7 Plot of canonical variates 2 versus 3, giving a better separation of groups D, E, F, and G than obtained in Fig. 6.

TABLE 9

DNA-base ratios within the numerically defined groups
of the genus Bacillus

Group	C	A$_1$	A$_2$	A$_1$	B	D	E	G	F	H
Species	*cereus*	*sphaericus*	*firmus*	*brevis*	*megaterium*	*pumilus*	*subtilis*	*polymyxa*	*licheniformis*	*circulans*
No. strains	10	7	4	11	8	11	13	4	14	6
% G+C range	36	37	40	39	42	42	44	46	48	50
	38	42	41	45	46	45	47	51	50	55
Mean	37	40	41	42	43	44	46	48	49	53

Formation and resistance to antibiotics

Bacillus strains form several, mostly peptide, antibiotics, some of which are applicable in therapy. The
antibiotics formed are generally specific for the groups
[Bonde, 1975]. On the other hand, the author is not in
agreement with Denis *et al.* [1975] regarding the diagnostic
value of the resistance patterns. Neither is this of any
importance in the separation of strains from marine habitats.

Grading of tests and characters

The tests can be graded according to their importance
for classification by the magnitude of eigenvectors. The
following order is then established (numbers in brackets
referring to Table 7): 1) breakdown of glucose (34), 2)
Voges-Proskauer test (32), 3) breakdown of arabinose (37),
4) hydrolysis of starch (33), 5) ONPG-test (47), 6) reserve fat (2), 7) dimension of spore (4), 8) Gram-stain
(10), 9) anaerobic growth (27, 30), 10) lecithinase test
(54), 11) formation of gas (31), 12) phenylalanine deaminase (63), 13) motility (13), 14) breakdown of mannitol (35), 15) breakdown of rhamnose (39), 16) diameter
of rod (5), 17) lysine decarboxylase (61), 18) reduction
of nitrate (52), 19) urease-test (56), 20) reduction of
methylene blue (53), 21) growth in propionate (49), 22)
growth in NaCl (25), 23) formation of ring and pellicle
(21), 24) growth in Sabouraud medium (26), 25) shape of
spore.
The remaining 51 characters were either of minor impor-

tance to classification or were redundant.

The tally

The output also gives the distribution of results on the individual groups as percentages. The most useful tests, of course, are those which are group-specific. Other useful tests are those where all positive or negative results, although not typical of the group as a whole, are found in only 1 (or 2) groups. The tally supports the final choice of tests for classification (key-characters) and for the description of the groups.

Allocation of strains from marine habitats to phena using the revised set of characters

It was possible to construct a dichotomous key for division of the remaining strains into groups taking into account the principles of Gower and Barnett [1971] and of Pankhurst [1970]. The reactions of typical strains in the defined groups and the per cent reactions of the groups are shown in Table 10.

Allocations of strains to phena A_1-I The first stage of the allocation of strains to phena can be done with only a few tests:

Test	Phena
	A_1A_2BCDEFGHI
No.34, 29 Breakdown of glucose: (- = weak or missing) (pH 6.2)	$-$ + A_1A_2 BCDEFGHI
No.32 Voges-Proskauer test	$-$ $-$ + A_1A_2 BI CDEFGH
No.37 Breakdown of arabinose	$-$ + $-$ + A_1A_2 BI CH DEFG

But since the allocation is less than unambiguous more tests are needed in a second stage in the procedure:

Groups A_1A_2 Tests: (4) spores swollen, A_1+, A_2; (33) hydrolysis of starch, A_1-, A_2+; (53) reduction of methylene blue, A_1+, A_2-; (60) ornithine decarboxylase, A_1+, A_2-.

In group A_1 70% of strains have round spores = *B.sphaericus*, 30% oval spores = *B.brevis*, the latter also have higher G+C base ratios.

TABLE 10

Distribution of results on the chosen characters for typical strains (T) and as a percentage for the whole group

Group character	A1 T	A1 %	A2 T	A2 %	B T	B %	I T	I %	C T	C %	D T	D %	E T	E %	F T	F %	G T	G %	H T	H %
2. Reserve fat	−	0	−	0	+	88	−	0	+	100	−	0	−	4	−	0	−	0	−	0
3. Spore round	+	70	−	0	−	0	−	0	−	3	−	0	−	4	−	0	−	0	−	0
4. Spore non-swelling	+	12	+	52	+	100	+	75	+	97	+	100	+	100	+	94	−	0	−	0
5. Diameter >0.9 µm	+	79	+	12	+	94	+	100	+	97	−	3	+	44	−	8	−	0	−	0
10. Gram + and variable	+	42	+	75	+	100	+	100	+	100	+	100	+	100	+	100	−	0	−	0
13. Motile	+	100	+	100	+	94	+	100	+	94	+	100	+	100	+	100	+	100	+	100
21. Ring/pellicle	−	27	−	0	−	6	−	50	−	88	+	93	+	100	−	98	−	0	−	22
25. 10% NaCl	−	12	−	50	−	37	−	0	−	3	+	86	+	76	−	92	+	88	−	0
26. Growth in Sabouraud's medium	−	3	−	6	−	31	−	0	+	94	+	93	+	96	+	100	+	100	−	22
27. Anaerobiosis	−	3	−	12	−	0	+	0	+	94	+	54	−	0	+	100	+	100	+	100
29. Acid from glucose	−	88	+	12	+	94	+	0	+	97	+	100	+	76	+	92	+	100	+	100
31. Gas	−	0	−	0	−	0	−	0	−	0	−	3	−	8	+	48	+	88	−	11
32. VP	−	0	−	0	−	0	+	0	+	97	+	100	+	100	+	100	+	100	+	100
33. Starch	−	0	+	75	+	94	+	100	+	91	−	0	+	100	+	100	+	100	+	100
34. Glucose	−	0	−	0	+	100	+	100	+	100	+	100	+	100	+	100	+	100	+	100
35. Mannitol	−	0	+	0	+	100	+	100	−	12	+	100	+	100	+	100	+	94	+	78
37. Arabinose	−	0	−	0	+	100	−	0	−	0	+	100	+	100	+	86	+	100	+	78
39. Rhamnose	−	0	−	0	−	6	−	75	−	3	−	0	+	8	+	100	+	35	+	56
47. ONPG	−	0	+	63	+	88	+	0	+	3	+	100	+	96	+	100	+	94	+	100
49. Propionate	−	9	−	0	−	6	+	75	+	67	−	11	−	4	−	96	−	0	−	0
52. Reduction of nitrate	+	15	−	50	−	12	−	0	+	97	−	0	+	100	+	100	+	59	−	11
53. Reduction of methylene blue	+	91	+	25	+	88	−	0	+	45	+	96	+	100	+	100	+	88	−	44
54. Lecithinase	−	3	−	0	−	0	+	0	+	88	−	0	−	0	−	0	−	18	−	0
56. Urease	+	48	−	6	−	0	−	0	−	0	−	0	−	8	−	34	−	0	−	0
58. Tyrosinase	−	9	+	12	+	63	+	100	+	97	−	0	−	0	−	0	−	0	−	0
60. Ornithine	+	100	−	37	+	81	+	100	−	3	−	4	−	24	−	6	−	0	−	44
61. Lysine	+	97	−	63	+	25	+	75	−	15	+	11	+	64	+	100	−	0	−	11

Groups

Groups C (B.cereus) *H* (B.circulans) Tests: (2) fat inclusions, C+, H-; (4) spores swollen, C-, H+; (10) Gram
stain, C+ or variable, H-; (54) lecithinase, C+, H-.

Groups B (B.megaterium) *I* (B.badius) Tests: (2) reserve
fat, B+, I-; (10) Gram stain, B+ or variable, I-; (13)
motility, all I strains are motile, B strains are non-
motile; (29) acidity, B+, I-; (33) starch hydrolysis, B+,
I-; (47) ONPG, B+, I-; (53) reduction of methylene blue,
B+, I-; (63) phenylalanine deaminase, some B+, all I-.

Groups D (B.pumilus), *E* (B.subtilis), *F* (B.licheniformis),
G (B.polymyxa) Tests: (53) no hydrolysis of starch, and
(52) no reduction of NO$_3$, D; (27,30) no anaerobic growth,
E; (4) rods swollen, G+, F-; (10) Gram negative, G, Gram
positive or variable, F; (25) growth in 10% NaCl, F+, G-;
growth in propionate, F+, G-; (31) formation of gas frequent and abundant, G, less so, F.

This set of characters is much like that of Geslin
[1975], of Gordon *et al.* [1973], in their "stepping stone"
procedure, and also related to that of Gibson and Topping
[1939]; formation of acetoin was specifically emphasized
by Boeyé and Aerts [1976].

Using the revised set of characters an attempt was made
to allocate the 182 intermediate strains and all the 271
strains from marine sources to the 10 established phena.
The way in which all strains from marine sources were
distributed to the groups, according to origin of sample,
is given in Table 11. This distribution is in good agreement with that achieved earlier for the 97 strains from
marine sources (Table 8).

The strains from the marine sources did not have any
special characteristics that distinguished them from
strains with other origins. Of the 413 strains examined
only 29 (7%) were able to grow at 5°C in media containing
5% and 10% NaCl, and only 11 (3%) of these were of marine
origin. These halophilic and psychrophilic strains were
equally represented in the groups equated with *B.sphaeri-
cus, B.megaterium, B.cereus, B.firmus, B.circulans* and
B.licheniformis.

Mesophilic strains, however, which can grow in 10% NaCl
are dominant in phena F (*B.licheniformis*), E (*B.subtilis*),
and D (*B.pumilus*) (Table 10). The ability to grow in high
concentrations of NaCl is a typical character of all the
strains in these groups and is not specially related to a
marine origin. Of the A$_2$ strains about 50% can grow in
10% NaCl but within this phenon the halophilic strains are
from marine sources.

In samples of sea water *B.licheniformis* (F) is dominant
(55%), followed by *B.subtilis* (E), irrespective of the
degree of pollution, and by *B.pumilus* (10%) in the non-
polluted samples.

In the sediment samples *B.licheniformis* is found quite

TABLE 11

*Allocation of 271 strains from marine habitats to the
defined phena on the basis of their origin*

Group / Species	A₁ sphaericus	brevis	firmus	"M"	A₂ megaterium	B cereus	C pumilus	D subtilis	E licheniformis	F polymyxa	G circulans	H badius	I	Number of strains
1. Coastal, polluted	0	4	2	9	3	3	8	8	0	0	0	0		37
2. Coastal, non-polluted	3	1	4	1	7	9	5	6	0	3	0	2		41
3. Sea sediment	15	5	0	3	1	7	1	0	1	0	0	0		33
4. Oceanic sediment	1	4	7	12	3	5	13	18	9	0	0	0		72
5. Irish Sea	1	5	7	8	1	1	3	3	8	0	0	0		37
6. Water, polluted	1	0	0	0	0	1	0	4	15	0	0	0		21
7. Water, non-polluted	1	3	1	1	0	1	5	5	13	0	0	0		30
	22	22	21	34	15	27	35	44	46	3	0	2		271

frequently in those from the Irish Sea, but this species
is not common in sediments whereas *B.subtilis* and *B.pumilus*
formed 18 to 25% of the polluted coastal samples, and in
the samples taken off Peru and in the Sargasso Sea. The
diversity of species is much greater in sediments than in
water. In unpolluted samples from coastal areas *B.sphae-
ricus* (45%), *B.cereus* (22%), and *B.megaterium* (17%) are
frequent. The presence in sediments of species which are
also frequent in sea water is easily explained by sedimen-
tation. The transmission of other species could take
place in several ways. Sewage or fresh water origin is a
possibility in some areas, but also dust storms might
carry strains, which generally are supposed to belong in
soil (*B.sphaericus, B.cereus* and *B.megaterium*). Another
possible mode of transmission is the faecal pellets of sea
animals [Jannasch and Wirsen, 1977].

*Allocation of culture collection strains
from marine sources*

Many of the "marine" species described by the authors
mentioned in the introduction have been reclassified with
well-established species. Included in the author's
collection are 4 strains received from the National
Collection of Marine Bacteria (NCMB, Aberdeen) which

have been reallocated: *B.catenula* (Duclaux) (NCMB 1522)
to *B.pumilus* (phenon D), *B.idosus* (Burchard) (NCMB 1523)
and *B.filaris* (Migula) (NCMB 1525) to *B.subtilis* (phenon
E), *B.glutinosus* (Kern) (NCMB 1526) to *B.cereus* (phenon C).
B.lubinskii (Kruse) (NCMB 1524) was unassigned.

Smith *et al*. [1952] and Gordon *et al*. [1973] have also
reclassified strains NCMB 1522 and NCMB 1526 to similar
effect but have also reallocated strains of some marine
species [ZoBell and Upham, 1944; Kriss, 1963], not included in the author's collection: *B.submarinus* [ZoBell and
Upham] to *B.subtilis*, *B.thalassokoites* [ZoBell and Upham]
to *B.subtilis*, *B.borborokoites* [ZoBell and Upham] to *B.
subtilis*, *B.abysseus* [ZoBell and Upham] to *B.licheniformis*,
B.imomarinus [ZoBell and Upham] to *B.firmus* and *B.thermo-
liquefaciens* [Kriss] to *B.stearothermophilus*.

Kriss [1963] listed 63 strains grouped in 29 species of
which 10 species (24 strains) could be equated with the
accepted species: *B.alvei*, *B.cereus*, *B.circulans*, *B.
megaterium*, *B.mycoides*, *B.pumilus*, *B.stearothermophilus*,
and *B.subtilis*. The allocation of the 19 remaining groups,
related to the descriptive system of Krasil'nikov, is not
possible at present.

Special characteristics of strains from marine sources

As stressed repeatedly, most strains from marine sources
fit into established species and have no special characteristics. In the section on quantitative examination it is
also stressed that it is very difficult to find evidence of
multiplication and metabolic activity of *Bacillus* in the
sea. It does, however, seem to be possible to define
strains found in the marine environment but not in any
other habitat. Such strains are found in the phenon A_2,
subdivision "M". These strains are related to *B.firmus*,
clustering in the same phenon and are also frequently of
marine origin. They often form yellow, pink or ochre pigments; however, the "M" strains are distinguished by a
certain plasticity of form. They very often grow in pigmented, butyrous, glistening colonies with a "cogwheel"
appearance (Fig. 8), and within the same culture, bulging
and non-bulging spores may be found (Fig. 9), as well as a
variation between round and oval bulging spores in the
same culture (Fig. 10). To this group belong *B.lubinskii*
and possibly also strains of *B.epiphytus*, *B.cirroflagello-
sus*, *B.filicolonicus*, and *B.limnophilus* described by
ZoBell and Upham [1944] and examined by Gordon *et al*.
[1973]. Strains described by a number of workers [Dias *et
al*., 1968; Boeyé and Aerts, 1976; Geslin, 1975] and some
B.sphaericus and *B."pantothenticus"* variants described by
Gordon *et al*. [1973], probably also belong in this poorly
studied group, which seems to be at a borderline between
B.firmus, *B.badius*, *B.coagulans* and *B.circulans*. However,
at present there are insufficient data to combine such

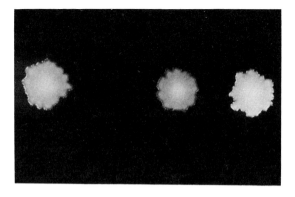

Fig. 8 Yellow and ochre colony of cogwheel appearance. Phenon A_2.

Fig. 9 Bulging and non-bulging spores in the same culture. Phenon A_2.

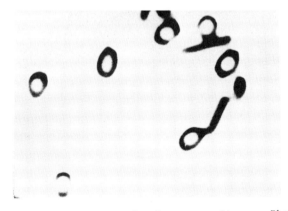

Fig. 10 Round and oval spores in the same culture. Phenon A_2.

strains into 1 group of specific rank. *Bacillus pacificus* [Delaporte, 1967] must according to the original paper, be considered to be of soil origin whereas many of the yellow pigmented strains of Turner and Jervis [1968] originating in salt marshes were examined by the author and allocated to phenon A_2.

Summary and Conclusion

A total of 271 *Bacillus* strains from marine habitats were isolated from samples of Danish sea water and samples of sediments collected from many estuarine and oceanic areas. Counts have also been made of *Bacillus* spores from these samples. *Bacillus* strains can be found in very high numbers in sediments being dominant in samples of low organic content and at greater depths where the numbers of strains of other species are declining. In water an accumulation of *Bacillus* species is found in and just above thermoclines.

After preliminary investigations, the non-metric multi-dimensional scaling of Kruskal was chosen on which to base a numerical classification and a collection of 413 strains was examined by means of 75 tests; 231 strains, 97 of marine origin, formed 10 clusters in a 3-dimensional approach. The great majority (70%) of isolates of marine origin were distributed at random among strains from other sources, and they did not exhibit specific characteristics. However, strains in phenon A_2, which contains about 20% of strains, could not be assigned to any recognized *Bacillus* species. By analysis using the canonical variate technique, the 75 tests were graded into 51 less useful and 24 useful characters, of which 10 to 15 were sufficient for grouping. By means of this revised set of tests all 271 isolates were reclassified and related to the origin of the sample. About 25% of strains were related to phenon A_2, were mainly of marine origin and had many traits in common such as morphological variability, differences in pigmentation, a negative VP test, and a weak ability or inability to breakdown glucose.

Acknowledgements

This work was supported by grants 512-2166, 5787, and 7311 from the Danish Medical Research Council, and by the International Service System Research Council.

References

Anderson, J.I.W. (1962). "Heterotrophic bacteria of North Sea water". Ph.D. Thesis. University of Glasgow.

Bartholomew, D.J. (1959). A test of homogeneity for ordered alternatives. *Biometrika* **46**, 36-48.

Bartholomew, J.W. and Paik, G. (1966). Isolation and identification

of obligate thermophilic spore-forming bacilli from ocean basin
cores. *Journal of Bacteriology* **92**, 635-638.

Baumann, L., Baumann, P., Mandel, M. and Allen, R.D. (1972). Taxonomy
of aerobic marine eubacteria. *Journal of Bacteriology* **110**, 402-429.

Boeyé, A. and Aerts, M. (1976). Numerical taxonomy of *Bacillus* isolates from North Sea sediments. *International Journal of Systematic Bacteriology* **26**, 427-441.

Bonde, G.J. (1962). "Bacterial Indicators of Water Pollution".
Thesis. Copenhagen: Teknisk Forlag.

Bonde, G.J. (1965). Classification of *Bacillus* spp. from marine sediments. *The Journal of General Microbiology* **41**, xxii.

Bonde, G.J. (1967). Bacteriology of water, sediments, and fish. In
"Report of the Investigations of the Swedo-Danish Committee on
Pollution of the Sound 1959-64" (ed. G.J. Bonde), pp.133-198.
Copenhagen: Statens Trykningskontor.

Bonde, G.J. (1968). Studies on the dispersion and disappearance
phenomena of enteric bacteria in the marine environment. *Revue
Internationale Océanographie Médicale* **IX**, 17-43.

Bonde, G.J. (1971). Examination of sediments in marine areas. In
"Reports of the Investigations of the Swedo-Danish Committee on
Pollution of the Sound 1965-70" (ed. T. Dackman), pp.288-291.
Copenhagen: Statens Trykningskontor.

Bonde, G.J. (1975). The genus *Bacillus*. *Danish Medical Bulletin* **22**,
41-61.

Bonde, G.J. (1976a). The marine *Bacillus*. *Journal of Applied
Bacteriology* **41**, vii-viii.

Bonde, G.J. (1976b). Kruskal's non-metric multidimensional scaling.
In "Compstat 1976" (eds. J. Gordesch and P. Naeve), pp.443-449.
Wien: Physica Verlag.

Brisou, J. (1955). "La Microbiologie du Milieu Marin". Editions
Médicin. Paris: Flammarion.

Delabre, M., Bianci, A. and Veron, M. (1973). Etude critique des
méthodes de taxonomie numerique. *Annales de Microbiologie
(Institut Pasteur)* **124 A**, 489-506.

Delaporte, B. (1967). Une Bactérie nouvelle de l'ocean Pacifique:
Bacillus pacificus n.sp. *Academie de Science. Comptes Rendues,
Série D* **264**, 3068-3070.

Denis, F. (1971a). "Contribution à l'étude des bactéries hétéro-
trophes du milieu marin: Inventaire de 2700 souches". Thèse de
Doctorat d'Etat ès Sciences. Poitiers No.C.N.R.S.A.O. 5720.

Denis, F. (1971b). Les *Bacillus* du milieu marin: Etude de 120
souches. *Comptes Rendue, Société Biologique* **165**, 2404-2406.

Denis, F., Geslin, M. and Nivet, A. (1975). Sensibilité aux antibio-
tique de différentes espèces de *Bacillus*. *Ouest Médical* **28**, 1629-
1633.

Dias, F.F., Agate, A.D. and Bhat, J.V. (1968). Coloured *Bacillus*
species resembling *B.firmus*. *Proceedings of the National Institute
of Science. India* **34B**, 9-16.

Edwards, P.R. and Ewing, W.H. (1972). "Identification of Enterobacteriaceae". 3rd ed. Minnesota: Burgess.

Geslin, M. (1975). "Place du genre *Bacillus* en microbiologie
comparée, étude de 409 souches de provenance variée". Thèse pour
le Doctorat en Pharmacie. Faculté de Médecine et de Pharmacie de

Poitiers.

Gianelli, F., Cabassi, E. and Ricci, B. (1970). Distribuzione qualitative dei batreri organotrofi del mare Adriatico. *L'igiene moderna 3-4* (cited after Geslin, 1975).

Gibson, T. and Topping, L.E. (1938). Further studies of the aerobic spore-forming bacilli. *Society of Agriculture and Bacteriology. Proceedings' Abstracts*, 43-4.

Gordon, R.E., Haynes, W.C. and Pang, C.H.-N. (1973). "The genus *Bacillus*". Washington, D.C.: United States Department of Agriculture.

Gower, J.C. and Barnett, J.A. (1971). Selecting tests in diagnostic keys with unknown response. *Nature* **232**, 491-493.

Hankin, L., Zucker, L.M. and Sands, D.C. (1971). Improved solid medium for the detection and enumeration of pectolytic bacteria. *Applied Microbiology* **22**, 205-209.

Huddleston, M. (1955). Marine bacteria of Cardigan Bay. I. *Journal of Applied Bacteriology* **18**, 22-28.

Jackson, D.K. and Bonde, G.J. (1971). DNA-base ratios of *Bacillus* strains related to numerical and classical taxonomy. *The Journal of General Microbiology* **63**, VII.

Jannasch, H.W. and Wirsen, O. (1977). Microbial life in the deep sea. *Scientific American* **236**, 42-52.

Jardine, N. and Sibson, R. (1970). "Mathematical Taxonomy". London, New York: Wiley.

Johnson, R.M., Katarski, M.E. and Weisrock, W.P. (1968). Correlation of taxonomic criteria for a collection of marine bacteria. *Applied Microbiology* **16**, 708-713.

Kriss, A.E. (1963). "Marine Microbiology (Deep Sea)" (translated by J.M. Shewan). Edinburgh: Oliver and Boyd.

Kruskal, J.B. (1964a). Multidimensional scaling by optimizing goodness of fit to a non-metric hypothesis. *Psychometrika* **29**, 1-27.

Kruskal, J.B. (1964b). A numerical method. *Psychometrika* **29**, 115-129.

Lemille, F., de Barjac, H. and Bonnefoi, A. (1969). Essai sur la classification biochimique de 97 souches de *Bacillus* du groupe I appartenant à 9 espèces differentes. *Annales de l'Institut Pasteur* **116**, 808-819.

MacLeod, R.A. (1965). The question of the existence of specific marine bacteria. *Bacteriological Reviews* **29**, 9-23.

Marmur, J. and Doty, P. (1962). Determination of the base composition of deoxyribonucleic acid from its thermal denaturation temperature. *Journal of Molecular Biology* **5**, 109-118.

Martin, S.M. and Skerman, V.B.D. (1972). "World Directory of Collections of Cultures of Microorganisms". New York and London: Wiley-Interscience.

Mayr, E. (1965). Numerical phenetics and taxonomic theory. *Systematic Zoology* **14**, 73-97.

Morita, R.Y. (1976). Survival of bacteria in cold and moderate hydrostatic pressure environments with special reference to psychrophilic and barophilic bacteria. In "The Survival of Vegetative Microbes" (eds. T.G.R. Gray and J.R. Postgate), pp.279-298. Cambridge: Cambridge University Press.

Murchelano, R.A. and Brown, C. (1970). Heterotrophic bacteria of Long Island Sound. *Marine Biology* **71**, 1 (cited after Geslin, 1975).

Nelson, J.D. and Colwell, R.R. (1975). The ecology of mercury-resistant bacteria in Chesapeake Bay. *Microbial Ecology* 1, 191-218.

Newton, B. (1924). Marine sporeforming bacteria. *Central Canadian Biology and Fishery* 1, 377-400.

Pankhurst, R.J. (1970). A computer program for generating diagnostic keys. *The Computer Journal* 13, 145-151.

Quigley, M.M. and Colwell, R.R. (1968). Properties of bacteria isolated from deep sea sediments. *Journal of Bacteriology* 95, 211-220.

Russel, N.L. (1891). Untersuchungen über im Golf von Neapel lebende Bakterien. *Zeitschrift für Hygiene* 11, 165-206.

Rüger, H.-J. (1973). Zum Vorkommen aerober sporenbildender Bakterien in nordostatlantischen Tiefseesedimenten. In "Meteor Forschungs-Ergebnisse", Reihe D 16, 60-64. Berlin and Stuttgart.

Rüger, H.-J. (1975). Bakteriensporen in marinen Sedimenten (Nord-atlantik, Skagerrak, Biskaya und Auftriebsgebiet von Nordwestafrika) - quantitative Untersuchungen. *Veröffentlichungen des Instituts für Meeresforschung, Bremerhaven* 15, 227-236.

Schubert, R.H.W. (1975). Der Nachweis von Sporen der *Bacillus*-species im Rahmen der hygienischen Wasserbeurteilung. *Zeitblatt für Bakteriologie und Hygiene, I Abteilung, Original B* 160, 155-162.

Schwarz, J.R. and Colwell, R.R. (1975). Heterotrophic activity of deep-sea sediment bacteria. *Applied Microbiology* 30, 639-649.

Schwarz, J.R., Walker, J.D. and Colwell, R.R. (1974). Growth of deep-sea bacteria on hydrocarbons at ambient and *in situ* pressure. In "Developments of Industrial Microbiology", Vol. 15, pp.239-249. Washington, D.C.: American Institute of Biological Sciences.

Shewan, J.M., Hobbs, G. and Hodgkiss, W. (1960). A determinative scheme for the identification of certain genera of Gram-negative bacteria, with special reference to the Pseudomonadaceae. *Journal of Applied Bacteriology* 23, 379-390.

Sieburth, J.Mc.N. (1964). Polymorphism of a marine bacterium (*Arthrobacter*). *Proceedings of Symposia of Experimental Marine Ecology* 2, 11-16.

Sierra, G. (1957). A simple method for the detection of lipolytic activity of micro-organisms and some observations on the influence of the contact between cells and fatty substrates. *Antonie van Leeuwenhoek* 23, 15-22.

Simidu, W. and Oisi, K. (1952). Studies on the putrefaction of aquatic perduals V. *Bulletin of the Japanese Society for Science and Fishery* 16, 547-549 (cited after Ferguson Wood, 1967).

Smith, L.D.S. (1968). The clostridial flora of marine sediments from a productive and from a non-productive area. *Canadian Journal of Microbiology* 14, 1301-1304.

Smith, N.R., Gordon, R.E. and Clark, F.E. (1952). "Aerobic spore-forming bacteria". Washington, D.C.: United States Department of Agriculture.

Sreenivasan, A. (1956). New species of marine bacteria tolerating high concentration of copper. *Current Science* (India) 25, 92-93 (cited after Ferguson Wood, 1967).

Stanley, S.O. and Morita, R.Y. (1968). Salinity effect on the maximal growth temperature of some bacteria isolated from marine environments. *Journal of Bacteriology* 95, 169-173.

Stenson, H.H. and Knoll, R.L. (1969). Goodness of fit for random rankings in Kruskal's non-metric scaling procedure. *Psychological Bulletin*, 122-126.

Sverdrup, H.U., Johnson, M.W. and Fleming, R.N. (1949). "The Oceans, their Physics, Chemistry, and General Biology". New York: Prentice Hall.

Thibault, P. (1971). "Cours de microbiologie systematique". Institut Pasteur Paris (cited after Geslin, 1975).

Trimble, R.B. and Ehrlich, H.L. (1970). Bacteriology of manganese nodules. *Applied Microbiology* **19**, 966-972.

Turner, M. and Jervis, D.I. (1968). Salt tolerance in pigmented and non-pigmented strains of *Bacillus* species isolated from soil. *Journal of Applied Bacteriology* **31**, 373-377.

Venkataraman, R. and Sreenivasan, A. (1954). Bacterial flora of sea water and mackerels off Tellicherry. *Proceedings of the National Academy of Sciences (India) Sect. A 20*, 651-655 (cited after Ferguson Wood, 1967).

Wood, E.J.F. (1953). Heterotrophic bacteria in marine environments in Australia. *Australian Journal of Marine and Freshwater Research* **4**, 160-200.

Wood, E.J.F. (1967). "Microbiology of Oceans and Estuaries". Amsterdam, London and New York: Elsevier.

Willemoes, M. (1964). A ball-stoppered quantitative sampler for the microbenthos. *Ophelia* **1**, 235-240.

ZoBell, C.E. and Upham, H.C. (1944). A list of marine bacteria including descriptions of sixty new species. *Bulletin of the Scripps Institute of Oceanography, University of California* **5**, 239-292.

ZoBell, C.E. (1946). "Marine Microbiology". Waltham, Massachusetts: Cronica Botanica.

ZoBell, C.E. and Johnson, F.H. (1949). The influence of hydrostatic pressure on the growth of terrestrial and marine bacteria. *Journal of Bacteriology* **57**, 179-189.

ZoBell, C.E. and Morita, R.Y. (1959). Deep sea bacteria. In "Galathea Reports", Copenhagen **1**, 139-154.

ZoBell, C.E. (1968). Bacterial life in the deep sea. *Bulletin of the Misaki Marine Biology Institute, Kyoto University* **12**, 77-96 (cited after Rüger, 1975).

Chapter 9

TAXONOMIC STUDIES ON *BACILLUS MEGATERIUM* AND ON AGAROLYTIC *BACILLUS* STRAINS

W. HUNGER and D. CLAUS

Deutsche Sammlung von Mikroorganismen, Gesellschaft für Biotechnologische Forschung mbH, Federal Republic of Germany

Introduction

The species concept currently adopted for the definition of *Bacillus* species is generally based on only a few phenotypic properties. It is not surprising, therefore, that such narrow definitions lead to some "complex" species and to species which comprise a spectrum of strains that cannot be clearly separated from related species. *Bacillus subtilis* may be cited as a typical example. According to Gordon *et al.* [1973] it is not possible to differentiate between *B.amyloliquefaciens* and *B.subtilis* strains using the test methods normally applied in the identification of *Bacillus* strains. DNA-DNA reassociation studies, however, have clearly shown that *B.subtilis*, as defined by Gordon *et al.* [1973], is composed of 2 genetically unrelated groups, with the neotype strain of *B.subtilis* in one group and organisms originally described as *B.amyloliquefaciens* in the second group [Seki *et al.*, 1975]. The 2 species may also be differentiated by pyrolysis gas-liquid chromatography and by API systems, but not by classical tests used in the identification of *Bacillus* species [O'Donnell *et al.*, 1980].

Since the classical studies of Smith *et al.* [1946, 1952] on the taxonomy of the genus, *B.megaterium* has usually been considered to be one of the better defined species of *Bacillus*. According to their species definition, slightly extended by Gordon *et al.* [1973] and Gibson and Gordon [1974], about 20 other aerobic spore formers (originally bearing different species names) are sufficiently closely related to *B.megaterium* to be considered as synonyms (Table 1).

Gordon *et al.* [1973], however, noted that within the species *B.megaterium* 2 merging aggregates of strains, connected by intermediate strains which blurred the division of the 2 groups, could be observed. As with other bacteria, *B.megaterium* appears to be composed of a series of strains forming a broad and continuous spectrum of properties. In a set of 60 strains studied by Gordon *et*

TABLE 1

Bacillus *names considered to be synonyms of*
B.megaterium *de Bary*
[*Gordon et al., 1973; Gibson and Gordon, 1974*]

B.agrestis	Werner [1933]
B.capri	Stapp [1920]
B.carotarum	Koch [1888]
B.cobayae	Stapp [1920]
B.cohaerens	Gottheil [1901]
B.danicus	Löhnis and Westermann [1909]
B.endoparasiticus	Benedek [1938]
B.flexus	Batchelor [1919]
B.fructosus	Ueda *et al.* [1967]
B.graveolens	Gottheil [1901]
B.immobilis	Steinhaus [1941]
B.malabarensis	Löhnis and Pillai [1907]
B.musculi	Stapp [1920]
B.oxalaticus	Migula [1894]
B.petasites	Gottheil [1901]
B.ruminatus	Gottheil [1901]
B.silvaticus	Neide [1904]
B.simplex	Gottheil [1901]
B.teres	Neide [1904]
B.tumescens	Zopf [1883]
Myxogeotrichum filarioides	Castellani [1965]

al. [1973] only 8 out of 22 properties were alike for all strains. Other positive characters have been found to vary in the range of 35 to 98% of the strains studied (Table 2).

According to Baumann-Grace and Tomscik [1957] 2 types differing in morphology and certain biochemical properties may be distinguished among strains of *B.megaterium*. Similar observations have been reported by Gibson and Gordon [1974] who described strains, classified as *B.megaterium*, as being closely related to *B.carotarum* of Koch [1888].

Other authors have also detected 2 different groups of bacteria within *B.megaterium*. Berger [1962] observed 2 groups with respect to the formation of capsules by cells growing with citrate as the sole carbon source and other workers [Fitz-James and Young, 1959; Rode, 1968; Watanabe and Takesna, 1970] described 2 groups on the basis of differences in the structure of their spore surface and in the conditions necessary for spore germination. Somerville and Jones [1972] found little homology between 2 strains of *B.megaterium* in the course of their DNA-DNA pairing studies.

The G+C content of the DNA of some *B.megaterium* strains

TABLE 2

Properties of 60 strains of B.megaterium
according to Gordon et al. *[1973]*

Property	% of positive strains
Motility	68
Catalase	100
Anaerobic growth	0
Voges-Proskauer reaction	0
Voges-Proskauer medium, pH <6.0	80
Temperature of growth	
Maximum 45°C	58
Maximum <40°C	47
Minimum <10°C	35
Egg-yolk reaction	0
Growth in:	
Lysozyme (0.001%)	0
Media, pH 5.7	88
NaCl (7%)	85
Acid formation from:	
Glucose	100
Arabinose	75
Xylose	62
Mannitol	85
Hydrolysis of starch	100
Use of citrate	98
Reduction of NO_3^- to NO_2^-	43
Deamination of phenylalanine	57
Decomposition of	
Casein	100
Tyrosine	67

has been found to fall within the range of 36 to 45% [Normore, 1973; Candeli *et al.*, 1978] suggesting that the species is heterogeneous. As different strains have been used in these experiments it is not certain whether the differences observed in each case correspond to the 2 main groups of *B.megaterium*.

Similar problems exist within many other *Bacillus* species. Values for the G+C content of the DNA very clearly show that some of the currently recognized *Bacillus* species are composed of genetically very unrelated strains (see Chapter 3). It may, therefore, be concluded that the conventional methods used for the recognition of aerobic endospore forming bacteria have only been partly successful in distinguishing between *Bacillus* species.

Modern species concepts necessarily should include data on the range of the DNA base composition found within taxospecies. Strains belonging to the same species gene-

rally differ only slightly in the mean G+C content of
their DNA [Jones and Sneath, 1970] though those with a
similar G+C content of their DNA may belong to different
species. The genetic relatedness of such strains may be
established by DNA-DNA reassociation studies (see Chapter
3). Strains belonging to a single species should not only
show a high degree of homology of their DNA but can be expected to have biochemical properties in common, thereby
allowing their phenotypical differentiation from strains
which have a similar G+C content but belong to a different
species.

For reasons such as those mentioned above the authors
have started a programme to study the genetic relationship
of the *Bacillus* strains grouped by Gordon *et al.* [1973] or
by others within a single species. In this Chapter
strains of the relatively well defined species *B.mega-
terium* are considered together with new isolates of agarolytic spore formers which, according to Hunger and Claus
[1978], are related to the less well known *B.agarexedens*
[Wieringa, 1941].

Studies on *Bacillus megaterium*

The 21 strains examined (Table 3) included cultures of
the 2 main groups of *B.megaterium* described by Gordon *et
al.* [1973] and Gibson and Gordon [1974] as well as some
intermediates. One of the strains (DSM* 1324) described
as a "*B.megaterium/cereus* intermediate" by Knight and
Proom [1950], differed from typical *B.megaterium* strains
primarily by having a positive Voges-Proskauer reaction.

DNA base composition

The mean base composition of the DNA was estimated from
the thermal denaturation temperature (T_m) by the method of
Marmur and Doty [1962]. Thermal denaturation was carried
out in a Pye Unicam SP 1800 equipped with an automatic
cuvette changer. The temperature of the cell block was
raised at a rate of $30°C$ h^{-1} with a Lauda U3-S15 thermos-
tat controlled by a Lauda P120 linear temperature pro-
grammer. The change in extinction value (ΔE_{260}) and the
temperature of the DNA solutions, measured directly with
Pt 100 resistance sensors connected to a Doric digital
thermometer DS 100-T5 in the cuvettes, were recorded automatically by a PA BCD-Moduprint. DNA base composition was
calculated from the T_m determined in standard saline
citrate buffer (1xSSC) from 3 separate determinations by
the equation: $mol\%GC=2.44(T_m-69.4)$ [De Ley, 1970] and ex-

---

*Deutsche Sammlung von Mikroorganismen - German Collection of Micro-
organisms.

pressed relative to the reference DNA of *Escherichia coli* K 12 (mol%GC=51.7; Gillis *et al.*, 1970).
The DNA base composition of the 21 strains examined fell between the values 34.8 to 47.0 %GC but within this range 2 narrower groups were apparent. Group I comprised 12 strains (37.0 to 38.1 %GC) and group II 6 strains (40.1 to 40.9 %GC), the 2 groups being connected by a single strain (DSM 1320) having a GC content of 38.8%. Two strains (DSM 1324 and 1488) fell outside of the 2 groups and possibly belong to different species (Fig. 1).

TABLE 3

Sources of Bacillus megaterium *strains*

Strain (DSM No.)	Source
32	ATCC 14581 ← R.E. Gordon ← T. Gibson ← W.W. Ford. Neotype strain.
90	Institut f. Mikrobiologie, Göttingen, ← G. Bohlken 3.Soil, Germany
319	Institut f. Mikrobiologie, Göttingen.
321	Institut f. Mikrobiologie, Göttingen.
322	C. Schaab, Garden soil, Göttingen.
333	D. Claus, Field soil, Göttingen.
337	Institut f. Mikrobiologie, Göttingen.
339	Institut f. Mikrobiologie, Göttingen.
344	Institut f. Mikrobiologie, Göttingen ← G. Bohlken 2.Soil, Germany
509	H.J. Somerville ← NCIB 8508 ← NRRL B-938.
510	H.J. Somerville ← J. Wolf FS 96.
1316	R.E. Gordon, NRS 602 ← J.R. Porter ← G. Bredemann; *B.agrestis* Werner [1933].
1317	R.E. Gordon, NRS 607 ← J.R. Porter ← C. Stapp; *B.capri* Stapp [1920].
1318	R.E. Gordon, NRS 610 ← J.R. Porter ← C. Stapp; *B.cobayae* Stapp [1920].
1319	R.E. Gordon, NRS 623 ← J.R. Porter ← C. Stapp; *B.musculi* Stapp [1923].
1320	R.E. Gordon, NRS 665 ← B.S. Henry 131; *B.flexus* Batchelor [1919].
1321	R.E. Gordon, NRS 960 ← J.R. Porter ← O.F. Edwards ← H.J. Cohn; *B.simplex* Gottheil [1901].
1322	R.E. Gordon, NRS 961 ← J.R. Porter ← NCTC 2597 ← W.W. Ford; *B.simplex* Gottheil [1901].
1323	R.E. Gordon, NRS 986 ← J.R. Porter ← C. Stapp; *B.teres* Neide [1904].
1324	Wellcome Res.Lab. ← H. Proom CN 736; *B.megaterium/cereus* intermediate.
1488	P.J. White 16.

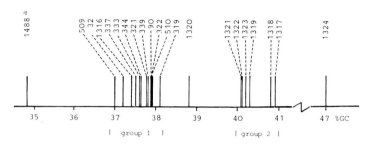

Fig. 1 DNA base composition of *Bacillus megaterium* strains; a, DSM number of strains.

For most of the strains the maximum difference in the DNA base composition is slightly less than 4% GC making it difficult to define subgroups without first considering results of DNA-DNA homology studies on the strains.

Interstrain DNA-DNA binding

Measurements of DNA reassociation have been done spectrophotometrically after De Ley *et al.* [1970]. This method, although used in only a few taxonomic studies, compares favourably with other reassociation techniques and has the advantage that it needs no isotopically labelled DNA [Gibbins and Gregory, 1972; Bradley, 1973; Martini and Phaff, 1973; Crombach, 1974; Swings and De Ley, 1975; Owen *et al.*, 1978; see also Chapter 3]. DNA was sheared with the aid of a French press, denatured in 2xSSC buffer in tightly stoppered cuvettes at 105°C for 15 mins and then quickly transferred to the spectrophotometer cell compartment held at the theoretically calculated optimum temperature of renaturation (T_{OR}). This temperature was arrived at using the equation: $T_{OR} = 47.0 + (0.51 \times \%GC)$ [De Ley *et al.*, 1970]. The renaturation of DNA fragments was monitored by the decrease in extinction during the initial 40 mins of the reaction. The degree of binding between DNA from different strains was calculated using the equation:

$$D = 100 \ \{4v_M - (v_A + v_B)\} \ / \ 2\sqrt{v_A \cdot v_B},$$

where v_A, v_B and v_M are the renaturation rates (decrease of extinction at 260 nm min^{-1}) of the single DNA samples (A,B) and of the mixture (M) [De Ley *et al.*, 1970]. In accordance with Johnson [1973] strains showing a DNA binding level in the range of D=60 to 100% were considered to belong to a single species. Using this method, D values of 10 to about 35% were observed with DNA of strains of different *Bacillus* species.

DNA homology groups

Eleven of the 12 strains from the GC group I formed DNA homology group A on the basis of the high levels of binding (D >78%) shown relative to the DNA reference strains *B.megaterium* DSM 333 and DSM 322 (Table 4). The DNA of the remaining strain of group I (DSM 1316) showed only a low level of binding (D = 30%). Strains DSM 1320 and 1488 also showed relatively little genetic similarity to these reference systems (Table 4). Within GC group II a rather high level of binding (D = 67-71%) was observed (DNA homology group B) using DNA from reference strains DSM 1317 and 1319. However, one of the 6 strains, DSM 1323, showed a relatively low homology value (D = 57%) with reference DNA from DSM 1317. Strain DSM 1320, which has a %GC content between the 2 GC groups, is not very closely related to DNA homology group B giving the low binding level of D = 24%.

It can be concluded, therefore, that the *B.megaterium* strains studied can be divided into at least 3 DNA homology groups. Further studies are required to determine strains DSM 1316 and DSM 1320, originally described as *B.agrestis* and *B.flexus* can be grouped within the same DNA homology group.

Phenotypic properties of B.megaterium *strains*

The standard methods of the ICSB Subcommittee on *Bacillus* were adopted for the phenotypic characterization of the test strains. These methods corresponded mainly to those described by Gordon *et al*. [1973]. Additional tests carried out were: urease formation [Christensen, 1946; as a control agar base without the addition of urea was used), hydrolysis of aesculin [Cowan and Steel, 1974], degradation of chitin [Skerman, 1967], splitting of amylose azure [Hunger, 1978] and formation of poly-β-hydroxy-butyrate [Jüttner *et al*., 1975].

It can be seen from Table 5 that the strains in the 2 main DNA homology groups can be distinguished clearly on the basis of morphological and biochemical properties. Thus, while group A is composed of strains having a mean cell diameter of 1.5 μm or more when grown in Difco nutrient broth, group B cells measure only 0.8 to 1.0 μm in diameter with a few broader cells occasionally being observed. However, the cell diameter of group B strains may be somewhat higher in media containing glucose.

With the biochemical properties usually tested in the characterization of *Bacillus* species, all strains of the DNA homology groups A and B gave positive results for catalase, the utilization of citrate, the decomposition of casein, the hydrolysis of starch and the acid formation from glucose. Negative results were observed in the following tests: Voges-Proskauer reaction, growth in the

TABLE 4

Degree of DNA homology between B.megaterium strains

DNA source (DSM strain no.)	Reference DNA from strain:											
	DSM 333			DSM 322			DSM 1317			DSM 1319		
	%D[a]	±SD[b]	ΔGC[c]	%D	±SD	ΔGC	%D	±SD	ΔGC	%D	±SD	ΔGC
32	87	5	0.4									
90				95	4	0.0						
319	95	5	0.5									
321	80	3	0.2	88	11	0.1						
322	79	3	0.3	100		0.0	31	3	3.0	24	4	2.4
333	100		0.0	79	3	0.3	22	5	3.3	30	3	2.7
337	85	5	0.1									
339	87	7	0.2									
344	86	1	0.0									
509	78	3	0.6									
510	84	4	0.3	95	1	0.0						
1317	22	6	3.3	31	3	3.0	100		0.0	70	9	0.6
1318										71	4	0.5
1319	30	3	2.7				70	9	0.6	100		0.0
1321	34	3	2.5				71	1	0.8			
1322							67	6	0.8			
1323	28	4	2.6				57	6	0.7	64	3	0.1
1316	30	1	0.2									
1320	33	7	1.2				30	4	2.1	24	4	1.5
1488	42	4	3.4									

a, %D Degree of binding; b, SD Standard deviation;
c, ΔGC Difference in the mean %GC contents of the DNAs in the reassociation mixture.

TABLE 5

Properties of B. megaterium strains

Strains (DSM number)	mol %GC	±SD[a]	DNA homology group	Mean cell width µm	Formation of poly-β-hydroxybutyrate	Deamination of phenylalanine	Hydrolysis of aesculin	Formation of urease	Yellow colonies on casein agar	Splitting of amylose azure	Hydrolysis of starch	Reduction of nitrate to nitrite	Brownish colonies on tyrosine agar	Decomposition of chitin	Growth in lysozyme	Voges-Proskauer reaction
1488	34.2	0.3	-	1.2	-	-	+	-	-	+	+	+	-	+	+	-
32	37.2	0.0	A	1.5	+	+	+	+	+	+	+	-	-	-	-	-
90	37.9	0.2	A	1.6	+	+	+	+	+	+	+	-	-	-	-	-
319	38.1	0.0	A	1.7	+	+	+	+	+	+	+	-	-	-	-	-
321	37.8	0.1	A	1.7	+	+	+	+	+	+	+	-	-	-	-	-
322	37.9	0.2	A	1.7	+	+	+	+	+	+	+	-	-	-	-	-
333	37.6	0.0	A	1.5	+	+	+	+	+	+	+	-	-	-	-	-
337	37.5	0.1	A	1.8	+	+	+	+	+	+	+	-	-	-	-	-
339	37.8	0.2	A	1.7	+	+	+	+	+	+	+	-	-	-	-	-
344	37.6	0.2	A	1.7	+	+	+	+	+	+	+	-	-	-	-	-
509	37.0	0.3	A	1.5	+	+	+	+	+	+	+	-	-	-	-	-
510	37.9	0.2	A	1.7	+	+	+	+	+	+	+	-	-	-	-	-
1316	37.4	0.1	C?	1.0	-	-	-	+	+	+	+	-	-	-	-	-
1320	38.8	0.2	C	0.9	-	-	-	+	-	+	+	-	-	-	-	-
1317	40.9	0.1	B	1.0	-	-	-	-	-	-	+	+	+	-	-	-
1318	40.8	0.1	B	1.0	-	-	-	-	-	-	+	+	+	-	-	-
1319	40.3	0.1	B	1.0	-	-	-	-	-	-	+	+	+	-	-	-
1321	40.1	0.1	B	0.8	-	-	-	-	-	-	+	+	+	-	-	-
1322	40.1	0.1	B	1.0	-	-	-	-	-	-	+	+	+	-	-	-
1323	40.2	0.1	B	0.8	-	-	-	-	-	-	+	+	+	-	-	-
1324	47.0	0.2	-	1.0	+	+	+	-	-	+	+	+	-	-	+	+

a, SD Standard deviation

presence of lysozyme, the egg-yolk and anaerobic growth
tests. The decomposition of tyrosine and the formation of
acid from mannitol, xylose and arabinose was found to be
variable in one or both of the DNA homology groups.

The 2 main homology groups can, however, be separated
using some of the tests listed in Table 5. Strains of DNA
homology group A clearly differ from those of group B in 8
out of 13 properties listed.

Strain DSM 1488 having a DNA base composition significantly lower than that shown by strains in the 2 main DNA
homology groups can also be distinguished from the other
test strains on the basis of phenotypic properties.
Similarly strains DSM 1316 and DSM 1320, which may form a
third DNA homology group, differ from strains in the 2
homology groups by up to 5 properties (Table 5).

The *B.megaterium/cereus* intermediate (DSM 1324) with
the highest %GC content recorded (47%), was the only
strain to give a positive Voges-Proskauer reaction. In
addition, it differs from bacilli classified in homology
groups A and B by up to 6 characters and from strains DSM
1316 and DSM 1320 in 4 properties (Table 5). The organisms
in DNA homology group A correspond closely to the morphological description of *B.megaterium* of de Bary [1884] and
include the type strain of the species. In contrast,
those in DNA homology group B have been considered [Gibson
and Gordon, 1974] to be similar to *B.carotarum* [Koch,
1888] and include strains originally carrying the names
B.capri, *B.cobayae*, *B.simplex* and *B.teres*.

From the results presented here it is evident that *B.
megaterium* as defined by Gordon *et al.* [1973] can be
classified into 4 genetically less related species (DNA
homology groups A,B,C and strain DSM 1488), which can be
differentiated also by phenotypic properties. One has to
conclude further that the strain named as a *B.megaterium/
cereus* intermediate (DSM 1324) is neither related to *B.
megaterium* nor to *B.cereus*, the latter has a DNA base composition of about 35 %GC (see Chapter 3).

The 60 *B.megaterium* strains studied by Gordon *et al.*
[1973] were analyzed in light of their own results and in
respect of the data derived from the present DNA homology
studies (Table 6). It is evident that strains in the 2
DNA homology groups can be differentiated by 2 tests: the
reduction of nitrate to nitrite and the deamination of
phenylalanine. According to these 2 tests it seems likely
that 31 out of the 60 strains belong to DNA homology group
A, 21 to DNA homology group B, 3 to the provisional DNA
homology group C and 3 strains, giving positive reactions
in both tests, to a further group not included in the present experiments (Table 6).

Neglecting the properties which have been found to be
variable within a certain DNA homology group, it is interesting that the 31 strains related to DNA homology group A
are nearly 100% alike in their properties as listed by

TABLE 6

Properties of the 60 strains of B.megaterium *listed
by Gordon* et al. *(1973) and grouped according
to their differences in respect to the formation of
nitrite from nitrate and the deamination of phenylalanine*

Group	"Complex"	A	B	C	D
Number of strains	60	31	23	3	3
	% of strains positive				
Property:					
Motility	68	90	65	33	66
Catalase	100	100	100	100	100
Voges-Proskauer reaction	0	0	0	0	0
Voges-Proskauer medium, pH <6.0	80	100	39	66	100
Temperature of growth					
Maximum 45°C	58	100	4	66	33
Maximum <40°C	47	0	97	33	66
Minimum <10°C	35	0	87	0	0
Egg yolk reaction	0	0	0	0	0
Growth in:					
Lysozyme	0	0	0	0	0
Media, pH 5.7	88	100	87	100	66
NaCl (7%)	87	100	70	100	100
Acid formation from:					
Glucose	100	100	100	100	100
Arabinose	75	100	(43)[a]	0	100
Xylose	62	(90)	(26)	0	100
Mannitol	85	100	(70)	100	33
Hydrolysis of starch	100	100	100	100	100
Use of citrate	98	97	100	100	100
Reduction of nitrate to nitrite	43	0	100	0	100
Deamination of phenylalanine	57	100	0	0	100
Decomposition of:					
Casein	100	100	100	100	100
Tyrosine	67	(61)	(83)	33	33

a, Results in brackets are variable within the DNA homology groups.

Gordon *et al.* [1973]. The only exceptions are the 3
strains recorded as non-motile and the single strain unable to utilize citrate as the sole source of carbon. In
contrast strains of *B.megaterium* related to DNA homology
group B are much more heterogeneous in their properties.
This heterogeneity may be related to the lower DNA binding
levels observed within this group as compared to the more
homogeneous group A. However, further DNA pairing studies
have to be done with more strains especially of the latter
group and such studies should also include representatives
of groups C and D.

TABLE 7

Sources of agarolytic Bacillus *strains*

Strain (DSM No.)	Source
34	ATCC 14176 (*B.circulans*) ← M. Shaw (*B.palustris* var. *gelaticus*). Decayed tree stump.
1327	I. Miehlmann 10. Meadow soil, Göttingen, 1972.
1352	D. Claus 39. Garden soil, Karlsruhe, 1970.
1353	D. Claus 45. Garden soil, Hannover, 1970.
1354	I. Miehlmann 34. Forest soil, Celle, 1972.
1355	W. Hunger 65. Volcanic soil, Volcano Paricutin, Mexico, 1975.
1356	I. Miehlmann 35. Forest soil, Göttingen, 1972.
1358	I. Miehlmann 19. Forest soil, Göttingen, 1972.
1472	I. Miehlmann L. Air contamination, Göttingen, 1972.
1473	D. Claus 4. Compost, Göttingen, 1970.
1474	I. Miehlmann 8. Lawn soil, Hannover, 1972.
1475	I. Miehlmann 12. Ditch mud, Ossendorf, 1972.
1477	D. Claus 26b. Field soil, Yugoslavia, 1970.
1478	I. Miehlmann 31a. Forest soil, Lüneburg, 1972.
1479	I. Miehlmann 31b. Forest soil, Lüneburg, 1972.
1480	I. Miehlmann 33. Forest soil, Dransfeld, 1972.
1481	D. Claus 44. Field soil, Rothamsted, England, 1970.
1482	K.A. Malik M. Contaminated agar plate, Göttingen, 1972.
1483	W. Hunger 62. Garden soil, Göttingen, 1974.
1484	W. Hunger 64. Forest soil, Göttingen, 1974.
1485	W. Hunger 68. Savanna soil, Tuxtla, Mexico, 1975.
1486	W. Hunger 69. Garden soil, Göttingen, 1976.
1487	W. Hunger 70. Lawn Soil, Göttingen, 1976.

Studies on Agarolytic Sporeformers

In the past few years the authors have isolated more than 20 agarolytic *Bacillus* strains (Table 7) which have properties corresponding to the description of *B.agar-exedens* [Wieringa, 1941; Hunger and Claus, 1978]. Few of these isolates show a significant similarity to the agar decomposing *B.palustris* var. *gelaticus* [Sickles and Shaw, 1934] which has been classified in the *B.circulans* complex [Gordon *et al.*, 1973]. According to Wieringa [1941], strains of *B.agar-exedens* differ from other agarolytic spore formers and from other *Bacillus* species mainly due to the inhibition of their growth by peptones in neutral or slightly acid, mineral-glucose media. This inhibitory effect is not observed in alkaline media [Hunger and Claus, 1978].

Preliminary studies on the biochemical properties of the isolates have shown that they form a compact group.

Unlike the agarolytic strains described by Wieringa [1941] half of the new isolates failed to hydrolyze starch [Hunger and Claus, 1978]. As this property is usually constant within *Bacillus* species, representatives of the starch positive and negative groups were examined for their DNA base composition using the DNA reassociation method described earlier.

DNA base composition

The mean base composition of the DNA of the agarolytic isolates was determined as before and the strains found to fall within the range of 34.4 to 52.7 %GC (Table 8). However, several subgroups can be recognized on the basis of the G+C data. This is especially true for the strains not inhibited by peptone as they form at least 2 groups, one within the range of 34.4 to 34.7 %GC (2 strains) and the second with values between 49.3 to 51.1 %GC (4 strains). The strains which are inhibited by peptone formed 1 sub-group comprising 3 strains with about 44.5 %GC, which can be separated from the bulk of strains that form a rather continuous spectrum within the range of 47.3 to 52.7 %GC.

According to their DNA base composition most of the agarolytic *Bacillus* strains are unrelated to the majority of *Bacillus* species since the DNA base composition of the latter is generally below 47%. However, higher values have been observed for *B.acidocaldarius, B.macerans* and with certain strains of *B.circulans, B.coagulans, B.poly-myxa* and *B.stearothermophilus* (see Chapter 3).

Interstrain DNA-DNA binding: DNA homology groups

It was not possible to estimate the DNA binding levels between all of the test strains so 7 DNA reference strains having DNA base compositions of 34.7, 44.5, 49.0, 49.5, 51.1, 51.3 and 52.7 %GC were studied. The reference strains did, however, include some organisms that were inhibited by peptones at neutral pH values and some which were not.

The DNA binding levels estimated between the reference strains and the other agarolytic *Bacillus* strains are listed in Table 9. High levels of DNA binding were found only with certain strains differing in their DNA base composition by less than 2 %GC. Other strains, however, differing by less than 1 %GC from the reference strains are genetically unrelated since binding levels of D <30% often were observed. From the results given in Table 9 it is clear that within the agarolytic strains studied 8 DNA homology groups can be distinguished: group 1 (6 strains: DSM 1327, DSM 1473, DSM 1475, DSM 1477, DSM 1478 and DSM 1479); group 2 (5 strains: DSM 1355, DSM 1483, DSM 1484, DSM 1485 and DSM 1486); group 3 (3 strains: DSM 1354, DSM 1474 and DSM 1480); group 4 (2 strains: DSM 1356 and

TABLE 8

DNA base content of agarolytic Bacillus *strains*

Strain (DSM No.)	Mean mol%GC	±SD[a]	Number of experiments	Inhibition by peptones
1472	34.4	0.4	3	-
1353	34.7	0.6	6	-
1352	44.4	0.7	3	+
1356	44.4	0.6	3	+
1482	44.5	0.3	3	+
1327	47.3	0.2	5	+
1479	47.8	0.1	3	+
1473	47.9	0.3	3	+
1478	48.2	0.0	2	+
1475	48.3	1.7	3	+
1477	49.0	0.6	4	+
1481	49.3	0.4	3	-
1358	49.5	0.5	4	-
1486	50.2	0.6	3	+
1485	50.8	0.4	3	+
1487	50.9	0.4	5	-
1484	51.0	0.5	5	+
34	51.1	0.8	4	-
1355	51.3	0.9	5	+
1483	51.7	0.4	5	+
1474	52.2	0.7	4	+
1480	52.6	0.6	3	+
1354	52.7	0.6	3	+
Reference strain B.*circulans* (type strain)[b]				
11	35.6	1.0	6	-

a, SD Standard deviation; b, Non-agarolytic.

DSM 1482); group 5 (1 strain: DSM 1352); group 6 (2 strains: DSM 1358 and DSM 1481); group 7 (2 strains: DSM 34 and DSM 1487) and group 8 (2 strains: DSM 1353 and DSM 1472). The homology groups show a high intra- but a low inter-group level of DNA binding (Table 10).

Within the strains inhibited by peptone 5 DNA homology groups (1 - 5) were recognized, each consisting of strains with a narrow DNA base range. The 3 strains with a rather low %GC content of about 44% fell into 2 groups (4 and 5) which shared only low DNA binding values (D = 14%). Due to the large difference in the %GC content of their DNA, reassociation experiments with the DNA of the latter strains have not been done with strains of DNA homology group 3 (%GC 52.2 - 52.7).

The strains not inhibited by peptone were recovered in 3 DNA homology groups (6 - 8) each composed of only 2 strains. The strains of groups 6 and 7 are not related to those inhibited by peptone even though they have an almost identical DNA base composition (compare group 6 with group 1 or 2, group 7 with 2) (Table 9).

The strains of DNA homology group 8, which can be separated from all other DNA homology groups by their low DNA base composition, share many phenotypic properties with strains of *B.circulans*. However, in the reassociation experiments little homology was found between these organisms and DNA from the type strain of *B.circulans*.

The classification of the agarolytic *Bacillus* strains into 8 DNA homology groups may be considered to be surprising, however, there is no reason to assume that the agarolytic property should be species specific.

Morphological, physiological and biochemical properties of agarolytic sporeformers

The methods described by Gordon *et al.* [1973] were used for the characterization of the agarolytic *Bacillus* strains. Since most of the isolates were inhibited by peptones in neutral or slightly acid media, the pH of the media was adjusted to 7.7. To fulfil the growth demands of the strains glucose (1% w/v) was added to liquid media when necessary and morphological observations made from cultures grown on nutrient agar supplemented with glucose (1% w/v) and urea (0.1% w/v). In addition the following tests were carried out with the test strains: hydrolysis of pectin [Wieringa, 1949], chitin [Skerman, 1967], dextran [Mencier, 1972], aesculin [Cowan and Steel, 1974], urea [Smith *et al.*, 1952], and DNA [Jeffries *et al.*, 1957].

In terms of morphological properties the strains differed mainly in the diameter of their vegetative cells which fell within the range of 0.5-0.7 to 1.2-1.5 μm. There was no correlation between cell diameter and clustering on the basis of the DNA homology data. The length of the straight rod-shaped cells fell within the range of 2-5 to 4-9 μm. In young cultures all strains were Gram positive although the staining was very uneven. In most strains bulging and non-bulging sporangia with oval spores, located at different positions within the mother cell, were observed. In only 3 cases were more than 50% of the sporangia swollen. All other strains were motile but the type of flagellation was not recorded.

The minimum and maximum temperature for growth of all but one strain lie in the range of 10-20 and 35-40°C respectively; the exception has a maximum temperature for growth of 45°C.

The agarolytic strains behave very similarly in respect of the tests usually used for the differentiation of *Bacillus* species. All of the strains were somewhat inactive but as far as positive results are concerned all of the strains formed catalase, hydrolysed hippurate and produced acid from glucose. The strains were negative for:

TABLE 9

DNA binding levels (%D) within groups of agarolytic
Bacillus *strains with similar DNA %GC content*

Range of mol%GC	Strains		Reference DNA	%GC	ΔGC[c]	%D[a]	±SD[b]	RE[d]
34.4 -35.6	8* {	1472 f	1353	34.4	0.3	106	4	4
		1353 f		34.7	0.0	100		
		11 e	1353	35.6	0.9	22	3	4
44.4 -44.5	5* {	1352	1482	44.4	0.1	14	6	4
		1356	1482	44.4	0.1	97	3	5
		1482		44.5	0.0	100		
44.4 -51.3		1352	1477	44.4	4.6	22	3	5
		1482	1477	44.5	4.5	29	4	3
		1327	1477	47.3	1.7	95	11	5
		1479	1477	47.8	1.2	100	1	5
	1* {	1473	1477	47.9	1.1	89	3	4
		1478	1477	48.2	0.8	80	9	5
		1475	1477	48.3	0.7	104	3	3
		1477		49.0	0.0	100		
		1481 f	1477	49.3	0.3	20	5	5
		1358 f	1477	49.5	0.5	11	3	4
		1355	1477	51.3	1.3	31	7	6
44.4 -52.7		1352	1355	44.4	5.9	21	8	6
		1482	1355	44.5	5.8	24	5	3
		1478	1355	48.2	3.1	32	7	5
		1475	1355	48.3	3.0	31	3	4
		1477	1355	49.0	2.3	31	7	6
		1481 f	1355	49.3	2.0	23	2	4
		1358 f	1355	49.5	1.8	31	5	5
		1487 f	1355	50.9	0.4	15	5	5
		1486	1355	50.2	1.1	63	10	5
		1485	1355	50.8	0.5	96	7	4
	2* {	1484	1355	51.0	0.3	102	5	5
		1355		51.3	0.0	100		
		1483	1355	51.7	0.4	101	1	5
		1474	1355	52.2	0.9	34	3	4
		1354	1355	52.7	1.4	23	8	5
48.3 -52.7		1475	1354	48.3	4.3	32	5	4
		1477	1354	49.0	3.7	30	3	3
		1358 f	1354	49.5	3.2	22	4	4
	3* {	1474	1354	52.2	0.5	97	2	4
		1480	1354	52.6	0.1	104	8	4
		1354		52.7	0.0	100		
44.4 -51.1		1352	1358	44.4	5.0	21	5	3
		1356	1358	44.4	5.0	26	3	3
	6* {	1481 f	1358	49.3	0.2	97	6	5
		1358 f		49.5	0.0	100		
		34 f	1358	51.1	0.6	24	4	5
50.9 -51.1	7* {	1487 f	34	50.9	0.2	98	4	5
		34 f		51.1	0.0	100		

TABLE 10

Intra- and inter-group DNA binding levels of DNA homology groups of agarolytic Bacillus *strains*

DNA homology group	No. of strains	DNA binding level (%D) to the reference strain of group:							
		1	2	3	4	5	6	7	8
1	6	>80							
2	5	<32 (3)[a]	>63						
3	3	<32 (2)	<34 (2)	>97					
4	2	29 (1)	24 (1)	nd[b]	97				
5	1	22 (1)	21 (1)	nd	14 (1)	100			
6	2	<20 (2)	<31 (2)	22 (1)	26 (1)	21 (1)	97		
7	2	nd	15 (1)	nd	nd	nd	24 (1)	98	
8	2	nd	nd	nd	nd	nd	nd	nd	100

a, numbers in brackets represent the number of strains selected from the group for DNA reassociation experiments with the DNA reference strain of the other DNA homology group; b, no determination.

the Voges-Proskauer reaction, growth in the presence of lysozyme, utilization of citrate, anaerobic growth, growth in 5% (w/v) NaCl, growth at pH 5.7, formation of indole, dihydroxyacetone or crystalline dextrins, decomposition of casein, gas formation from glucose, reduction of nitrate to nitrite and for the egg yolk reaction.

Some of the strains gave positive results for the degradation of tyrosine, the hydrolysis of starch and for the deamination of phenylalanine. Tyrosine was decomposed by 4 out of 6 strains of DNA homology group 1 and weakly by the 2 strains of DNA homology group 4. Hydrolysis of starch was observed with all strains of DNA homology groups 1, 5, 6, 7 and 8, whereas phenylalanine was deaminated by all of the strains of homology groups 1, 3 and 4 (Table 11).

Footnote to Table 9 (opposite)

For a, b and c see Table 4; d, RE: number of reassociation experiments; e, *B.circulans* neotype; f, strains not inhibited by peptones. * DNA homology group.

TABLE 11

Properties of agarolytic Bacillus *strains corresponding to the DNA homology groups*

Strains (DSM No.)	mol %GC	DNA homology group	pH 7.7	pH 7.7 plus 1% glucose	pH 6.8 plus 1% glucose	Deamination of phenylalanine	Starch	Tyrosine	Pectin	Dextran	Chitin	Aesculin	Urea	DNA
1327	47.3	1	−	+	−	+	+	+	−	−	−	+	+	−
1479	47.8	1	−	+	−	+	+	−	−	−	−	+	+	−
1473	47.9	1	−	+	−	+	+	+	−	−	−	+	+	−
1478	48.2	1	−	+	−	+	+	−	−	−	−	+	+	−
1475	48.3	1	−	+	−	+	+	+	−	−	−	+	+	−
1477	49.0	1	−	+	−	+	+	+	−	−	−	+	+	−
1486	50.2	2	−	+	−	−	−	−	−	+	−	+	+	−
1485	50.8	2	−	+	−	−	−	−	−	+	−	+	+	−
1484	51.0	2	−	+	−	−	−	−	−	+	−	+	+	−
1355	51.3	2	−	+	−	−	−	−	−	+	−	+	+	−
1483	51.7	2	−	+	−	−	−	−	−	+	−	+	+	−
1474	52.2	3	−	+	−	+	−	−	+	−	−	+	+	−
1480	52.6	3	−	+	−	+	−	−	+	−	−	+	+	−
1354	52.7	3	−	+	−	+	−	−	+	−	−	+	+	−
1356	44.4	4	−	+	−	+	−	(+)	+	−	−	+	+	+
1482	44.5	4	−	+	−	+	−	(+)	+	−	−	+	+	+
1352	44.4	5	−	+	−	−	+	−	−	+	+	+	+	+
1481	49.3	6	−	+	+	−	+	−	−	+	−	+	+	+
1358	49.5	6	−	+	+	−	+	−	−	+	−	+	+	+
1487	50.9	7	−	+	+	−	+	−	+	−	−	+	−	−
34	51.1	7	−	+	+	−	+	−	+	−	−	+	−	−
1472	34.4	8	−	+	+	−	+	−	−	+	−	+	+	+
1353	34.7	8	−	+	+	−	+	−	−	+	+	+	+	+

As with other biochemically "inactive" *Bacillus* species the standard tests usually applied within the genus *Bacillus* are not adequate for a species differentiation within the agarolytic sporeformers. A search for additional tests is needed. Some of the properties which seem to be specific for one or more of the DNA homology groups and which may be useful for their differentiation are shown in Table 11. However, more agarolytic *Bacillus* strains belonging to the defined DNA homology groups need to be examined to determine the value of these putative diagnostic tests.

The tests listed in Table 11 separate all of the homology groups except groups 6 and 8. Two groups (5 and 8) differ only in 1 property, 3 groups in 2, 6 groups in 3, 7 groups in 4, 3 groups in 5 and 6 groups in 7 of the properties examined. It should be noted that strains in homology group 8 share phenotypic properties in common to those of groups 5 and 6, even though these strains show marked differences in their DNA base composition (ΔGC 10 - 15%).

Before the DNA-DNA homology studies were performed on the agarolytic *Bacillus* strains, the authors reported the reisolation of *B.agar-exedens* [Hunger and Claus, 1978]. However, which of the fresh isolates should be grouped with that species?

Wieringa [1941] separated *B.agar-exedens* from some other agarolytic sporeformers isolated by him solely on the basis of the inhibition of growth by peptones, a property which could be relieved by the addition of urea. In this respect strains in the DNA homology groups 1 - 5 may belong to *B.agar-exedens*. However, since all the strains designated as *B.agar-exedens* by Wieringa hydrolysed starch only strains in DNA homology groups 1 and 5 correspond to the original description of the species. The 6 strains of DNA homology group 1 cannot be differentiated from the single strain of DNA homology group 5 using the properties tested by Wieringa [1941]. The single strain produces a yellowish pigment when grown in mineral glucose medium supplemented with yeast extract (0.01% w/v). Since Wieringa [1941], who grew his strains in a similar medium, did not record the formation of pigment it seems most likely that the strains of DNA homology group 1 correspond to *B.agar-exedens*, a species name which according to Rule 12a of the International Code of Nomenclature of Bacteria [Lapage *et al.*, 1975] should now be written as *B.agarexedens*.

Finally it should be mentioned that all of the strains included in the DNA homology group 2 were isolated by a technique [Hunger and Claus, 1978] slightly modified from the original method described for the isolation of agarolytic sporeformers by Wieringa [1941].

Concluding Remarks

From the results presented above and from some other studies it is obvious that taxospecies of the genus *Bacillus* as defined by Gordon *et al.* [1973] may be composed of strains which according to DNA base composition or DNA pairing studies are genetically different and thus should be considered as strains of separate species. The reason for this heterogeneity within *Bacillus* species as currently defined is the relatively small number of phenotypic

properties which have been used in the characterization of
species and in the unknown stability of certain properties
used as species characters.

Stanier *et al.* [1976] have pointed out that

> most bacteria can be identified only by finding out what they can
> do, not simply how they look. This confronts the bacterial taxo-
> nomist with an additional problem. To find out what a bacterium
> can do, he has to perform experiments with it. The number of
> possible experiments that can be performed is extremely large,
> and although all will reveal facts, the facts so revealed will
> not necessarily be taxonomically significant ones, in the sense
> of contributing to a differentiation of the organism under study
> from related assemblages. Consequently, the bacterial taxonomist
> can never be sure that he has performed the right experiments for
> taxonomic purposes; he may well have failed to perform certain
> experiments that would have shown him significant clustering in a
> collection of strains, and therefore erroneously conclude that he
> is dealing with a continuous series. There is no obvious way to
> get around this difficulty, except to make phenotypic character-
> izations as exhaustive as possible.

Such an exhaustive description of phenotypes is far
from being achieved within the genus *Bacillus*. However,
instead of testing more and more phenotypic properties,
not knowing whether these include characters of value for
differentiation it seems to be more useful to apply DNA
pairing studies to the checking of the homogeneity of
species. Such studies may lead not only to the recognition of genetically different subgroups within hitherto
designated taxospecies but are also very useful as a basis
for the interpretation of phenotypic properties; mainly
in determining which characters correlate best with appropriate subgroups or which phenotypic properties tend to be
more conserved than others and thus may be successfully
applied in the identification of new isolates.

It may well be that 2 or more DNA homology groups found
within a "species" cannot be differentiated by tests pre-
sently applied for the characterization of *Bacillus*
species. In such cases a search for additional phenotypic
properties is needed in order to allow the identification
of strains by simple routine methods.

References

Batchelor, M.D. (1919). Aerobic spore-bearing bacteria in the intestinal tract of children. *Journal of Bacteriology* 4, 23-34.

Baumann-Grace, J.B. and Tomscik, J. (1957). The surface structure and
serological typing of *Bacillus megaterium*. *Journal of General
Microbiology* 17, 227-237.

Benedek, T. (1938). Further investigations on *Bacillus endoparasiti-
cus*. *Mycopathologia* 1, 26-40.

Berger, W. (1962). Bedingungen für die Kapselbildung der Spezies
Bacillus megaterium. *Pathologia et Microbiologia* 25, 871-884.

Bradley, S.G. (1973). Relationship among mycobacteria and nocardiae based upon deoxyribonucleic acid reassociation. *Journal of General Microbiology* **113**, 645-651.

Candeli, A., Mastrandrae, V., Cenci, G. and De Bartolomeo, A. (1978). Sensitivity to lytic agents and DNA base composition of several aerobic spore-bearing bacilli. *Zentralblatt für Bakteriologie, Parasitenkunde, Infektionskrankheiten und Hygiene, II. Abteilung,* **133**, 250-260.

Castellani, A. (1965). Further observations on *Myxomicrobium multiplex* and a preliminary report on another peculiar pleomorphic slime organism isolated from human lesions. *Mycopathologia et Mycologia Applicata* **26**, 359-372.

Christensen, W.B. (1946). Urea decomposition as a means of differentiating *Proteus* and para-colon cultures from each other and from *Salmonella* and *Shigella*. *Journal of Bacteriology* **52**, 461-470.

Cowan, S.T. and Steel, K.J. (1974). "Manual for the Identification of Medical Bacteria". Cambridge: Cambridge University Press.

Crombach, W.H.J. (1974). Relationships among coryneform bacteria from soil, cheese and sea fish. *Antonie van Leeuwenhoek* **40**, 347-359.

De Bary, A. (1884). "Vergleichende Morphologie und Biologie der Pilze, Mycetozoen und Bakterien". Leipzig: Wilhelm Engelmann.

De Ley, J. (1970). Reexamination of the association between melting point, buoyant density, and chemical base composition of deoxyribonucleic acids. *Journal of Bacteriology* **101**, 738-754.

De Ley, J., Cattoir, H. and Reynaerts, A. (1970). The quantitative measurements of DNA hybridization rates. *European Journal of Biochemistry* **12**, 133-142.

Fitz-James, P.C. and Young, J.E. (1959). Cytological comparison of spores of different strains of *Bacillus megaterium*. *Journal of Bacteriology* **78**, 755-764.

Gibbins, A.M. and Gregory, K.F. (1972). Relatedness among *Rhizobium* and *Agrobacterium* species determined by three methods of nucleic acid hybridization. *Journal of Bacteriology* **111**, 129-141.

Gibson, T. and Gordon, R.E. (1974). *Bacillus*. In "Bergey's Manual of Determinative Bacteriology" (eds. R.E. Buchanan and N.E. Gibbons), pp.529-550. Baltimore: The Williams and Wilkins Co.

Gillis, M., De Ley, J. and De Cleene, M. (1970). The determination of molecular weight of bacterial genome DNA from renaturation rates. *European Journal of Biochemistry* **12**, 143-153.

Gordon, R.E., Haynes, W.C. and Pang, C.H.-N. (1973). "The Genus *Bacillus*". Washington, D.C.: United States Department of Agriculture.

Gottheil, O. (1901). Botanische Beschreibung einiger Bodenbakterien. *Zentralblatt für Bakteriologie, Parasitenkunde, Infektionskrankheiten und Hygiene, II. Abteilung* **7**, 481-497; 529-544; 680-691.

Hunger, W. (1978). Phänotypische und genotypische Untersuchungen an *Bacillus megaterium* und agarolytischen *Bacillus*-Stämmen. Thesis. University of Göttingen.

Hunger, W. and Claus, D. (1978). Reisolation and growth conditions of *Bacillus agar-exedens*. *Antonie van Leeuwenhoek* **44**, 105-113.

Jeffries, C.D., Holtman, D.F. and Guse, D.G. (1957). Rapid method for determining the activity of microorganisms on nucleic acids. *Journal of Bacteriology* **73**, 590.

238   W. HUNGER and D. CLAUS

Johnson, J.L. (1973). Use of nucleic-acid homologies in the taxonomy of anaerobic bacteria. *International Journal of Systematic Bacteriology* **23**, 308-315.

Jones, D. and Sneath, P.H.A. (1970). Genetic transfer and bacterial taxonomy. *Bacteriological Reviews* **34**, 40-81.

Jüttner, R.-R., Lafferty, R.M. and Knackmuss, H.-J. (1975). A simple method for the determination of poly-β-hydroxybutyric acid in microbial bio-mass. *European Journal of Applied Microbiology* **1**, 233-237.

Knight, B.C.J.G. and Proom, H. (1950). A comparative survey of the nutrition and physiology of mesophilic species in the genus *Bacillus*. *Journal of General Microbiology* **4**, 508-538.

Koch, A. (1888). Ueber Morphologie und Entwicklungsgeschichte einiger endosporer Bakterienformen. *Botanische Zeitung* **46**, 227-287.

Lapage, S.P., Sneath, P.H.A., Lessel, E.F., Skerman, V.B.D., Seeliger, H.P.R. and Clark, W.A. (1975). "International Code of Nomenclature of Bacteria". Washington, D.C.: American Society of Microbiology.

Löhnis, F. and Pillai, N.K. (1907). Ueber stickstoffixierende Bakterien. II. *Zentralblatt für Bakteriologie, Parasitenkunde, Infektionskrankheiten und Hygiene, II. Abteilung* **19**, 87-96.

Löhnis, F. and Westermann, T. (1909). Ueber stickstoffixierende Bakterien. IV. *Zentralblatt für Bakteriologie, Parasitenkunde, Infektionskrankheiten und Hygiene, II. Abteilung* **22**, 234-254.

Marmur, J. and Doty, P. (1962). Determination of the base composition of deoxyribonucleic acid from its thermal denaturation temperature. *Journal of Molecular Biology* **5**, 109-118.

Martini, A. and Phaff, H.J. (1973). The optical determination of DNA-DNA homologies in yeasts. *Annali de Microbiologia ed Enzimologia*, **23**, 59-68.

Mencier, F. (1972). Méthode simple et rapide de mise en évidencen des microorganismes producteurs de dextranase. *Annales de l'Institut Pasteur* **122**, 153-157.

Migula, W. (1894). Über den Zellinhalt von *Bacillus oxalaticus* Zopf. *Arbeiten aus dem bakteriologischen Institut der Technischen Hochschule zu Karlsruhe* **1**, 137-147.

Normore, W.D. (1973). Guanine-plus-cytosine (GC) composition of the DNA of bacteria, fungi, algae and protozoa. In "Handbook of Microbiology", Vol. 2 (eds. A.I. Laskin and H.A. Lechevalier), pp.558-740. Cleveland: CRC Press.

Neide, E. (1904). Botanische Beschreibung einiger sporenbildenden Bakterien. *Zentralblatt für Bakteriologie, Parasitenkunde, Infektionskrankheiten und Hygiene, II. Abteilung* **12**, 1-32.

O'Donnell, A.G., Norris, J.R., Berkeley, R.C.W., Claus, D., Kaneko, T. Logan, N.A. and Nazaki, R. (1980). Characterisation of *Bacillus subtilis*, *Bacillus pumilus*, *Bacillus licheniformis* and *Bacillus amyloliquefaciens* by pyrolysis gas-liquid chromatography: characterisation tested using DNA-DNA hybridisation; biochemical tests; and API systems. *International Journal of Systematic Bacteriology* **30**, 458-459.

Owen, R.J., Legros, R.M. and Lapage, S.P. (1978). Base composition, size and sequence similarities of genome deoxyribonucleic acids from clinical isolates of *Pseudomonas putrefaciens*. *Journal of General Microbiology* **104**, 127-138.

Rode, L.J. (1968). Correlation between structure and spore properties in *Bacillus megaterium*. *Journal of Bacteriology* **95**, 1979-1986.

Seki, T., Oshima, T. and Oshima, Y. (1975). Taxonomic study of *Bacillus* by deoxyribonucleic acid-hybridization and interspecific transformation. *International Journal of Systematic Bacteriology* **25**, 258-270.

Sickles, G.M. and Shaw, M. (1934). A systematic study of microorganisms which decompose the specific carbohydrates of the *Pneumococcus*. *Journal of Bacteriology* **28**, 415-431.

Skerman, V.B.D. (1967). "A Guide to the Identification of the Genera of Bacteria". Baltimore: Williams and Wilkins Co.

Smith, N.R., Gordon, R.E. and Clark, F.E. (1946). "Aerobic mesophilic sporeforming bacteria". United States Department of Agriculture Miscellaneous Publications 559.

Smith, N.R., Gordon, R.E. and Clark, F.E. (1952). "Aerobic spore-forming bacteria". Washington, D.C.: United States Department of Agriculture.

Somerville, H.J. and Jones, M.L. (1972). DNA competition studies within the *Bacillus cereus* group of bacilli. *Journal of General Microbiology* **73**, 257-265.

Stanier, R.Y., Adelberg, E.A. and Ingraham, J.L. (1976). "The Microbial World". Engelwood Cliffs, New Jersey: Prentice Hall Inc.

Stapp, C. (1920). Botanische Untersuchung einiger neuer Bakterienspezies, welche mit reiner Harnsäure oder Hippursäure als alleinigem organischen Nährstoff auskommen. *Zentralblatt für Bakteriologie, Parasitenkunde, Infektionskrankheiten und Hygiene, II. Abteilung* **51**, 1-71.

Steinhaus, E.A. (1941). A study of the bacteria associated with thirty species of insects. *Journal of Bacteriology* **42**, 757-790.

Swings, J. and De Ley, J. (1975). Genome deoxyribonucleic acid of the genus *Zymomonas* Kluyver and van Niel 1936: base composition, size and similarities. *International Journal of Systematic Bacteriology* **25**, 324-328.

Ueda, K., Higashi, S. and Origuchi, K. (1967). Studies on production of D-fructose from D-sorbitol by fermentation. (Part II). Isolation of fructose-producing bacteria. *Journal of Fermentation Technology* **45**, 541-549.

Watanabe, K. and Takesna, S. (1970). Surface changes of two types of *Bacillus megaterium* spores differing in their response to n-butane. *Journal of General Microbiology* **63**, 375-377.

Werner, W. (1933). Botanische Beschreibung häufiger am Buttersäureabbau beteiligter sporenbildender Bakterienspezies. *Zentralblatt für Bakteriologie, Parasitenkunde, Infektionskrankheiten und Hygiene, II. Abteilung* **87**, 446-475.

Wieringa, K.T. (1941). *Bacillus agar-exedens*, a new species, decomposing agar. *Antonie van Leeuwenhoek* **7**, 121-127.

Wieringa, K.T. (1949). A method for isolating and counting pectolytic microbes. In "Fourth International Congress for Microbiology. Copenhagen 1947". Report of Proceedings. Copenhagen: Rosenkilde and Bagger.

Zopf, W. (1883). "Die Spaltpilze". Breslau: Eduard Trewendt.

Chapter 10

INSECT PATHOGENS IN THE GENUS *BACILLUS*

H. DE BARJAC

Institut Pasteur, Paris, France

Introduction

The insect pathogens of the genus *Bacillus* can be allocated to 1 of 5 species: *B.thuringiensis, B.larvae, B.len-
timorbus, B.popilliae* and *B.sphaericus*. Amongst them are
found representatives of all the 3 morphological groups of
Smith *et al.* [1952]. *B.thuringiensis* is in the first
group and is characterized by non-swollen sporangia. The
next 3 species fall into the second group whose members
produce oval or ellipsoidal spores, causing swelling of
the sporangia. *B.sphaericus* occurs in the third group
which has round spores, also swelling the sporangia. This
mode of grouping has received much criticism [for example,
Wolf and Barker, 1968]. Nevertheless, in the absence of
any better criteria it is felt that this arrangement
should be retained as a basis for a practical identification of the *Bacillus* species.

All these 5 species are well known as insect pathogens
except *B.sphaericus* which is usually regarded as a sapro-
phyte; the discovery of its pathogenicity for mosquito
larvae is quite recent [Kellen *et al.*, 1965; Singer,
1973].

Comparing the 5 species from the points of view of
fundamental research and of their practical use in the
biological control of insects, *B.thuringiensis*, is by far
the most important. Next is *B.popilliae* which has been
successfully used in the USA against *Popillia japonica*;
but data concerning this bacterium is still lacking.
B.lentimorbus comes third in this list although it has not
received much attention and, finally, there is *B.sphaeri-
cus* which is becoming more and more interesting from a
practical point of view. In contrast, *B.larvae* is a
pathogen for useful insects and not for noxious ones,
being toxic for honey bees.

From the taxonomic point of view the best known species
are *B.thuringiensis* and *B.sphaericus*. Strains of these 2
species are easy to grow *in vitro*, in laboratory culture
media; many strains of each species are known and they
have been extensively studied.

TABLE 1

Main characters of the five entomopathogenic
Bacillus *species*

A. Large rods (diameter of vegetative cells >1 μm), forming
elliptical spores contained in unswollen sporangia.
- Presence of parasporal bodies (= crystals) which
appear as losangic, or square, round or indefinite
in shape.
- Pathogens for Lepidoptera or Diptera larvae
(Mosquitoes, Simuliidae).
= B.*thuringiensis*

B. Smaller rods (diameter of vegetative cells <1 μm)

   1. Forming elliptical spores in swollen sporangia.
No growth on ordinary media.
     a. Pathogens for some Coleoptera larvae
(Popillia, Melolontha): "Milky disease"
     b. Parasporal bodies, with various shapes
= B.*popilliae*
     bb. No parasporal bodies
= B.*lentimorbus*
     aa. Pathogens for honey bees larvae: "American
foul-brood" - No parasporal bodies
= B.*larvae*

   2. Forming spherical spores in swollen sporangia.
Growth on ordinary media, even at pH 6
- Not reactive on many carbohydrate substrates
- Some strains are pathogenic for Mosquito larvae
= B.*sphaericus*

The main characters of the 5 species are given in Table
1. In addition, for B.*thuringiensis* (Table 2) and for
B.*sphaericus* (Table 4) are given general phenotypic
patterns based on 80 cultural and biochemical reactions.
All *Bacillus* strains received or isolated in this laboratory are characterized with these tests.

The taxonomy of each species will now be considered
separately.

Taxonomy of the Insect Pathogens

B.*thuringiensis*

Acting as the International Reference Centre for B.*thuringiensis* since 1965, this laboratory now holds a culture
collection of 745 strains of this species. The classification of these B.*thuringiensis* isolates is based on biochemical (see Table 2) and serological criteria. The de-

TABLE 2

General phenotypic pattern of B.thuringiensis *strains*

A. *Common characters*

- Facultative anaerobiosis
- Growth on nutrient agar at pH 6, in nutrient broth + NaCl 7%,
  w/v and on KCN medium [Brown]
- No growth on citrate medium [Simmons]
- Acid production in anaerobic broth + glucose 1%, w/v
- Gas production from nitrates in anaerobic conditions
- Nitrites produced from nitrates in aerobic broth
  (type A nitrate-reductase)
- Positive methyl red, haemolysis and RNAse tests
- Fermentative reaction on Hugh-Leifson's medium + glucose
- Acid production in media + ribose, or glucose, fructose,
  maltose, trehalose, glycerol or soluble starch
- No production of indole, H_2S, oxidase, tetrathionate-reductase,
  lysine decarboxylase, phenylalanine deaminase, tryptophane
  deaminase, ornithine decarboxylase, β-xylosidase, α-fucosidase,
  α-mannosidase
- No attack on pectin, no acid production in media containing
  arabinose, or xylose, galactose, rhamnose, sorbose, lactose,
  raffinose, melibiose, erythritol, adonitol, mannitol, dulcitol,
  sorbitol, m-inositol, amygdalin, inulin or α-methylglucoside.

B. *Differential characters*

According to the strains: differences of reaction in:
- production of acetylmethylcarbinol, lecithinase, urease,
  β-galactosidase, DNAse, lipase, arginine dihydrolase
- action on aesculin, chitin, starch-agar, casein-agar,
  gelatin-agar, coagulated serum, litmus milk
- formation of surface-pellicle in broth and pigment on
  potatoes
- acid production in media + mannose, or sucrose, cellobiose
  or salicin
- production of heat-stable exotoxin
- nature of H-antigen.

---

termination of the H-antigen by flagellar agglutination in
1962 by de Barjac and Bonnefoi has allowed the recognition
of different serotypes which correspond with different
patterns of biochemical characters and entomopathogenic
specificities. In other words, each serotype has a specific H-antigen, gives specific biochemical reactions and
has a typical spectrum of activity on insects. Some serotypes can be subdivided according to the presence of H-
antigenic subfactors and/or the presence of biochemical
differences. Finally, *B.thuringiensis* strains can be
allocated to different subspecies or varieties on the

basis of the serotype and biotype of the particular strain.
At present 19 such subspecies are recognized and these are
listed in Table 3.

TABLE 3

Different subspecies of Bacillus thuringiensis

Serotype H	Biotype	Subspecies
1	I	*thuringiensis*
2	II	*finitimus*
3a	III 1	*alesti*
3a, 3b	III 2	*kurstaki*
4a, 4b	IV 1	*sotto*
4a, 4b	IV 1'	*dendrolimus*
4a, 4c	IV 2	*kenyae*
5a, 5b	V 1	*galleriae*
5a, 5c	V 2	*canadensis*
6	VI	*subtoxicus*
6	VI'	*entomocidus*
7	VII	*aizawai*
8	VIII	*morrisoni*
9	IX	*tolworthi*
10	X	*darmstadiensis*
11	XI	*toumanoffi*
12	XII	*thompsoni*
13	XIII	*pakistani*
14	XIV	*israelensis*

Roughly speaking, the first 13 subspecies are specific
pathogens of lepidopterous larvae with greater or lesser
virulence for the different species, according to the serotype concerned. The fourteenth subspecies is specifically
pathogenic to mosquito and *Simulidae* larvae. In both
cases, the insecticidal power appears to be mainly related
to the endotoxin of the crystals or parasporal inclusions
[de Barjac, 1978].
Other approaches have been made to the classification
of *B. thuringiensis*. For example, the esterase patterns of
vegetative cells have been determined and good congruence
between the patterns obtained and the classification based
on serotypes shown to exist [Norris, 1964]. Another method,
which also produced results in agreement with H-serotyping,
is that based on the crystal antigens [Krywienczyk and
Angus, 1967]. In most cases, each H-serotype is characterized by a specific crystal antigen pattern.
Phage-typing was, however, unsuccessful when tried with
12 different bacteriophages [de Barjac *et al.*, 1974]. The
subspecies of *B. thuringiensis*, as recognized on the basis
of H-serotyping and biochemical pattern, can be neither
differentiated nor subdivided on the basis of their suscep-

tibility to these bacteriophages.

In conclusion, as far as *B.thuringiensis* is concerned, the most sensitive, reliable and specific way of characterizing subspecies is the H-agglutination method. The use of anti-H sera, it is also the easiest and quickest technique for identifying *B.thuringiensis* isolates.

B.sphaericus

In contrast to *B.thuringiensis*, *B.sphaericus* is inactive on many substrates *in vitro* (see Table 4) and different biochemical patterns enabling the grouping of strains are not found. In this laboratory the auxanograms method has been tried but the results have been largely unhelpful.

TABLE 4

General phenotypic pattern of B.sphaericus *strains*

A. *Common characters*

-   Strict aerobiosis
-   Growth on nutrient agar at pH 6, and in KCN medium
-   No growth on anaerobic broth + glucose or + nitrates, and on citrate medium
-   Production of RNAse
-   No production of nitrate-reductase, lecithinase, lysine-decarboxylase, phenylalanine-deaminase, tryptophan-deaminase, arginine dihydrolase, β-galactosidase, β-xylosidase, α-fucosidase and α-mannosidase
-   No production of acetylmethylcarbinol, indole, H_2S, and surface pellicle in broth, methyl red test negative
-   No action on aesculin and starch agar. No reaction in Hugh-Leifson's medium + glucose
-   No acid production in media containing arabinose, xylose, ribose, glucose, galactose, mannose, rhamnose, sorbose, fructose, sucrose, maltose, lactose, trehalose, cellobiose, melibiose, raffinose, melezitose, erythritol, adonitol, mannitol, dulcitol, sorbitol, m-inositol, soluble starch, inulin, salicin, amygdalin or α-methylglycoside.

B. *Differential characters*

According to the strains, differences of reaction in:
-   growth in broth + NaCl 7%, w/v
-   production of urease, tetrathionate-reductase, DNAse, oxidase, lipase, haemolysis and pigment
-   action on casein-agar, gelatin-agar, coagulated serum, litmus milk, chitin and pectin
-   acid production in medium + glycerol

TABLE 5

Serological classification of some B.sphaericus strains

	Serotype	1	20 Q	25 K	21 SSII₁	23 1404	53 1930	H-suspensions 6 C6	9 SC2	11 G2	7 6530	22 14577	24 1593	33 1593-4	34 1881	35 1691
5125 CIP = 1F	1a, 1b	25600	400	800												
Q Singer = 20F*	}	6400	19200	12800												
K Singer = 25F*	1a	1600	12800	25600												
SSI1 Singer=21F*	2				25600	19200	25600									
1404 Singer=23F*					25600	25600	25600									
1930 Singer=53F*					25600	25600	25600									
C₆ = 6F	3							19200	25600	12800						
SC₂ = 9F								12800	12800	12800						
G₂ = 11F								25600	6400	19200						
6530 CIP = 7F	4										25600	25600				
14577 Singer=22F (ATCC)											25600	25600				
1593 Singer=24F*	5												25600	12800	12800	12800
1593-4 WHO=33F*													25600	19200	25600	25600
1881 Singer=34F*													25600	25600	25600	25600
1691 Singer=35F*													25600	25600	25600	25600
C₃ = 13F	6															
5288 CIP = 2F	7															
1537 Singer=26F	8															
7054 Singer=19F	9															
A2R = 5F	10															
5289 CIP = 3F	11															
5116 CIP = 8F	12															
3P = 10F	13															

*Strains pathogenic to mosquitoes.

Thus, faced with the problem of differentiating *B.sphae-ricus* strains which are pathogens for mosquito larvae, all isolates have been subjected to an H-serotyping method, similar to that previously developed for *B.thuringiensis*. Here again, with *B.sphaericus*, the H-antigen determination succeeded in characterizing different strains. Table 5 shows the data obtained on a group of *B.sphaericus* including saprophytic and entomopathogenic strains.

Other methods have been applied to the taxonomy of *B. sphaericus* species, principally DNA-DNA homology and phage-typing [Krych *et al*., 1980]. The DNA-DNA homology determinations on 46 strains have led to the recognition of 5 groups. It is interesting that all the isolates ento-motoxic for mosquito larvae fall into group IIa.

The results of phage-typing experiments are even more remarkable as the grouping achieved is in perfect agree-ment with that obtained by H-serotyping. The strains toxic for mosquitoes are divided into 3 different serotypes (H_{1a}, H_2 and H_5), and each one of these serological groups reacts specifically with a number of bacteriophages (Table 6). So, the results from the 2 techniques reinforce each other and the high similarity of the subdivisions resul-ting from H-serotyping and from phage-serotyping empha-sizes the value of this basic differentiation.

TABLE 6

Parallelism of serotype and bacteriophage type of B.sphaericus *strains pathogenic for mosquitoes*

B.sphaericus strains	Sero-types H	Sensitivity to bacteriophages											DNA homology group*
		13	11	12	1A	1B	2	3	63	4	6	SS T	
Kellen K	H_{1a}	+	+	+	-	-	-	-	-	-	-	+	
Kellen Q		+	+	+	-	-	-	-	-	-	-	+	
SS II.1	H_2		+	+	+	+	+	+	+	-	-	-	
1404-924 B		+	+	+	+	+	+	+	+	-	-	-	II a
1593	H_5	+	-	-	-	-	-	-	-	+	+	-	
1881		+	-	-	-	-	-	-	-	+	+	-	
1691	°	-	-	-	-	-	-	-	-	+	+	-	

*According to Krych *et al*. [1978, 1980].

According to the studies which have been made on the mosquito pathogenic strains of *B.sphaericus*, their toxicity may be related to an unknown toxin, presumably located in the cell wall [Davidson *et al*., 1975; Myers and Yousten, 1980].

B.popilliae and B.lentimorbus

These 2 species cause the milky diseases of Japanese
beetles (*Popillia japonica*). The name of the diseases
comes from the fact that the larvae turn milky white because of the heavy production of *Bacillus* spores in the
haemolymph. In this case the main expression of the pathogenicity looks more like a septicaemia than a toxaemia.

Usually, sporulation of strains of both these species
occurs readily *in vivo* when injected into or fed to
susceptible larvae, and the spores are infective *per os*.
But, *in vitro*, on special artificial media, only selected
strains have been reported to sporulate and these spores
cause milky disease only by injection in the haemolymph.

B.popilliae and *B.lentimorbus* do not grow on ordinary
media such as those supporting the growth of *B.thuringien-
sis* or *B.sphaericus*. They need complex media, supplemented
with vitamins and growth factors; *B.lentimorbus* being even
more fastidious in its nutritional requirements. All these
difficulties account partly for the lack of fundamental
knowledge about these two species.

The common characteristics of *B.popilliae* and *B.lenti-
morbus* can be listed as:
-   Facultative anaerobiosis and growth in complex media
-   No growth on serial transfer in nutrient broth, or in
    broth + NaCl 7%, w/v or Simmons citrate medium
-   Acid production from media + trehalose, but no acid
    production from arabinose, or xylose, mannitol, or
    soluble starch
-   No action on casein-agar, gelatin-agar or starch-agar
-   No production of indole, catalase, nitrate reductase or
    phenylalanine deaminase.

The main difference between the 2 species consists in
the presence of parasporal bodies in the sporangia of *B.
popilliae*. These parasporal bodies are of various shapes.
They remain usually associated with the spores after the
final autolysis of the sporangia and their role in entomo-
pathogenicity is not clear.

B.larvae

B.larvae is also a species requiring complex media, enriched with vitamins and amino-acids, for growth. It will
not survive serial transfer in nutrient broth but strains
from culture collections can be grown on usual media to
which yeast extract has been added.

B.larvae has sometimes been confused with *B.alvei*. It
is, however, very easy to differentiate *B.alvei* because of
its production of indole. Moreover, *B.alvei* is a saprophytic species, unlike *B.larvae* which is pathogenic for
honey bees.

The most common characteristics of *B.larvae* strains are
the following:

- Facultative anaerobiosis and growth on media + yeast extract 1%, w/v even at pH 6
- No growth in broth + yeast extract + NaCl 7%, w/v
- Acid production from ribose, glucose, trehalose, mannose and glycerol
- No acid production from arabinose, xylose, fructose, sorbose, rhamnose, sucrose, maltose, lactose, salicin, inulin, soluble starch and different polyols
- No production of catalase, nitrate reductase, urease, arginine dihydrolase, ornithine decarboxylase, phenyl-alanine deaminase or DNAse
- No production of indole and acetylmethylcarbinol
- Positive reactions on casein-agar and gelatin-agar.

Conclusions

Because of their importance in the control of insects *B. thuringiensis* and *B. sphaericus* are the most intensively studied species among the insect pathogens. There is also considerable interest in the fundamental problem of the biogenesis of *B. thuringiensis* crystals.

From the taxonomic point of view, the 5 insect patho-genic species are relatively easy to diagnose on the basis of phenotypic characters. More work is, however, required on genotypic features with a view to establishing a sound basis for intraspecific differentiation.

References

de Barjac, H. and Bonnefoi, A. (1962). Essai de classification bio-chimique et serologique de 24 souches de *Bacillus* de type *B. thu-ringiensis*. *Entomophaga* **8**, 223-229.

de Barjac, H., Sisman, J. and Cosmao Dumanoir, V. (1974). Descrip-tion de 12 bacteriophages isoles a partir de *Bacillus thuringiensis*. *Comptes Rendus Hebdominaires des Séances de l'Academie des Sciences Serie D* **279**, 1939-1942.

de Barjac, H. (1978). Un nouveau candidat a la lutte biologique contre les Moustinques: *Bacillus thuringiensis* var. *israelensis*. *Entomophaga* **23**, 309-319.

Davidson, E., Singer, S. and Briggs, J.D. (1975). Pathogenesis of *Bacillus sphaericus* SS II-1. Infections in *Culex pipiens quinque-fasciatus* (= *C. pipiens fatigans*) larvae. *Journal of Invertebrate Pathology* **25**, 179-184.

Kellen, W.R., Clark, T.B., Lindegren, J.E., Ho, B.X., Rogoff, W.N. and Singer, S. (1965). *Bacillus sphaericus* (Neide) as a pathogen of Mosquitoes. *Journal of Invertebrate Pathology* **7**, 442-448.

Krych, V.K., Johnson, J.L., Hedrick, J.C. and Yousten, A.A. (1978). Taxonomy and identification of Mosquito-pathogenic strains of *Bacillus sphaericus*. In "Progress in Invertebrate Pathology - Proceedings of the International Colloquium on Invertebrate Pathology" (ed. J. Weiser), pp.99-100. Prague.

Krych, V.K., Johnson, J.L. and Yousten, A.A. (1980). Deoxyribonucleic acid homologies among strains of *B. sphaericus*. *International*

Journal of Systematic Bacteriology **30**, 476-484.

Krywienczyk, J. and Angus, T.A. (1967). A serological comparison of several crystalliferous insect pathogens. *Journal of Invertebrate Pathology* **9**, 126-128.

Myers, P.S. and Yousten, A.A. (1980). Localization of a mosquito-larvae toxin of *Bacillus sphaericus* 1593. *Applied and Environmental Microbiology* **39**, 1205-1211.

Norris, J.R. (1964). The classification of *Bacillus thuringiensis*. *Journal of Applied Bacteriology* **27**, 439-447.

Singer, S. (1973). Insecticidal activity of recent bacterial isolates and their toxins against Mosquito larvae. *Nature* **244**, 110-111.

Smith, N.R., Gordon, R.E. and Clark, J.B. (1952). "Aerobic Spore-forming Bacteria". Washington, D.C.: United States Department of Agriculture.

Wolf, J. and Barker, A.N. (1968). The genus *Bacillus*: aids to the identification of its species. In "Identification Methods for Microbiologists, Part B" (eds. B.M. Gibbs and D.A. Shapton), pp.93-109. London and New York: Academic Press.

Chapter 11

TAXONOMIC AND RELATED ASPECTS OF THERMOPHILES WITHIN THE GENUS *BACILLUS*

J. WOLF* and R.J. SHARP[†]

*Department of Microbiology, The University of Leeds, Leeds, UK
†The P.H.L.S. Centre for Applied Microbiology and Research, Porton, Salisbury, UK

Historical Aspects

The occurrence in microorganisms of uniquely heat resistant spores was demonstrated by Ferdinand Cohn in 1876 in the mesophile *Bacillus subtilis*. This discovery was followed within a few years by the startling isolation from the cool waters of the Seine of a thermophilic aerobic spore-former by Miquel [1888]; it grew at 73°C. During the next 50 years curiosity maintained an interest in aerobic spore-forming *Bacillus* strains capable of growth at temperatures at which denaturation of proteins, inactivation of enzymes and destruction of microorganisms readily occur.

This early period is well summarized in the review of Allen [1953]. The early reports record the ubiquitous occurrence of thermophiles and the ease of their isolation from virtually any sample of water, soil, or mud derived not only from tropical regions but also from sea water [MacFadyen and Blaxall, 1894] and even freshly fallen snow [Golikowa, 1926]. Thermophilic members of the genus *Bacillus* predominated in these isolations; non-spore-formers were less common. The activity of these strains at high temperature led to the early demonstration of their thermostable enzymes [Oprescu, 1898; Pringsheim, 1913] while their nuisance value in sugar refining, resulting from their ability to grow in hot (60°C) sucrose (50%) syrups, was clearly described by Laxa in 1900. The ability of these organisms to actively multiply during the process of milk pasteurization and thus spoil the product instead of improving its keeping quality was observed by Jacobsen [1918] within only a few years of the experimental intro-duction of the subsequent universally adopted process.

The comparative lack of interest in thermophiles during the past 50 years and in problems in enzymology and mole-cular biology which they pose is at last being rapidly re-dressed. Thus, the last few years have seen the publica-tion of 8 symposia dealing with these bacteria and it is clear that their heat stable and rapidly acting enzymes, active at temperatures (60°C+) at which the effect of

contaminants can be virtually ignored, are likely to prove
valuable agents of commercially exploitable value for the
hydrolysis of polysaccharides and proteins refractive to
or unsuitable to conventional reagents and as biological
detergents. The following abbreviated list of references
to some recent publications may be of interest: Brock
[1978]; Fields [1970]; Friedman [1978]; Heinrich [1976];
Kushner [1978]; Loginova and Egorova [1977]; Ljungdahl
[1979]; Shilo [1979]; Williams [1975]; Zeikus [1979];
Zuber [1976]. Only those of Brock, Fields and of Loginova
and Egorova deal with taxonomic aspects.

Taxonomic studies and descriptions of *Bacillus* thermophiles until about 1949 are rather inadequate. The exceptions which represent distinct milestones among the thermophiles of *Bacillus* deserve mention. These comprise
Bacillus coagulans [Hammer, 1915], *Bacillus stearothermophilus* [Donk, 1920] and the rediscovered *Bacillus thermodenitrificans* of Ambrož by Mishustin [1950]. Original data
and taxonomic studies on the strict thermophiles are almost
entirely due to the patient and persistent efforts of Ruth
Gordon and her colleagues spanning 25 years (1949 to 1974).

Because of the uncommonly long period of gestation of
the current edition of "Bergey's Manual of Determinative
Bacteriology" the chapter on *Bacillus* [Gibson and Gordon,
1974] failed to incorporate several important recent relevant studies on thermophiles. It is these expanding recent
studies viewed against the classic work of the Gordon
school which have engendered this chapter.

Mention must, however, also be made of the pioneering
efforts of Cameron and Esty [1926] at the National Canners
Association (NCA), Washington D.C. on the causative
organisms associated with spoilt canned foods. During 1920
to 1926 they examined some 6000 samples which yielded over
200 *Bacillus* cultures. Some 60 proved to be mesophiles
(55°C-); almost 100 were facultative thermophiles (37°C+
and 55°C+) and 55 isolates qualified as obligate thermophiles (37°C- and 55°C+). These latter were further subdivided into 4 groups one of which (Group 100) consisted of
cultures, many showing bulging sporangia, which actively
hydrolysed starch, were non-proteolytic on serum or gelatin, grew optimally at 55 to 60°C and even at 72.5°C. In
fact these historic isolates deserve re-examination as a
potential source of extreme thermophiles. This group of
55 isolates has provided many of the authentic cultures of
B.stearothermophilus included in the classic work of Gordon
and Smith [1949]. Thus, cultures of *B.stearothermophilus*
ATCC 7953, 7954 and 12977 correspond to the original NCA
strains 1518, 1503 and 1492, respectively. Gordon and
Smith identified the original Donk strain, now ATCC 12980,
with this group of isolates. In fact, except for positive
motility this group can be most readily identified with
recent descriptions of this species in the last 2 editions
of "Bergey's Manual of Determinative Bacteriology" [Breed

et al., 1957; Gibson and Gordon, 1974]. Developments and progress in the fields of taxonomy and nomenclature during the past 50 years can conveniently be considered in a number of sections.

The early years (1926-1948)

During these 2 decades progress in the physiology and systematics of *Bacillus* thermophiles was very limited. The period starts with the second edition of "Bergey's Manual of Determinative Bacteriology" [Bergey *et al.*, 1925] with the listing of some sporing thermophiles and with *B.coagulans* [Hammer, 1915] appearing amongst the mesophiles. In fact, because *B.coagulans* sometimes fails to grow at 65°C (many strains *do*, at an *acid* pH) its thermophilic tendency has but rarely been utilized as a critical characteristic in diagnostic keys.

In 1928 Hussong and Hammer proposed *Bacillus calidolactis* to accommodate extreme thermophiles causing the coagulation of pasteurized milk. These grew at 75°C, coagulated and reduced litmus milk and would not grow on standard nutrient agar unless glucose was added. During more detailed tests on several of these isolates, Smith *et al.* [1952] concluded that some were strains of *B.coagulans*, whilst the others could be easily typed as members of *B. stearothermophilus*.

The same year witnessed the isolation and adequate description of *Bacillus kaustophilus* Prickett [1928], an important strict thermophile from milk. Isolates showed distinct bulging of the sporangium, produced reduction and coagulation of litmus milk at 63°C and on prolonged incubation at 56°C showed weak peptonization, liquified gelatin, hydrolysed starch and reduced nitrate to nitrite and gas. They grew at 73 to 75°C but not at 80°C, and qualify as extreme thermophiles. In more recent studies Walker and Wolf [1971] found the ability to denitrify nitrate to gas, presumably nitrogen, an important diagnostic indicator of a particular type or subgroup within *B.stearothermophilus*. The meticulous description of Prickett has not received adequate recognition.

Bacillus coagulans was re-examined by Sarles and Hammer [1932] who amplified its characteristics and description. Its failure to reduce nitrate or to liquify gelatin, its pronounced fermentative and acidophilic properties and growth at 55°C were all duly emphasized.

In the sixth edition of "Bergey's Manual of Determinative Bacteriology" Smith [1948] lists descriptions of 19 mesophilic species of *Bacillus* and of 20 thermophiles capable of growth at 55°C. The thermophiles were classified largely on sporangial morphology and several species were recognized including those mentioned above. Smith emphasized the difficulty of presenting a more rational scheme given the meagre data available on the thermophiles.

This inadequate state of affairs must have acted as a powerful stimulus for the more detailed study presented shortly thereafter by Gordon and Smith [1949].

Gordon and Smith (1949-1974)

In 1946 Smith, Gordon and Clark published the first monograph on the mesophilic species of *Bacillus*. The monograph contained detailed descriptions of species based on a thorough examination of over 600 cultures and introduced order into a hitherto chaotic situation. The original monograph was restricted almost entirely to mesophilic species but included an amended description of *B.coagulans* based on a study of 8 strains. The authors extended the monograph in 1952 and this work has in turn been updated [Gordon *et al.*, 1973]. A fundamental study of *Bacillus* thermophiles [Gordon and Smith, 1949] followed the publication of the original monograph and the work reported in this paper formed the basis of subsequent entries in the seventh and eighth editions of "Bergey's Manual of Determinative Bacteriology" [Breed *et al.*, 1957; Gibson and Gordon, 1974]. This classical paper has also served as a stimulus and mine of information for all subsequent studies on *Bacillus* thermophiles.

In their seminal study Gordon and Smith [1949] examined 216 thermophilic cultures derived from a great variety of sources including laboratories in the U.S.A. and Europe as well as their own isolations (44) from dairy products, soil and silage. These test strains were subjected to a large number of physiological, cultural and microscopic tests. Temperature growth tests showed that cultures from at least 5 *Bacillus* species grew at 55°C, many representing facultative thermophilic variants of mesophilic strains of *B.subtilis*, *B.brevis* and *B.circulans*. Results at 60°C, 65°C and 70°C indicated that all the strains of *B.stearothermophilus* would grow at 60°C and 65°C but at 70°C only 45 out of 87 strains grew. Strains of the other thermophilic species, *B.coagulans*, all grew at 45°C, 1 failed at 50°C while 7 and 50 strains out of 73 did not grow at 55°C and 60°C respectively. The study thus finally resolved itself into a comparison of 87 and 73 strains of these 2 respective thermophilic species.

A detailed examination of the results indicated some fundamental differences between *B.coagulans* and *B.stearothermophilus* and virtually suggested that all previously described *Bacillus* thermophiles including new isolates from natural sources, conveniently fell into one or other of these 2 species. This attractive, and convenient classification was treated with caution by the authors who guardedly stated in their Discussion that: "the writers are reluctant to accept the collection of cultures described here as representing the entire group of aerobic thermophilic spore bearing bacteria". For the next 20 years and more,

however, their tentative conclusions readily satisfied
most requirements.

Definition of Bacillus *thermophiles*

This account is deliberately limited to the better
known thermophilic *Bacillus* species and types for which
considerable information is scattered in the literature.
Much of the available data are on *B.stearothermophilus* and
a variety of *Bacillus* isolates which have been identified
with it. *Bacillus stearothermophilus* is an obligate thermophile with optimum and maximum temperatures of growth of
about 55°C to 60°C and about 65°C respectively. In addition the 3 caldoactive species of Heinen and Heinen [1972],
B.caldolyticus, *B.caldotenax* and *B.caldovelox*, are considered. These 4 species contain the better known neutro-
philic obligate thermophiles. At the other extreme there
are the acidophilic thermophilic strains which fall into
at least 2 distinct species: *B.coagulans*, a facultative
thermophile which grows optimally at about 50°C and at a
pH of 5-6.5 with a maximum growth temperature about 60 to
65°C, and *B.acidocaldarius*, the more extreme obligate acid
thermophile with an optimum pH of about 3, an optimum temperature of growth at about 60°C and a maximum growth temperature of about 65°C. These pragmatic and self-imposed
limits which bring in organisms growing actively at about
60°C, permit the exclusion of virtually all other *Bacillus*
species except for several strains of *B.brevis* which grow
at 60°C [Gordon *et al.*, 1973] and which clearly invite
further study.

The constancy of the minimum and maximum growth temperatures is unresolved. Gordon and Smith [1949] observed a
drop in the minimum growth temperature after strains had
been stored for 4 years at 28°C, whilst Long and Williams
[1959] confirmed earlier observations that growth at the
lower end of the temperature range is materially affected
both by the medium and nature of the inoculum, spores
proving more responsive to growth at 37°C than vegetative
cells. Similarly even the optimum growth temperature is
subject to modification [Williams, 1975; Schenk and
Aragno, 1979] and hence it is the temperature limits within which these can be affected which may be of greatest
significance.

Taxonomic and ecological studies (1951-1980)

Gyllenberg [1951] examined a collection of *Bacillus*
thermophiles, 15 of which grew at 65°C. Eleven of the 15
strains had oval or cylindrical spores and swollen sporangia but 1 strain exhibited spherical spores while reactions such as gelatin liquefaction, starch hydrolysis,
acetoin production or nitrate reduction were variable.
The strains reduced and coagulated litmus milk. Many of

these strict thermophiles can now be identified with *B. stearothermophilus* Groups 2 and 3 [Walker and Wolf, 1971]. Here, therefore, at least 1 strain produced spherical spores not easily identifiable with existing descriptions of *Bacillus* thermophiles where sporangia were characterized by oval or cylindrical spores. Similarly, Stark and Tetrault [1952] described 35 isolates, all showing distinct bulging of the sporangia, with 24 of them able to reduce nitrate to gas. These isolates can also be allotted to the *B. stearothermophilus* groups of Walker and Wolf [1971].

An extensive study by Allen [1953] on thermophiles of the genus *Bacillus* contains a substantial amount of original observation. Although the experimental parts lack essential details, and have not been subsequently pursued, her findings cannot be ignored because they find confirmatory echoes in subsequent work.

In her search for thermophiles Allen used an impressively wide selection of enrichment media and applied these to a great variety of source material, finally obtaining a collection of 105 cultures all of which grew at 65°C and were classified into 4 groups:

Group 1 Organisms with distinct oval, bulging sporangia, morphologically and biochemically similar to *B. circulans*. These organisms bear a strong resemblance to a number of *B. stearothermophilus* strains which Walker and Wolf [1971] identified with their Group 3.

Group 2 These organisms showed slight bulging of the sporangium, grew in acidified media, and closely resembled *B. coagulans*. These isolates do not appear to have been examined in detail by Allen or subsequent workers.

Group 3 These organisms resembled *B. subtilis* on biochemical and morphological grounds; their sporangia were totally non-swollen. Some reduced nitrate to gas and may be more closely related to *B. licheniformis*.

Group 4 The isolates in this group produced distinctly swollen sporangia and spherical spores situated terminally. They resembled *B. sphaericus*.

Although Smith and Gordon [1955a, b] have expressed reservations about the authenticity of some of these isolates, nonetheless these results of Allen are important. Firstly, she describes the occurrence of *Bacillus* isolates, capable of growth at 65°C, which produce cylindrical spores showing no swelling whatever (the *B. subtilis* type) and also terminal spherical spores occurring in distinctly swollen or distended sporangia (the *B. sphaericus* type). Secondly, similar reports of the occurrence of obligate (65°C) thermophiles resembling *B. subtilis* or *B. sphaericus* in morphological features have been subsequently made by a number of workers [Golovacheva *et al.*, 1965; Klaushofer and Hollaus, 1970; Schenk and Aragno, 1979]. Since these workers examined materials from totally different environments it is evident that these thermophiles exist in different habitats. Even though the isolates of Allen may no longer be available, her observations merit recognition.

Golovacheva *et al.* [1965] studied the microflora of 2 localities renowned for their thermal springs and situated some thousands of miles apart; namely mount Yangan-Tau in the Southern Urals and Kunashir Isle (Kuril chain) which is situated on the northern Pacific coast. An outline of the general microflora of the soil and water samples collected in these regions is given in Loginova *et al.* [1962], whilst the properties of the strictly thermophilic *Bacillus* isolates are contained in an abbreviated account [Golovacheva *et al.*, 1965].

The Urals locality was represented by 110 samples and the Kunashir Isle by 24. Of the 71 isolates from Kunashir Isle only 12 grew at 65 to 70°C while 49 strict *Bacillus* thermophiles were isolated from the Urals samples. These 61 isolates were allocated to 7 species. The first 3 are the oft quoted *B.stearothermophilus, B.thermodenitrificans* (which according to Walker and Wolf [1971] represent a subgroup of the former) and the thermoacidophilic species, *B.coagulans*; the remaining 4 consisting of *B.brevis, B. circulans, B.lentus* and *B.megaterium*. The latter are normally regarded as mesophilic species, but many of the isolates of the Russian workers were classified in these taxa, mostly on morphological criteria, since they fitted rather uncomfortably into any of the 3 thermophilic species mentioned above.

The 12 Kunashir isolates did not include any strains of *B.stearothermophilus*, whilst *B.coagulans* was represented by a solitary culture. The isolates from the Urals on the other hand yielded 8 strains identified as *B.stearothermophilus* and 1 as *B.thermodenitrificans*. In this extensive study, therefore, only a single isolate of *B.stearothermophilus* was able to reduce nitrate to gas while 11 of the 12 isolates from Kunashir Isle seem to be taxonomically atypical thermophiles not easily identifiable with either of the 2 established thermophilic species of *Bacillus*. Furthermore, the predominant thermophilic organisms in each of these localities resemble *B.circulans* (26 out of 49 from the Urals and 5 out of 12 from the Pacific). Since 13 of the 31 isolates produced gas from nitrate there are good grounds for considering them as varieties or types of *B.stearothermophilus*.

One of the 71 isolates from the Urals was at first typed as a thermophilic variant of *B.megaterium*. In a subsequent paper Golovacheva *et al.* [1975] noted that this strain exhibited some unusual features and proposed a new species, *Bacillus thermocatenulatus*, for it. The isolate produces yellowish colonies, reduces nitrate to gas and it should be emphasized is a facultative anaerobe. One of the

characteristic features of *B.megaterium* strains, distinguishing them from representatives of related species, is
their strict aerobic requirements. *Bacillus megaterium* is
heterogeneous and can be classified into 2 clearly defined
and separate units of species rank each with a characteristic GC value [Hunger and Claus, Chapter 9; Priest *et
al.*, Chapter 5]; the organisms in each taxon are strict
aerobes. Strains in 1 of the constituent groups (Type A)
give yellow colonies on casein agar but this is hardly
sufficient for considering such strains as even distinctly
related to *B.stearothermophilus* or *B.thermocatenulatus*.
Similarly the results and claims of Stahl and Olsson [1977]
require confirmation. Their findings imply that the
thermophilic variant of a typical *B.megaterium* culture
simultaneously changes its basic metabolism from a strict
aerobe to facultative anaerobe. Admittedly this possibility arises from the work of Jung *et al.* [1974] where
several classical strains of *B.stearothermophilus* grew at
the surface of the medium at 37°C, whilst at 55°C growth
was homogeneous throughout the medium. This observation
clearly requires confirmation.

Klaushofer and Hollaus (1970)

In a taxonomic study of aerobic thermophiles associated
with the process of sugar beet extraction Klaushofer and
Hollaus [1970] examined material derived from several
Austrian plants during the 1965, 1966 and 1967 seasons.
Some 80 *Bacillus* isolates were obtained, all capable of
growth at 65°C. A few additional strains from other
sources were also included and for reference purposes some
authenticated strains of *B.stearothermophilus*, namely ATCC
12980 (NCA 26, the original strain of Donk); NCA 1503
(ATCC 7954) and ATCC 8005 (the Prickett strain of *B.kausto-
philus*).
    The 87 cultures were subjected to the physiological and
biochemical tests of Smith *et al.* [1952] and their media
were closely adhered to. The 68 characters recorded were
examined by computer using the Simple Matching Coefficient
[S_{SM}; Sokal and Sneath, 1963] and 4 groups were obtained,
some of which could be further subdivided (Fig. 1). The
essential characters demarcating the 4 defined groups deserve description since in many respects they substantiate
the classification of Walker and Wolf [1971].

Group I The 14 strains in this group produced sporangia
which were distinctly swollen. The organisms failed to
grow anaerobically, grew in NaCl (3%, w/v) broth, produced
no visible changes in litmus milk media and were characterized by considerable inertness and inability to ferment a
great variety of sugars or to hydrolyse gelatin, casein or
starch. This group corresponds to *B.stearothermophilus*
Group 2 of Walker and Wolf [1971]. The subdivision of

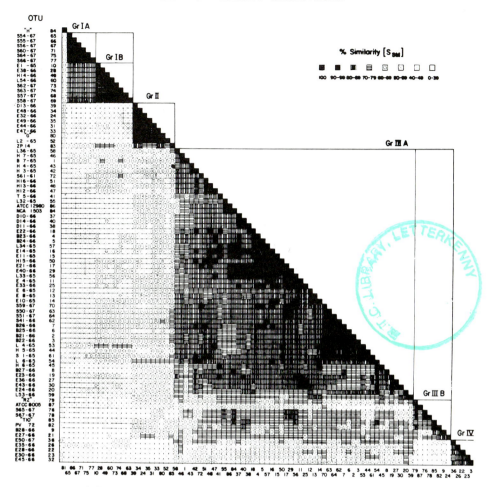

Fig. 1 Similarity matrix (coefficient of Sokal and Michener). Reproduced with permission from Klaushofer and Hollaus [1970].

this group into 2 distinct subgroups is hardly justifiable since the essential difference between subgroups IA and IB was the ability of the 8 strains of the latter to grow at a slightly higher maximum and a corresponding higher minimum growth temperature (68°C and 37°C) when compared with strains in subgroup IA (64°C and 28°C).

Group II The 8 strains in this group exhibited a definite but slightly swollen sporangium. In contrast to the Group I strains these isolates were physiologically active, they grew well in anaerobic media, for example, glucose broth and in nitrate peptone water, hydrolysed starch and fermented sugars. They were also able to totally reduce nitrate to gas and were classified as *B.thermodenitrificans*. This pattern of reactions is consistent with those being classified as *B.stearothermophilus* Group I of Walker and Wolf [1971] which also contains *B.kaustophilus*.

Group III This group comprised 56 of the 80 isolates derived from the sugar beet extracts and the reference
strains *B.stearothermophilus* ATCC 7954 and 12980. The
strains were allocated to 2 distinct subgroups. The majority of the strains (48 out of 58) were placed in subgroup
IIIA together with the 2 *B.stearothermophilus* reference
strains. The organisms in this subgroup produced distinctly swollen sporangia, grew anaerobically, were sensitive
to NaCl (3%, w/v), hydrolysed starch and formed no gas from
nitrate when the latter was reduced. Sugar reactions
varied considerably, some strains fermenting lactose but
many giving a characteristic acid clot and reduction reaction in litmus milk or in yeast milk. These organisms
are readily identifiable as typical strains of *B.stearo-*
thermophilus [Donk] and correspond to Group 3 of Walker
and Wolf [1971] which also probably contains strains described as *Bacillus calidolactis* [Galesloot and Labots,
1959].
    The remaining 8 strains form subgroup IIIB which differs
in many respects from Group IIIA and is heterogeneous.
This subgroup contains *B.stearothermophilus* ATCC 8005 which
was classified in Group 1 by Walker and Wolf [1971]. The
strains in subgroup IIIB require further study; the properties of some of them suggest a possible relationship to
the caldoactive strains of Heinen and Heinen [1972].

Group IV This group consists of 5 strains clearly different from those in the other 3 groups. The Group IV strains
were provisionally identified as *B.coagulans* and some of
their properties, especially the stimulatory effect on
growth of acidic media, favour this suggestion. If this
identification were confirmed these strains would represent
a highly thermophilic type of *B.coagulans*, since they grew
at 66 to 68°C, and below 28°C. This wide temperature range
(+40°C) is worth exploiting in comparative studies of the
effects of temperature on enzyme stability, end products
and nutritional requirements at the upper and lower limits
of growth.
    In a further study, Hollaus and Klaushofer [1970]
examined the immunological relationship of their sugar
isolates by comparing the reaction of vegetative cell pre-
cipitinogens of a high proportion of their cultures against
a few 'O' antisera from each group. The antiserum to
strain L 54 (IB) was the more suitable, since it also reacted to all the strains from groups IB and IA, but not
with the other groups. The reactions of the 2 group II
antisera were less satisfactory, since they failed to react
with many of their homologous strains but did react with
strains in unrelated groups IIIA and IIIB. The reactions
of 4 antisera in group IIIA were only partially specific
whilst the serum to strain RZ in group IIIB was perhaps

the most interesting since it reacted with virtually all
the strains in groups II, IIIA and IIIB. As a primary
industrial tool for rapid identification, this antiserum
might be as suitable as the one recommended by Klaushofer
et al. [1962].

Using the conventional vegetative 'O' agglutination
approach, Walker and Wolf [1971] did not observe cross reactions amongst strains of their 3 groups. Unquestionably
the spore agglutinins are considerably more specific and
the more promising for precise identification.

The Leeds School (1959-1980)

The interests of this centre have been the immunology
[Norris and Wolf, 1961] and taxonomy [Wolf and Barker,
1968] of the genus *Bacillus* and on the germination requirements of *Bacillus* and *Clostridium* species. The work on
the taxonomy of *Bacillus* thermophiles is largely that of
Walker [1959] and Walker and Wolf [1961, 1968, 1971];
amplification and confirmation of their results has been
recently completed by Chowdhury and Wolf (in preparation).

Walker and Wolf [1971] reported the result of an extensive study of the biochemical, physiological and serological properties of 230 strains of *B.stearothermophilus* including the 75 strains of Smith *et al.* [1952], 16 strains
from Galesloot and Labots [1959], previously described as
B.calidolactis or *B.thermoliquefaciens*, and 5 strains from
Grinsted and Clegg [1955]. Using essentially the methods
of Smith *et al.* [1952] they were able to divide the
strains into 3 distinct major groups, 2 of which were
further divided into minor subgroups (Table 1).

Group 1 These organisms were characterized by the ability
to form gas from nitrate under anaerobic conditions (except
3 strains in group 1b4) and the ability to weakly hydrolyse
starch within the immediate vicinity of the colony. The
strains did not grow, or grew poorly in anaerobic glucose
broth, and produced sporangia which were either slightly or
definitely swollen with oval to cylindrical spores. Group
1 which comprised the largest group of organisms (127) was
subdivided into 5 subgroups based on the maximum and minimum temperature for growth (organisms in group 1a did not
grow at 70°C), reduction of nitrate to gas and fermentative
ability. *Bacillus kaustophilus* isolated by Prickett [1928]
was found to correspond to group 1b4.

Group 2 Organisms in this group were relatively inert,
they did not hydrolyse starch, gelatin, or casein or reduce
nitrate but were the only strains to show growth in NaCl
(3%, w/v). In general, the temperature range for growth
was lower than for the other groups, none grew at 70°C and
out of the 40 strains in the group all but 5 grew at 30°C.
The starch negative strains isolated by Daron [1967],

TABLE 1

Reactions characterizing the 3 principal groups
of Bacillus stearothermophilus*

	Group 1	Group 2	Group 3
Morphology of spores	oval to cylindrical	oval	oval to cylindrical
Swelling of sporangium	slight to definite	definite	definite
Growth in 3%, w/v NaCl broth	-	+	-
Hydrolysis of:			
Gelatin	+-	-	+-
Casein	-	-	+-
Starch	R	-	++
NO_2 from NO_3	+	-	+-
Gas from NO_3	+	-	-
Tomato-yeast milk	unchanged or reduction, or reduction and weak curd	unchanged	acid, clot, reduction
Litmus milk	unchanged or slight reduction	unchanged	unchanged or acid, clot
Growth in glucose anaerobically	poor or -	-	+
Acid from:			
Arabinose	+-	-	-
Cellobiose	+-	+	+-
Lactose	-	-	+-
Mannitol	+-	+	-
Rhamnose	+-	+	+-
Xylose	+-	+	-

*Data from Walker and Wolf [1971]; +, positive; -, negative;
+-, variable; ++, strongly positive; R, restricted.

Epstein and Grossowicz [1969] and Sharp *et al*. [1979] all
appear to belong to this taxon since all show growth in
NaCl (5%, w/v) and no growth at 70°C. Strains in this
group produce oval spores positioned either centrally or
subterminally in swollen sporangia.

Group 3 This taxon was more homogeneous than group 1 and
contained 63 strains classified into 4 subgroups. Organisms in this group characteristically produce a strong amylase reaction, giving a zone of hydrolysis around the

colony, are unable to reduce nitrate under anaerobic conditions but grow in glucose anaerobically and form oval to
cylindrical spores in distinctly swollen sporangia. The
48 strains in subgroup 3a can be distinguished from those
in subgroups 3b1, 3b2 and 3c by their inability to change
litmus milk or to ferment lactose. Groups 3b1 and 3b2
contain the 10 strains of *B.calidolactis* isolated by Grinsted and Clegg [1955] and Galesloot and Labots [1959],
respectively. These strains showed close similarity to
the 5 group 3c cultures for, unlike the other test strains,
all of these organisms ferment lactose and produce an acid
clot and reduction in litmus milk. The 5 strains comprising group 3c were identified as *B.thermoliquefaciens* by
Galesloot and Labots [1959]. These strains differed from
groups 3b1 and 3b2 in showing growth at 37°C (3 strains
grew at 30°C) and they were the only organisms in group 3
showing only restricted starch hydrolysis. Table 2 groups
a number of strains according to the classification of
Walker and Wolf [1971].

Using the methods established by Norris and Wolf [1961],
Walker and Wolf [1971] followed their biochemical and
physiological study with an examination of agglutination
reactions with antisera prepared against spores from representatives of the major and minor groups which they had
identified. Antisera prepared to a representative of subgroup 1a, showed positive agglutination with the other 39
strains in that subgroup. Cross reaction occurred with 1
of the 70 strains included in group 1b1 and 3 of those in
group 1b4, no agglutination occurred with any strains from
subgroups 1b2, 1b3 or groups 2 and 3. Antisera prepared
against spores from 2 strains in subgroup 1b1 showed
varied reactions with other strains in this subgroup and
also cross reacted with strains from subgroup 1b2. No
cross reaction was found with subgroups 1a, 1b3 (1 strain
only agglutinated), 1b4 or groups 2 and 3. Agglutination
with antisera prepared from a member of group 2 showed
positive agglutination with all strains of group 2, but no
agglutination with members of groups 1 and 3. Three
strains from group 3 used to prepare spore agglutinins
failed to evoke an antibody response.

Attempts by Walker and Wolf [1971] to use 'O' antigens
from vegetative cells proved to be of limited application
since the reactions proved relatively strain specific, or
at most specific to only a small number of related strains.
An antiserum prepared to the vegetative cells of a representative of group 2 showed positive agglutination to 29
of the 40 strains within this group.

Although the inability to elicit spore agglutinins to
the strains in group 3 limited the value of sera agglutination reactions for classification, the fact that no cross
reactions were observed between the 3 major groups, using
either spore or vegetative agglutinins, adds considerable
weight to the individuality of these major subgroups based

on biochemical and physiological characterization. However, spore agglutinins to this group have since been successfully used [Chowdhury and Wolf, in preparation].

TABLE 2

Classification of recognized Bacillus *thermophiles after Walker and Wolf [1971]*

Group 1	Group 2	Group 3
B.kaustophilus ATCC 8005 [Prickett, 1928]	*B.stearothermophilus* [Daron, 1967]	*B.stearothermophilus* (NCA 1503, ATCC 7954)
B.thermodenitrificans	*B.stearothermophilus* RS93 [Sharp *et al.*, 1980]	*B.stearothermophilus* NCA 1518, ATCC 7953 [Donk, 1920]
B.stearothermophilus ATCC 12016	*B.stearothermophilus* [Epstein and Grossowicz, 1969]	*B.calidolactis* [Galesloot and Labots, 1959]
B.caldotenax [Heinen and Heinen, 1972]		*B.calidolactis* [Grinsted and Clegg, 1955]
B.caldovelox [Heinen and Heinen, 1972]		*B.thermoliquefaciens* [Galesloot and Labots, 1959]
		B.stearothermophilus NCA 1356
		B.stearothermophilus NCA 1492
		B.stearothermophilus NCA 26

ATCC, American Type Culture Collection, Rockville, Maryland, U.S.A.; NCA, National Canners Association, Washington D.C., U.S.A.; NCIB, National Collection of Industrial Bacteria, Torry Research Station, Aberdeen, Scotland, U.K.

Examination of esterase enzymes has provided valuable information for the classification of a wide range of microorganisms including *Corynebacterium* [Robinson, 1966], *Streptococcus* [Lund, 1965] and *B.thuringiensis* [Norris and Burgess, 1963; Norris, 1964]. Baillie and Walker [1968] applied this technique to study the collection of strains assembled by Walker and Wolf [1971]. Using starch gel electrophoresis they examined the esterase patterns of 217 strains and found that they could be divided into groups closely resembling those based on the physiological and serological reactions.

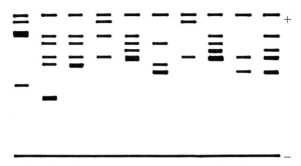

Fig. 2 Diagrammatic representation of the esterase patterns obtained by starch gel electrophoresis of extracts of the 217 strains of thermophilic aerobic sporeformers. Reproduced with permission from Baillie and Walker [1968].

In group 1 [Walker and Wolf, 1971] they established 6 subgroups (Fig. 2). Subgroup 1a (38 cultures) strains all had identical esterase patterns, while those in subgroup 1b1 (67 strains) exhibited 2 distinct patterns and could be further divided into 2 groups 1b1(A) and 1b1(B). Nine of the 10 strains in subgroup 1b2 had identical esterase patterns, the exception being identified with subgroup 1b1 (A). The 4 representatives of subgroup 1b3 showed a more divergent pattern, 2 had esterase patterns identical with subgroup 3a, 1 corresponded to subgroup 1b1(A) and the other to subgroup 1b1(B). The 3 cultures comprising subgroup 1b4 gave identical esterase patterns, as did 38 strains from group 2. The presence of esterase enzymes in this group proved difficult to detect once again illustrating the biochemical inertness of organisms in this group.

The 46 cultures comprising subgroup 3a all shared a common esterase pattern [Baillie and Walker, 1968]. The 2 strains of *B.calidolactis* of Grinsted and Clegg [1955] from group 3b1, however, differed in their esterase patterns from the 3 strains of *B.calidolactis* of Galesloot and Labots [1959] from group 3b2. This latter group shared an identical esterase pattern with subgroup 3c which comprised 4 of the strains of *B.thermoliquefaciens* from Galesloot and Labots [1959]. Examination of proteins from these strains by acrylamide gel electrophoresis again confirmed the 3 main groups as indicated by esterase analysis, but subdivisions within groups 1 and 3 were less evident [Baillie and Walker, 1968].

Analysis of polar lipids has further supported the subdivision of *B.stearothermophilus* into 3 distinct groups (Fig. 3). Thus Minnikin and his colleagues [1977] found that representatives of group 2 contained monoglycosyl and diglycosyl diacylglycerols which were absent in strains of groups 1 and 3. Conversely phosphatidylethanolamine (PE) was a characteristic component of groups 1 and 3 but totally absent in organisms of group 2.

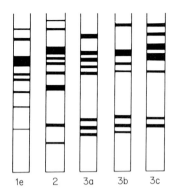

Fig. 3 Diagram of an acrylamide gel stained to detect protein. Inserts (left to right) were extracts of 1e (T107), 2 (T143), 3a (T188), 3b (T208) and 3c (T229). Reproduced with permission from Baillie and Walker [1968].

The caldoactive bacteria (1971-1980)

Following the isolation of the Gram-negative non-sporing thermophile *Thermus aquaticus* [Brock and Freeze, 1969] from a superheated pool in Yellowstone National Park, a sporulating thermophile was isolated from the same pool [Heinen, 1971]. The temperature at the sampling site was a constant 86°C and the pH 8.2. The new strain, designated YT-G, was reported to be a Gram-negative rod, which produced short filaments distinctively composed of a number of single cells and produced dull yellowish grey colonies. Two more thermophilic strains, YT-F and YT-P, were subsequently isolated from Yellowstone and compared with the original strain YT-G [Heinen and Heinen, 1972]. All 3 strains were found to be Gram-variable though young cells of strains YT-F and YT-P appeared Gram-negative. The physiological and morphological properties of the strains are shown in Tables 3 and 4.

Strain YT-G utilizes a variety of carbon sources such as glucose, succinate, pyruvate and acetate when grown at 70 to 75°C, but only pyruvate is used at 80°C. Growth does not occur, however, unless the basal medium is supplemented with Brain Heart Infusion (B.H.I.) for which there is an increased requirement when the growth temperature is increased from 70°C to 80°C. Strain YT-P utilizes glucose, sucrose and pyruvate and grows without the addition of B.H.I. while strain YT-F grows best with succinate as a carbon source. Growth at 60°C with sucrose as carbon source showed that YT-F had a requirement for methionine and that strains YT-G and YT-P were prototrophic [Sharp and

Atkinson, unpublished data].

Heinen and Heinen [1972] considered the production of extracellular proteases and amylases a distinguishing character of strain YT-P but Sharp *et al.* [1980] report strong amylase production from YT-P and restricted production from strains YT-G and YT-F (Table 4). The abundance of branched chain fatty acids was typical of all 3 strains although the amount of individual branched and straight chain fatty acids changed with growth temperature [Weerkamp and Heinen, 1972]. The fatty acid composition of the individual strains can be recognized under standard conditions. Logarithmic phase cells of YT-G are resistant to rupture by ultrasonics while strains YT-F and YT-P are relatively sensitive. Strain YT-F was motile with many flagellae.

TABLE 3

*Characteristics of the caldoactive strains**

	B.caldotenax	*B.caldolyticus*	*B.caldovelox*
Strain designation	YT-G	YT-P	YT-F
Optimum pH range	7.5 - 8.5	6.0 - 8.0	6.3 - 8.5
Optimum temperature for growth	80°C	72°C	60° - 70°C
Maximum temperature for growth	85°C	82°C	76°C
Cell diameter	0.48 μm	0.7 μm	0.62 μm
Position of spores	terminal	terminal	terminal
Swelling of sporangia	definite	definite	definite
Shape of spores	oval	cylindrical	oval

*Data compiled from Heinen and Heinen [1972].

Heinen and Heinen [1972] differentiate all 3 caldoactive strains from *B.stearothermophilus* on the basis of their temperature optima, fatty acid pattern and submicroscopical structure. The 3 strains can be distinguished by their readiness to sporulate, their different temperature optima, and the morphological differences in their cell walls and membranes. Heinen and Heinen proposed the names *B.caldotenax*, *B.caldolyticus* and *B.caldovelox* for strains YT-G, YT-P and YT-F, respectively, and coined the term caldoactive to describe extremely thermophilic bacteria which show no active metabolism at lower temperatures.

Sharp *et al.* [1980] compared the biochemical properties, DNA base composition (Table 3), bacteriophage and bacteriocin sensitivities (Table 5), esterases, constitutive enzyme production, and antibiotic resistance of 7 strains of *B.stearothermophilus* and the 3 caldoactive strains.

TABLE 4

Sugar fermentations and biochemical reactions of thermophilic Bacillus species

	B. stearothermophilus NCA 1503	*B. stearothermophilus* ATCC 12016	*B. stearothermophilus*[1] Strain 240	*B. stearothermophilus*[2] Strain 262	*B. stearothermophilus*[2] Strain R593	*B. stearothermophilus*[3] Strain 136	*B. stearothermophilus*[4] Strain 10	*B. caldotenax*[5]	*B. caldovelox*[5]	*B. caldolyticus*[5]	*B. coagulans* ATCC 8038	*B. coagulans* ATCC 12245
Adonitol	-	-	-	-	+	-	-	-	-	-	-	-
Arabinose	-	+	-	-	-	-	w	-	-	-	-	-
Dextrin	+	-	+	+	-	+	+	+	+	+	+	+
Erythritol	-	-	-	-	-	+	-	-	-	-	-	-
Galactose	w	+	+	+	-	-	w	+	-	-	+	+
Glucose	+	+	+	+	w	+	+	+	+	+	+	+
Glycerol	+	+	+	-	+	w	+	+	+	w	w	w
Glycogen	+	w	+	+	-	-	+	+	-	+	+	-
Inositol	-	-	-	+	-	w	+	-	-	-	-	-
Lactose	-	-	-	-	-	-	-	-	-	-	-	-
Laevulose	+	w	+	-	w	+	w	+	w	w	+	+
Maltose	+	+	+	w	+	+	+	+	w	+	+	+
Mannitol	+	w	-	w	+	+	-	w	+	w	w	-
Mannose	+	+	+	-	w	+	+	+	+	+	+	+
Raffinose	+	-	-	-	-	-	+	-	w	-	+	-
Rhamnose	-	-	-	-	-	w	-	-	-	-	-	-
Salicin	-	w	-	w	-	w	-	-	-	-	+	-
Starch	+	-	-	-	-	-	-	-	-	+	w	+
Sucrose	+	-	+	+	-	+	+	+	+	+	-	+

Sorbitol	−	−	−	−	W	+	−	−	−	+	−
Trehalose	+	−	+	+	W	+	+	+	+	+	+
Xylose	−	+	−	−	−	+	−	−	−	W	W
NaCl tolerance	3%	1%	3%	4%	5%	7%	3%	3%	2%	2%	2%	2%
Sodium azide	−	−	W	W	−	−	−	−	−	−	W
Sabouraud agar	+	−	+	+	+	−	+	+	−	+	+
NO3 → NO2	+	+	+	+	+	+	+	+	+	−	+
NO2 reduction*	+	+	−	+	+	−	+	+	−	−	−	−
Anaerobic reduction of NO3 to gas	−	+	+	+	+	+	−	−	−	−	−
Catalase production	−	−	−	−	+	+	−	−	+	+	+	+
Oxidase	−	+	+	+	−	W	+	+	+	+	−	−
Gelatin hydrolysis	−	W	−	+	−	W	+	+	+	+	−	−
Casein digestion	W	W	+	+	+	−	+	+	+	+	+	−
Starch digestion**	+	+	R	R	−	−	+	R	R	+	−	W
Growth in anaerobic dextrose broth	+	+	+	+	−	−	+	W	W	W	+	+
Maximum growth temperature °C	74	76	77	78	68	65	75	85	76	82	61	61
Minimum growth temperature °C	41	40	41	42	39	41	41	−	−	−	25***	25***
% G+C content	51.9	54.2	52.3	62.2	54.5	47.5	50.9	64.8	65.1	52.3	48.3	46.9

All results recorded after 48 h; further incubation resulted in no additional activity.
Weak positive reactions denoted by W.
No strain utilized aesculin, dulcitol, inulin or citrate or produced indole or acetoin; * + indicates no residual nitrite present; **digestion recorded as + for a large zone of hydrolysis, and R for a restricted zone; ***growth at 25°C following 3 days incubation.
1. B.stearothermophilus strain 240 was isolated from chalky soil in the Salisbury area; 2. B.stearothermophilus 262 from an Icelandic hot spring; B.stearothermophilus RS93 from a hot water supply [Sharp et al., 1979]; 3. B.stearothermophilus 136 by Epstein and Grossowicz [1969]; 4. B.stearothermophilus 10 was from Dr. N.E. Welker, North Western University, Ill., U.S.A.; 5. cultures of B.caldotenax, B.caldovelox and B.caldolyticus were from Dr. W. Heinen, University of Nijmegen, Holland.

Data compiled from Sharp et al. [1980].

On the basis of biochemical and physiological properties
B.caldotenax YT-G and *B.caldovelox* YT-F can be classified
in Walker and Wolf's [1971] group 1 while *B.caldolyticus*
YT-P appears to have characteristics of both groups 1 and
3 (Table 2). The antibiotic sensitivity patterns of the 3
caldoactive bacteria did not show any significant differences from those of *B.stearothermophilus* strains while
sensitivity to selected bacteriophages indicated that the
former were more sensitive to infection than the latter.

TABLE 5

Phage and bacteriocin sensitivity of thermophilic
Bacillus *cultures**

	JS017	JS020	JS025	TP1C	TP84	12016	Thermocin 93 (zone ϕ, mm)
B.*stearothermo-philus* NCA 1503	-	-	-	+	++	-	22
B.*stearothermo-philus* ATCC 12016	+++	-	-	-	++++	++++	20
B.*stearothermo-philus* 240	-	-	-	-	+++	++++	18
B.*stearothermo-philus* 262	++	-	-	+	++	+	21
B.*stearothermo-philus* RS93	++++	+++	++++	+	+	-	0
B.*stearothermo-philus* 136	-	-	-	-	-	-	0
B.*stearothermo-philus* 10	++	-	-	+	++	-	24
B.*caldotenax*	++++	++++	++++	++++	++++	++++	22
B.*caldovelox*	++++	++++	++++	++++	++++	++++	22
B.*caldolyticus*	+++	+++	+++	+++	++	++	22
B.*coagulans* ATCC 8038	-	-	-	-	-	-	0
B.*coagulans* ATCC 12245	-	-	-	-	-	-	0

*Bacteriophage suspension (10^6 p.f.u. ml^{-1}) and dilutions were plated
with 0.1 ml of 6 h bacterial cultures and incubated at 60°C for 18 h.
Thermocin 93 was added to wells cut into TSBA plates seeded with bacterial suspensions; plates were incubated at 60°C for 18 h and then
zone diameters were recorded.
++++, Plating efficiency 1 to 0.1; +++, plating efficiency 0.1 to
0.01; ++, plating efficiency 0.01 to 0.001; +, indicates
low level of phage lysis; -, no visible indication of phage infection and no evidence of phage lysis.
Data from Sharp *et al.* [1980].

Examination of esterase enzymes by starch gel electrophoresis using the method of Baillie and Walker [1968]
showed the presence of essentially identical bands from
B.caldotenax, *B.caldovelox*, *B.caldolyticus* and *B.stearo-
thermophilus* strain 262. Examination of the esterase
patterns permitted the grouping of strains which corresponded closely with groups based on biochemical and physiological data.

Estimations of the % G+C content indicated that *B.caldo-
tenax*, *B.caldovelox* and *B.stearothermophilus* strain 262
fell into the 62 to 65%GC range while *B.caldolyticus* and
the remaining 6 strains of *B.stearothermophilus* showed the
lower values of 47.5 to 54.5%GC. DNA:DNA pairing studies
revealed close DNA homology between *B.caldovelox* and *B.
caldotenax* and to a lesser extent with *B.stearothermo-
philus* strain 262. *Bacillus caldolyticus* had closer DNA
homology with *B.stearothermophilus* NCA 1503 than with *B.
caldotenax*.

Phenotypically the caldoactive bacteria appear to closely resemble other strains classifiable with *B.stearothermo-
philus*, a taxon which embraces a wide variety of different
strains. Growth temperatures of 75°C or above were reported for 2 aerobic sporeformers in the sixth edition of
"Bergey's Manual of Determinative Bacteriology" [Breed *et
al.*, 1948] and there are reports by a number of workers of
adaption to growth at higher temperatures. The membrane
lipid composition, reported to differ from that previously
described for *B.stearothermophilus*, may be a consequence
of the higher optimum growth temperature. There are
several reports [Daron, 1970; Heinen *et al.*, 1970; Weerkamp and MacElroy, 1972] of increases in the percentage of
saturated and branched chain fatty acids at increased
growth temperatures.

Bacillus Thermophiles - an Emerging Taxonomy

In the preceding review of the literature attention has
been drawn to descriptions of strict thermophiles (65°+)
giving rise to (a) cylindrical spores within non-swollen
sporangia and (b) spherical spores contained within distinctly swollen or distended sporangia as demonstrated in
the recent report on *B.schlegelii* [Schenk and Aragno,
1979]. Previous to the current edition of "Bergey's
Manual of Determinative Bacteriology" [Gibson and Gordon,
1974] such organisms could not be accommodated in taxonomic keys since the only obligate thermophilic species was
B.stearothermophilus, with the Donk isolate as the type
strain, which has ellipsoidal spores within swollen sporangia.

During the past 2 decades evidence has accumulated
making such a restrictive attitude no longer tenable.
Furthermore, the variability in sporangium morphology
among several mesophilic species argues against such an

approach. Thus, the simultaneous occurrence of ellipsoidal and spherical spores in the same strain of *B.pantothenticus* [Proom and Knight, 1950] is obvious in the photographic evidence of the authors. Again, Gordon *et al.* [1973] record that in *each* of the 11 strains of this species there were equal proportions of ellipsoidal and spherical spores in each preparation. Similarly Bonde [1975] in a classic paper on the genus *Bacillus* has emphasized the variability in the morphology of sporangial features in several species in which more than 1 spore type occur.

The primary subdivision of the genus on sporangial morphology was an important feature in the pioneering and taxonomic efforts of the late Tom Gibson [Gibson and Topping, 1938] and of Nathan R. Smith [see, for example, Smith *et al.*, 1946, 1952]. By the time the current classification on *Bacillus* was being composed, Tom Gibson was adamant that the earlier views on the constancy and reliability of these sporangial features could no longer be defended [Gordon, 1977].

In the current edition of "Bergey's Manual of Determinative Bacteriology" [Gibson and Gordon, 1974] these morphological characters are neither used for the subdivision of the genus into main groups or for the demarcation of species. Differentiation is now based almost entirely on physiological and biochemical characters. Tom Gibson's disapproval of the use of sporangial and spore morphology for taxonomic purposes has not, however, deterred Gordon *et al.* [1973] from continuing to offer it as a most useful aid in the identification of the genus *Bacillus*. In fact, in the latest monograph 2 distinct keys or pathways are presented; the hitherto conventional approach based primarily on morphology and applicable to *typical* strains and an alternative, based mainly on physiology.

Nevertheless, the usefulness of the traditional key based on morphology is worth emphasizing. Even if more than 1 morphological sporangium type is present, the occurrence of swollen or unswollen sporangia, of oval or spherical spores, even in small numbers, is highly significant and of considerable help in identification.

The apparent demotion in the diagnostic value of sporangial morphology in the genus as a whole and for the thermophiles in particular makes a reappraisal of the classifications of Walker and Wolf [1968, 1971] and Klaushofer and Hollaus [1970] on *B.stearothermophilus* both pertinent and instructive.

Group 1 of Klaushofer and Hollaus [1970] consists of 2 closely related subgroups IA and IB. Strains in subgroup 1B are slightly more thermophilic than those in IA. Since strains in both subgroups produce spherical spores within distended sporangia, they were considered to represent thermophilic types of *B.sphaericus*. Klaushofer and Hollaus, however, also noted the simultaneous presence of oval spores in their preparations and considered the possi-

bility that their organisms might correspond to Group 2 of
Walker and Wolf [1971]. One of us (J.W.) has therefore
recently examined representative strains of subgroups IA
and IB, namely DSM 461, 462, 463 and 464. The spore morphology and extent of swelling were typical of the Walker
and Wolf Group 2, with which can be associated biochemically the strain NRS 1511 of Daron [1967].

The second group described by Klaushofer and Hollaus
contains 9 strains and is readily equated with the 127
strains comprising the denitrifying Group 1 of Walker and
Wolf [1971]. These organisms closely correspond in their
reactions to those given by the original isolation of *B.
kaustophilus* by Prickett [1928]. An earlier description
of this type is that of Ambrož [1913]. However, his new
species *Denitrobacterium thermophilum* is not acceptable
and the claims of Prickett with his description of *B.
kaustophilus* will have to be seriously considered. The
latter has priority claims over the suggestion [Mishustin,
1950] of considering *B.thermodenitrificans* as corresponding to the isolate of Ambrož.

As recalled earlier, Golovacheva *et al.* [1965] isolated
1 strain only of this species from their Ural samples.
Essentially these organisms show slight to definite
swelling of the sporangium, are sensitive to NaCl (3%,
w/v), give a restricted reaction on starch, and either
fail completely, or grow poorly, anaerobically. Another
outstanding feature is that they reduce nitrate to gas.

The third group [Klaushofer and Hollaus, 1970] is composed of a large homogeneous subgroup, IIIA, of 50 strains
and a minor subgroup, IIIB, consisting of 8 strains which
display a heterogeneous set of reactions. Subgroup (IIIA)
is important since it comprises a high proportion (50/87)
of isolates from sugar beet extractions. Many of the
organisms conform well with the reactions of the Donk
strain of *B.stearothermophilus* and the inclusion of reference strains ATCC 12980 (NCA 26) and of ATCC 7954 (NCA
1503) clearly indicate a close relationship with Group 3
of Walker and Wolf [1971].

It is worth emphasizing that neither the 63 strains representing Group 3 in the study of Walker and Wolf [1971]
nor the 50 strains of Klaushofer and Hollaus [1970] indicate a particularly tight homogeneous group. The variability in many reactions is very marked but both the
similarity index of the latter and the serological tests
of Chowdhury and Wolf (unpublished data) clearly support
substantial affinities and the sharing of some basic properties.

It is now timely to summarize our views on the obligate
Bacillus thermophiles. The difficulties involved are
clearly stated by Gibson and Gordon [1974] in their concluding paragraph on *B.stearothermophilus*. Nonetheless,
the additional evidence of the past decade does help to
resolve part of the present complex dilemma. The results

of Walker and Wolf [1968, 1971], Klaushofer and Hollaus
[1970] and more recently of the Porton group [Sharp *et al.*,
1980] strongly suggest that *B.stearothermophilus* contains
some distinct and well-defined elements which merit promo-
tion to individual species rank.

1. There is an overwhelming case for distinguishing Group
   1 of Walker and Wolf [1971] and designating this type
   as *B.kaustophilus* [Prickett, 1928]. This species
   would include ATCC 12016 (NCA 2184) examined by Sharp
   et al. [1980]. Organisms in this species are distinguished from the other 2 groups recognized by Walker
   and Wolf by their ability to reduce nitrate to gas, by
   failure to grow anaerobically and their relative sensitivity to NaCl. Strains have a G+C ratio of about
   54. Sporangial swelling is variable. Within the
   species, biochemical and immunological variations are
   substantial.

2. Group 2 of Walker and Wolf [1971] is outstanding in
   its biochemical inertness, a largely negative and
   questionable taxonomic criterion. The organisms in
   this taxon are, however, strictly aerobic, relatively
   tolerant to NaCl and immunologically homogeneous and
   completely unrelated to strains in the other 2 groups
   described by Walker and Wolf.

3. Group 3 of Walker and Wolf [1971] is based on the best
   studied and most widely occurring obligate thermophile,
   B.stearothermophilus [Donk, 1920]. Although originally
   described as unable to reduce nitrate and as non-
   proteolytic, the examination of 63 strains by Walker
   and Wolf and the 50 strains (50/58) of Group IIIA of
   Klaushofer and Hollaus indicate that usually each of
   these reactions is positive. The organism is characterized by producing a sediment in broth, a distinctly
   wide zone of hydrolysis on starch agar (only 6/63 gave
   a restricted reaction) and acid clot and reduction in
   tomato-yeast-milk. Arabinose and mannitol are rarely
   fermented. Proteolysis of gelatin and casein is most
   frequently positive.

Of the 3 groups of Walker and Wolf, it is the only one
to give good growth anaerobically in glucose broth. It is
important, however, to stress the great variations in the
pattern of reactions and in the high number of variable
combinations within these reactions. The 63 strains yielded 21 different subgroups or patterns. Although immunologically Group 3 was initially refractive, subsequent
attempts based on disintegrated spores as antigens indicated full support for the delimitation boundaries of this
group [Chowdhury and Wolf, in preparation].

The extensive variability in the reactions of cultures
within this group would seem to suggest that it may well
contain several elements best segregated to themselves.
Recent criteria such as determination of G+C ratios and
especially DNA homologies have hardly been applied to mem-

bers of this ubiquitous group of thermophiles, which clearly deserves more detailed study.

This group also included 10 cultures classified by
Galesloot and Labots [1959] as *B.calidolactis* and *B.thermo-
liquefaciens*.

Gordon and Smith [1949] were unhappy with the inadequate
original description of *B.calidolactis* and *B.thermolique-
faciens* and with the authenticity of the cultures they received (NRS 15 and 16). These 2 strains, together with the
cultures examined by Galesloot and Labots [1959], can be
classified in Group 3. It is these strains (6/10) which
were responsible for the exceptional restricted reaction to
starch. Galesloot and Labots [1959] suggested that the
term *B.thermoliquefaciens* should be reserved for the non-
obligate (30°C+) strains. Gordon *et al*. [1973] similarly
found that the 6 strains of Galesloot and Labots [1959]
conformed well with the typical reactions given by the
selected strains they re-examined (Table 22, pp.214-215).
Present evidence for retaining these 2 species is not convincing.

Bacillus thermocatenulatus [Golovacheva *et al*., 1975]
is sufficiently different from previously described thermophilic rods to be considered as a distinct species. The
characters worth emphasizing are the negative reactions on
starch and gelatin, its ability to grow anaerobically *and*
to reduce nitrate to gas. The maximum growth temperature
is 78°C and the optimum 57°C in broth and 67°C on agar;
the G+C ratio is high (69%). Colonies are chromogenic
(yellow).

The most recent of the obligate thermophiles, *B.schle-
gelii* [Schenk and Aragno, 1979], is the first chemolitho-
autotroph to be comprehensively described. It is strictly
aerobic, oxidizes hydrogen in the presence of O_2 and CO_2,
also grows heterotrophically, produces a distinctly
swollen sporangium and spherical spores. Its optimum temperature is 70°C and its G+C ratio 67-68%.

The Thermoacidophiles

B.coagulans

The organism was first described by Hammer in 1915,
followed by more detailed tests in 1932 [Sarles and
Hammer]. A definitive study of the species based on an
examination of 77 strains derived from a variety of
sources was made by Gordon and Smith [1949].

Morphologically both swollen and unswollen sporangia
occur and from the studies of Wolf [1968] there are 2 distinct physiological types. Type A will grow at 65°C providing the pH is distinctly acid (6.2). Type B is the
more acidophilic and grows at pH 4.5 and at a lower temperature, 23°C, at pH 6.2. Essentially *B.coagulans* is a
facultative thermophile, growing well at 45 to 55°C. Many

strains grow well at 60°C. Secondly the organism is distinctly acidophilic, some strains grow poorly at pH 7.0 and
are stimulated to more profuse growth at pH 6.0 (Table 6).

TABLE 6

Differentiation of B.stearothermophilus
from B.coagulans

	B.stearothermophilus	*B.coagulans*
Growth at pH 5.7 (PPA)	-	+
Acidophily (pH)*	7>6	6>7
Growth in azide (0.2%)	-	+
Growth in sulphadiazine 5μ/ml	-	+
Growth temperature (°C):		
Maximum	65-75	55-60**
Minimum	30-45	15-25
Anaerobic growth	-	+
Spore H.R.D. value (min. at 120°C)	4-5	0.1
% G+C	44-53	47-56

PPA, proteose peptone acid agar pH 5.7; *acidophily (pH), comparative
stimulation or inhibition at the two pH values on nutrient agar slopes
at 50°C; **maximum growth temperature, some strains of *B.coagulans*
[type A, Wolf, 1968] grow at 65°C at pH 6.2 (these are AMC negative);
Group 3 of Walker and Wolf [1971] are (+); Gibson and Gordon [1974].

A recent isolate of Belly and Brock [1974] derived from
the hot springs of Yellowstone National Park indicates
that strains more acidophilic than hitherto recognized,
readily occur in these hot, acid environments. Their isolate grew over a wide acid pH range (2 to 6), had an optimum at about pH 3.5 and failed to grow at pH 7. The G+C%
ratio of this strain was (56%) higher than hitherto recorded in comparison with values of 47 to 48% [Gibson and
Gordon, 1974] and 44.4% [Bonde, 1975].
The present species probably comprises several distinct
elements as perhaps indicated by the wide range of G+C%
values. Thus, the 2 strains (318 and 343) examined by
Bonde [1975] appeared to belong to 2 distinct and markedly
different groups IIIB and IIAT, respectively. On the
other hand, 11 strains examined by Seki *et al.* [1978] presented a highly homologous group on the basis of DNA homology, in spite of the fact that their strains included
types showing both swollen and non-swollen sporangia.
The organism is primarily saccharolytic, non-proteolytic (some strains liquefy gelatin weakly) and microaerophilic. It was thus the most common cause of over-
acidification in canned fruits and vegetables, especially
tomato juice. Improvements in hygiene and an increase in
the heat treatment have largely overcome this fault. The

comparatively lower heat resistance of its spores in comparison with those of *B.stearothermophilus* has been a
further factor in the resolution of this problem.

Extreme thermoacidophiles (B.acidocaldarius)

Uchino and Doi [1967] described the isolation of hitherto unknown *Bacillus* thermophiles capable of growth at high
temperatures and extreme acidity. They were derived from
natural hot springs of Japan. Closely related types have
since been isolated from similar acid hot springs in
Yellowstone National Park, U.S.A. [Darland and Brock,
1971]; volcanic hot springs of Naples [De Rosa *et al.*,
1971-3]; hot springs and soil of the north Pacific island
of Kunashir, U.S.S.R. [Loginova *et al.*, 1978] and other
hot springs of Japan [Oshima *et al.*, 1977].

A most adequate description of this novel type of
Bacillus is given by Darland and Brock [1971] who have
proposed the new specific name of *Bacillus acidocaldarius*
for it. The organism produces terminal to subterminal
spores within sporangia which are slightly swollen.
Growth occurs aerobically at a temperature range of 45 to
70°C and over a pH range of 2 to 6. The Russian strains
grow even at 75°C. The organism is mildly proteolytic
(gelatin), ferments a variety of carbohydrates and hydro-
lyses starch, nitrates are not reduced, nor is acetylmethylcarbinol produced.

The G+C% ratio is high, 59 to 66% (61 to 62% for the
type strain of Darland and Brock). The spores are remarkably heat sensitive. Thus at 86°C, 50% lose viability
within 10-12 mins. Spores germinate in solutions of mineral and organic acids without the need of activation or
germination agents. Germination is rapid at pH 3 to 5,
slow at pH 6 and absent at pH 7 [Handley, 1977].

The Biochemical, Physiological and Genetic
Basis of Thermophily

A taxonomic study of *Bacillus* thermophiles would be incomplete without some consideration of the impressive progress which has been made particularly in the last decade
on the identity and role of the various cell components
which contribute to and control thermal stability. Firstly, only some of the components of these thermophiles are
associated with thermophily and it is these which may
prove particularly useful as markers of specific identity.
Secondly, since many enzymes and other cellular components
are identical with those of corresponding mesophiles these
would presumably prove unlikely as taxonomic discriminators. This section has been deliberately restricted to
evidence based on studies with members of the genus
Bacillus.

Proteins

Militzer *et al.* [1949] examined the thermal stability
of malate dehydrogenase in extracts from an obligate thermophile grown at 65°C and of the mesophile, *B.subtilis*.
The enzyme from the thermophile was stable for 120 mins at
65°C; in contrast, that from *B.subtilis* was rapidly inactivated.

Amelunxen and Lins [1968] compared the thermostability
of 11 enzymes in extracts from *B.stearothermophilus* and
from the mesophile *B.cereus*. Nine of the enzymes from the
former were more thermostable than the corresponding
enzymes from *B.cereus*. However, pyruvate kinase and glutamate oxaloacetate transaminase showed similar extents of
inactivation at 70°C for the thermophilic and mesophilic
strains.

Singleton and Amelunxen [1973] considered 3 distinct
mechanisms to account for the activity of thermophilic
microorganisms:

1.   Thermophiles may contain factors which increase the
     thermal stability of enzymes.
2.   Conversely, mesophiles may contain factors which in-
     crease the thermolability of their enzymes.
3.   Cellular components of thermophiles may have inherent
     stability independent of exogenous factors.

Attempts to transfer heat stability and heat lability
[Amelunxen and Lins, 1968; Koffler and Gale, 1957;
Koffler, 1957] by mixing cell-free extracts from thermophiles and mesophiles failed to change the original properties of individual enzymes. These studies did not, however, take into account the possible presence of stabilizing factors tightly bound to the enzyme and not readily
transferable. Amelunxen and Murdock [1978] point out that
thermal stability persists even after extensive purification and recrystallization of certain enzymes. Also,
since molecular weights of thermostable enzymes and their
homologous enzymes from mesophiles are essentially identical, any stabilizing factors present would be of very low
molecular weight.

Some enzymes of thermophiles proved unstable at the
optimum growth temperature of the organism. These enzymes
may be stabilized by normal cell constituents such as membranes, cofactors and substrates. Welker [1976] reported
the loss of thermal stability of alkaline phosphatase in
B.stearothermophilus when released from protoplasts, and
attributed the stabilization of the enzyme to the cell
membrane. Glutamine synthetase from *B.stearothermophilus*
is stabilized by binding of L-glutamate and NH_4^+, or L-
glutamate, Mn^+ and ATP, by increases in polar amino acids,
probably by formation of disulphide bridges and by enzyme
aggregation driven by energy released by hydrophobic interactions [Wedler, 1978].

Calcium ions confer stability on several extra-cellular

enzymes such as α-amylase [Yutani, 1976], thermolysin
[Tajima *et al.*, 1976] and proteinases [Sidler and Zuber,
1977]. Cobalt ions are involved in the thermal stability
of aminopeptidase 1 from *B.stearothermophilus* NCIB 8924
[Deranleau and Zuber, 1977].

The stabilization of proteins by aggregation has been
suggested by Wedler and Hoffman [1974]. With glutamine
synthetase loss of activity at 70°C was greater at a
protein concentration of 0.04 mg ml^{-1} as compared with one
of 4 mg ml^{-1}. Amelunxen and Singleton [1976] demonstrated
thermolabile glycerophosphate dehydrogenase in the facultative thermophile *B.coagulans*. The enzyme was stabilized
only at a critical ionic strength by the addition of salt.
They attributed stabilization to a highly charged intra-
cellular macromolecular environment.

The rapid resynthesis of macromolecules to counteract
thermal inactivation in thermophiles has been suggested,
but evidence for this is rather insufficient. Allen [1953]
reviews some of the earlier work relating to this hypothesis. Koffler [1957] has pointed out that assuming a
doubling in the rate of enzyme reactions for each 10°C
rise in temperature, thermophiles should exhibit 16 times
more activity at 70°C than at 30°C.

The only thermostable enzyme reported to differ considerably from its mesophilic counterpart was an α-amylase
from a strain of *B.stearothermophilus* [Manning and Campbell, 1961]. It exhibited a semi-random coil configuration existing in a semi-denatured state. The enzyme had a
molecular weight of 15,600, well below that of other amylases, it contained 4 cysteine residues and was stabilized
by 1 disulphide bridge. This enzyme appears to be an
exception. Other α-amylases isolated from *B.stearothermo-
philus* [Isono, 1970; Pfueller and Elliot, 1969] have
molecular weights in the region of 50,000 and contain
fewer cysteine residues.

Small conformational changes occur without denaturation
in some thermophile proteins when heated from room temperature to 55°C or 60°C. This is in contrast to most mesophile proteins which are denatured at this temperature.
Magsanaga and Nosoh [1974] demonstrated a conformational
change in the glutamine synthetase of *B.stearothermo-
philus* well below the temperature of enzyme inactivation;
at 55°C the enzyme became susceptible to thermolysin digestion.

Almost all *in vitro* studies, on the thermal stability
of proteins, suggest some inherent stability of the protein
structure which is dependent upon the amino acid sequence
and composition. Several proteins from thermophiles and
mesophiles have proved similar in molecular weight and in
structure and differences were associated with a very small
number of amino acids. Alterations in a small number of
amino acids in the polypeptide chain may result in changes
affecting the stability of the secondary and tertiary

structures of the molecule. These changes may involve formation of disulphide bridges, increased hydrogen bonding, increases in hydrophobic bonds and ionised group interactions.

The number of disulphide bridges does not correlate well with enzyme thermal stability. Ribonuclease has high thermal stability attributed to the presence of 4 disulphide bonds [Spackman *et al.*, 1960; Smith *et al.*, 1963]. Data from most thermophilic enzymes, however, suggest the occurrence of fewer disulphide bonds than in their mesophilic counterparts.

Several proteins have now been isolated and their amino acid sequence determined in a variety of organisms and it is now possible to compare amino acid sequences in thermophiles and mesophiles and also in higher organisms.

Examination of ferredoxins from the thermophiles *Clostridium tartarivorum* and *C.thermosaccharolyticum* and the mesophiles *C.pasteurianum*, *C.acidurici* and *C.butyricum* [Tanaka *et al.*, 1971; Devanathan *et al.*, 1969] indicated similarities in their molecular weights, iron, inorganic sulphide and cysteine content, adsorption spectra and the number of amino acid residues. The proteins from the thermophiles, however, showed greater thermal stability. Since ferredoxin is a small (molecular weight 15,000) protein having no secondary or tertiary structure, the increased stability is presumably associated with changes in the amino acid composition. Comparison of sequence data indicated that the thermophile ferredoxins were the only ones to contain histidine which replaced either serine or tyrosine. These may therefore serve as ligands for the tighter binding of atoms of iron or the differences in charge may increase the opportunities of hydrogen bonding. There is also some evidence for a greater number of basic amino acids in thermophile protein. Perutz and Raidt [1975] suggested this might permit the formation of additional ionic bonds. In the thermophile ferredoxins a maximum number of 4 ionic bonds are formed in *C.thermosaccharolyticum* and only 1 in the mesophilic *C.acidurici*. Increased thermal stability also correlates with increase in the number of glutamic acid residues. *Clostridium pasteurianum* and *C.acidurici* contain a total of 2 each, while the thermophile *C.tartarivorum* contains 5 and *C.thermosaccharolyticum* contains 7 residues.

Hase *et al.* [1976] reported a total of 6 glutamic acid residues in ferredoxin from *B.stearothermophilus*. They point out that glutamic acid residues are the best helix promoting amino acid [Robson and Pain, 1971]. The ferredoxin from *B.stearothermophilus* also had a lower number of cysteine residues than other ferredoxins examined.

Arginine content particularly in lactate dehydrogenase [Frank *et al.*, 1975] appears to increase in thermophile proteins [Ljungdahl *et al.*, 1976; O'Brien *et al.*, 1976] at the expense of lysine. In contrast to the ferredoxins

the overall number of basic amino acids remains constant
although the incorporation of arginine at specific points
at the expense of lysine in other positions results in
greater thermal stability.

Examination of the enzyme rhodanase (molecular weight
15,000) indicates the half-life of the enzyme from *Thio-
bacillus denitrificans* to be 0.5 mins at 65°C; from *B.
subtilis* the half-life was 4.5 mins and from *B.stearother-
mophilus* 36 mins. The only significant changes in amino
acid composition was in the aspartate:glutamate ratio,
from 8:2 in *T.denitrificans*, 5:5 in *B.subtilis* and 1:10 in
B.stearothermophilus [Atkinson, 1976]. No primary sequence
data are available on this enzyme although it is interesting to note that a change from aspartate to glutamate
requires only a single base change in the third base of the
triplet code.

Calculations indicate that to increase the melting temperature of a molecular weight 35,000 polypeptide chain,
from 35°C to 45°C requires only a 3% increase in hydrogen
bonding and that up to 10 K cal/mole differences in activation energy can be obtained by changes in 1 or 2 amino acid
residues [Atkinson, 1976]. A thirty-fold difference in
thermal stability at 60°C of triosephosphate isomerase from
rabbit muscle and *B.stearothermophilus* can, therefore, be
accounted for by a difference in activation energy of only
2.2 K cal/mole [Hocking and Harris, 1976]; similarly, the
differences in the thermal stability of the ferredoxins
could be due to differences of only 4.5 to 8.5 K cal/mole.
Studies on the inactivation of Haemoglobin A and an abnormal Haemoglobin A2 indicated the latter to have a 6-fold
greater thermal stability at 45°C. The abnormal protein
had only 10 amino acid changes out of a total of 670 amino
acid residues. Three-dimensional studies indicated that
only 3 of these contributed to increased interactions in
the form of 2 non-polar interactions and 1 hydrogen bond
[Perutz and Raidt, 1975].

Present evidence indicates and emphasizes the close
structural similarity of proteins from mesophiles and
thermophiles. Differences which occur are subtle and
appear to involve a small but significant number of amino
acid residues. These amino acid substitutions are responsible for changes in hydrogen bonding, hydrophobic bonding
and ionic bonding. Slight changes in bonding within the
protein molecule appear to account substantially if not
entirely for increases in the thermal stability of proteins
in thermophiles.

Cell walls

There is little evidence that the cell wall plays any
significant role in thermal stability. Forrester and
Wicken [1966] studied the walls of *B.coagulans* and *B.
stearothermophilus* grown at 37°C and 55°C. At the higher

temperature there was an increase in the proportion of
mucopeptide and a decrease in the proportion of teichoic
acid in the walls of both organisms. Each had a higher
wall lipid content than is usually found in Gram-positive
organisms.

Novitsky *et al*. [1974] compared the amino acid and amino
sugar composition of the wall peptidoglycan of a facultative thermophilic strain of *B.coagulans* grown at 37°C and
55°C. At 55°C, all of the wall components except alanine
were present in a higher proportion, the cells also showed
a lower level of amylolytic activity, bound more Mg^{++} and
contained less peptide cross bridging.

Membranes

Heilbrun [1924] and Belehradek [1931] observed that
membrane lipids of thermophiles had higher melting points
than those of mesophiles. They suggested that the melting
temperature of cell lipids might represent the upper temperature limit for cell growth. Examination of mesophilic
and thermophilic strains of *Bacillus* sp. [Cho and Salton,
1964; Shen *et al*., 1970] indicated that the cell membranes
of thermophiles generally contained a higher content of
saturated and branched-chain fatty acids. Examination of
facultative and obligate thermophilic strains [Ray *et al*.,
1971; Weerkamp and MacElroy, 1972] indicated that increasing the growth temperature resulted in a change in membrane lipids with a higher melting point. An increase in
growth temperature normally results in the production of
lipids containing a relatively lower proportion of
straight-chain saturated fatty acids and often in the production of membrane lipids with greater than average chain
length [McElhany, 1976].

Daron [1970] reported a 3- to 4-fold increase in
branched chain fatty acids when the growth temperature of
his strain of *B.stearothermophilus* (NRS 1511) was increased
from 40°C to 60°C. Heinen *et al*. [1970] found the level
of the branched chain fatty acid iso-C17 in the membrane
of *Thermus aquaticus* increased from 30% to 50% when the
growth temperature was raised from 50°C to 80°C. Conversely the level of the iso-C16 component was reduced from 30%
to 10% of the total fatty acid fraction.

Comparison of wild type cells of *B.stearothermophilus*
grown at 42°C and 65°C indicated that at the latter there
was an increase of 27% in fatty acids with melting points
above 55°C [Souza *et al*., 1974]. A temperature sensitive
mutant lacked the ability to make similar changes in its
fatty acid composition above 52°C. The authors proposed
that the lowest and highest boundary temperatures for the
growth of thermophiles were dependent on the lipid mixtures
which could be synthesized by the organism.

Most cell lipids are known to undergo phase transitions
and phase separations [Lee, 1977]. The temperature at

which these occur depends upon the chemical composition of the phospholipids, such as the length of fatty acid chains, the degree of unsaturation, the presence of methyl or cyclopropyl sidechains and the change and size of the head group, whilst the presence of neutral lipids such as cholesterol, proteins and ions would also have an effect.

Esser [1978] considered that phase separations and phase transitions might serve as a means of amplifying small signals into having a large effect on the membrane and may be determining factors in the outcome of immunological re-actions expressed at the cell surface.

McElhany [1976] found that cells of *Acholeplasma laid-lawii* could grow within as well as above the temperature range of phase transition.

Esser [1978], after freeze etching and fluorescent label spectroscopy, concluded that *B.stearothermophilus* was able to grow at a temperature when most, if not all, of the lipid hydrocarbon chains were in the fluid state. A further increase in temperature, above the lipid melting point, resulted in the weakening of the various physical interactive forces still present (van der Waals forces, for example) resulting in randomization of the membrane compo-nents and lack of function.

The regulating mechanism for this control in bacteria has not been fully elucidated although studies on *Tetra-hymena pyriformis* [Thompson and Nozawa, 1977] suggested that 'fatty acid desaturase', a membrane bound enzyme might be functioning as a thermometric regulator.

A study of the protein content of the cell membrane re-veals an increase in the level of membrane protein as growth temperature is increased [Wisdon and Welker, 1973]; thus, protoplasts of *B.stearothermophilus* showed an in-crease in the ratio of protein to lipid as the growth tem-perature was increased. This may in part explain the thermal stabilization of certain enzymes by association with the cell membrane.

DNA

Comparisons of DNA melting temperatures (Tm) of DNA iso-lated from thermophilic strains of *Bacillus* and from meso-philes, for example, *E.coli*, indicate no apparent correla-tion between Tm values and the ability to grow at higher temperatures [Marmur, 1960; Welker and Campbell, 1965; Saunders and Campbell, 1966]. Stenesh *et al.* [1968] studied the DNA of some mesophiles and thermophiles of the genus *Bacillus*. The DNA from the thermophiles showed a consistently higher guanine and cytosine (G+C) content (53% compared with 45% for the mesophiles). DNA with a higher G+C content is more stable at higher temperatures, since guanine and cytosine bind *via* 3 hydrogen bonds, while adenine and thymine bind *via* only 2. The extreme thermo-philes *B.caldovelox* and *B.caldotenax* [Sharp *et al.*, 1980]

and *B.acidocaldarius* [Gibson and Gordon, 1974] have G+C
contents in the region of 60%. Various strains of *B.
stearothermophilus* have a G+C content in the ranges 49 to
53% and 44 to 46% and *B.coagulans* 47 to 48% [Gibson and
Gordon, 1974]. Table 7 compares the optimum growth temperature and the percentage G+C content in various *Bacillus*
species. Although in general this indicates higher G+C
values with increasing growth temperature, there is no
direct evidence of a relationship between increased DNA
thermostability and the ability to grow at higher temperatures.
    Stenesh [1976] reported studies on the fidelity of DNA
replication in *B.licheniformis* and *B.stearothermophilus* at
a range of temperature between 37°C and 72°C. Using
nearest neighbour frequency analysis [Stenesh and Roe,
1972] they found that differences related to increase in
temperature. These were more pronounced in *B.stearothermo-
philus* and particularly in adenine-thymine dinucleotides
where bases are bonded by only 2 hydrogen bonds.
    Increasing the growth temperature of *B.stearothermo-
philus* NCA 1503 increased significantly the mutation frequency of several genes. The terminal methionyl trans-
ferase involved in methionine biosynthesis is a temperature
sensitive protein such that *B.stearothermophilus* NCA 1503
is auxotrophic for methionine only above 53°C. Segregation
and reversion frequencies at higher growth temperatures are
in the region of 10^{-4}. This is one order of magnitude
higher than that found for "hot spots" in genes from mesophilic organisms [Atkinson, personal communication].

RNA

    Ribosomal RNA from thermophiles appears to have a higher
G+C content than that from mesophiles. Pace and Campbell
[1967] studied ribosomes from 19 different microorganisms
whose optimum growth temperatures ranged from 20°C to 70°C.
There was a general increase in G+C content with increasing
growth temperature. Irwin *et al.* [1973] investigated the
ribosomes and ribosomal RNA from psychrophilic, mesophilic
and thermophilic *Clostridium* species. The thermal melting
temperatures (Tm) of ribosomes were 64.4, 63.8 and 68.7 to
59.6°C respectively, and Tm values of ribosomal RNA were
similar for all 3 groups, 66.2 to 67.9°C. Stenesh *et al.*
[1968] reported that the total RNA from thermophiles had a
G+C content of 61.4% compared with 56.9% from mesophilic
strains. These values are higher than those for ribosomal
RNA, indicating transfer RNA to be more stable than ribosomal RNA.
    Studies on the heat stability of ribosomes by Friedman
[1978] indicate a more orderly conformation in ribosomes
from *B.stearothermophilus* than from *E.coli*, which may contribute to increased thermal stability.
    Agris *et al.* [1973] reported that transfer RNA from *B.*

TABLE 7

*A comparison of the %G+C content and optimum growth
temperatures in representatives of some* Bacillus *species*

Species	Maximum growth temperature (°C)	%G+C content	References
B. *thermocatenulatus*	-	69	Golovacheva *et al.* [1975]
B. *thermoruber*	-	67	Guicciardi *et al.* [1968]
B. *schlegelii*	70 opt.	67-68	Schenk and Aragno [1979]
B. *caldovelox*	76	65.1	Sharp *et al.* [1980]
B. *caldotenax*	85	64.8	Sharp *et al.* [1980]
B. *acidocaldarius*	65-70	61-62	Darland and Brock [1971]
B. *caldolyticus*	82	52.3	Sharp *et al.* [1980]
B. *stearothermophilus*	65-75	44-46 49-53	Gibson and Gordon [1974]
B. *coagulans*	50-60	37-48 47-48	Belly and Brock [1974] Gibson and Gordon [1974]
B. *macerans*	40-50	49-51	
B. *licheniformis*	50-55	43-47	
B. *polymyxa*	35-45	43-46	
B. *circulans*	35-50	35,47	
B. *subtilis*	45-55	42-43	Gibson and Gordon [1974]
B. *pasteurii*	33-42	42	
B. *firmus*	40-45	41	
B. *sphaericus*	30-45	37,43	
B. *megaterium*	35-45	36-38	
B. *cereus*	35-45	32-33 33-37	

stearothermophilus grown at 70°C had 1.4 times more methyl groups than the transfer RNA cells grown at 50°C. It did not result in any increase in bonding since the thermal melting profiles were similar.

Saunders and Campbell [1966] showed, as expected, that the base composition of messenger RNA from B. *stearothermophilus* was almost identical to that of its DNA.

Studies on translation of the genetic code [Arca *et al.*, 1965] showed the mischarging of isoleucyl transfer RNA with valine by isoleucyl tRNA synthetase from B. *stearothermophilus* to occur at temperatures of 70° to 75°C but not at 50° to 60°C.

The incorporation of 1 amino acid in response to the codon for another has been studied by Friedman and Weinsten [1964] in B. *stearothermophilus*. They have shown that this type of error in translation (termed 'ambiguity') is affected by such conditions as temperature and ionic composition during amino acid incorporation into the grow-

ing polypeptide chain. In most cases the amino acids involved were those in which the codons contained at least 2
bases that were present in the copolymer used as a template. Stenesh [1976] reports the implication of the ribosome as the main cause of ambiguity in *B.stearothermo-
philus*. Schlanger and Friedman [1973] and Gorini [1971]
suggest the control of translation accuracy by the ribosome, involving the presence of a transfer RNA screening
site on the ribosome, which interacts with a portion of the
transfer RNA molecule preventing the incorporation of
particular amino acids into the polypeptide.

Studies by Stenesh [1976] into the degeneracy of the
genetic code and the possibility that thermophiles and
mesophiles differ in their relative utilization of synonym
codons suggested a preference of certain synonym codons to
be a function of incubation temperature.

Genetic basis of thermophily

Genetic transfer within *B.stearothermophilus* has not
advanced to the extent where it can identify the genetic
basis of thermophily. Isono [1970] reported the transformation of an amylase deficient mutant of *B.stearothermo-
philus* to the positive wild type. This transformation,
carried out at 37°C, has not apparently been followed by
further reports on this system. Attempts by Welker [1978]
over a number of years, to establish a system of genetic
transfer in *B.stearothermophilus*, has resulted in the
ability to transfect *B.stearothermophilus* strain 4S. This
was accomplished with DNA from bacteriophage TP1C or TP84,
and with the help of intact phage TP-12 which appears to
behave in the manner of a competence factor. Welker [1978]
also reports the transformation of rifampicin resistance in
B.stearothermophilus strain 4S, without the use of the
helper phage.

Studies on the genetic transfer of thermophily have been
mainly limited to the *B.subtilis* transformation system
developed by Spizizen [1959]. McDonald and Matney [1963]
reported the transformation of streptomycin resistant *B.
subtilis* 168 which was unable to grow at 55°C. The transformants which arose at a frequency of 10^{-4} grew at 55°C,
but only 10 to 20% of these high temperature transformants
still retained their high level resistance to streptomycin.
Lindsay and Creaser [1975] using DNA from *B.caldolyticus*
transformed *B.subtilis* to grow at 70°C. They also report
the cotransference of genes mapping close to those coding
for ribosomal and tRNA functions. Studies of the enzyme
L-histidinol dehydrogenase indicated that the transformed
cells possessed a thermostable enzyme similar to that
present in *B.caldolyticus*. They suggested that changes in
the synthesis of proteins by thermophiles at the ribosomal
or transfer RNA level give rise to translationally altered
enzymes which can function at higher temperatures. Fried-

man and Mojica-a [1978] similarly reported the transformation of *B.subtilis* with DNA from streptomycin resistant *B.
caldolyticus*. Their work implies the transfer of genes
coding for ribosomal proteins from donor to recipient
during transformation to thermophily.

Studies of homologous enzymes from mesophiles and
thermophiles show small changes in amino acid sequences
which can result in increases in thermal stability through
hydrogen bonding, hydrophobic bonding and ionic bonding.
It yet remains to be revealed if these small but significant amino acid changes are determined by the genetic code,
and if the genes for all thermophile proteins differ from
those of complementary mesophile proteins in having a small
number of different base triplets in the code.

The work of McDonald and Matney [1963], Lindsay and
Creaser [1975] and Friedman [1978] suggests that only a
small number of genes are involved in determining the
thermophily of an organism, and these operate at the level
of translation. This control would involve the controlled
mischarging of transfer RNA molecules resulting in selection of particular amino acids in preference to others.
Alternatively, the ribosome may possess a site which enables selective screening of incoming transfer RNA molecules, or a failure to recognize certain transfer RNA
anticodons.

Further comparative studies into the role of ribosomal
proteins, hybridization of DNA and messenger RNA from
thermophiles and mesophiles, and studies of genetic transfer between thermophiles and mesophiles should throw additional light on the extent of genetic basis of thermophily.

References

Agris, P.F., Koh, H. and Soll, D. (1973). The effects of growth temperatures on the *in vivo* ribose methylation of *Bacillus stearothermophilus* transfer RNA. *Archives of Biochemistry and Biophysics*
154, 227-282.

Allen, M.B. (1953). The thermophilic aerobic spore forming bacteria.
Bacteriological Reviews **17**, 125-173.

Ambrož, A. (1913). *Denitrobacterium thermophilum* spec. nova. Ein
Beitrag zur Biologie der thermophilen Bakterien. *Zentralblatt für
Bakteriologie und Parasitenkunde* **II**, 3-16.

Amelunxen, R.E. and Lins, M. (1968). Comparative thermostability of
enzymes from *Bacillus stearothermophilus* and *Bacillus cereus*.
Archives of Biochemistry and Biophysics **125**, 765-769.

Amelunxen, R.E. and Murdock, A.L. (1978). Life at high temperatures:
molecular aspects. In "Microbial Life in Extreme Environments"
(ed. D.J. Kushner), pp.217-278. London and New York: Academic
Press.

Amelunxen, R.E. and Singleton, R. (1976). Thermophilic glyceralde-
hyde-3-P dehydrogenases. In "Enzymes and Proteins from Thermophilic Microorganisms" (ed. H. Zuber), pp.107-120. Basel and
Stuttgart: Birkhauser Verlag.

Andersen, A.A. and Werkman, C.H. (1940). Description of a dextro-lactic acid forming organism of the genus *Bacillus*. *Iowa State College Journal of Science* 14, 187-194.

Arca, M., Frontali, L. and Tecce, G. (1965). Lack of specificity in the formation amino-acyl-s RNA as a possible source of coding errors. *Biochemica et Biophysica Acta* 108, 326-328.

Atkinson, A. (1976). Thermostable enzymes. *Journal of Applied Chemistry and Biotechnology* 26, 577-578.

Baillie, A. and Walker, P.D. (1968). Enzymes of thermophilic aerobic spore forming bacteria. *Journal of Applied Bacteriology* 31, 114-119.

Belehradek, J. (1931). Le mechanisme physico-chemique de l'adaptation thermique. *Protoplasma* 12, 406-434.

Belly, R.T. and Brock, T.D. (1974). Widespread occurrence of acido-philic strains of *Bacillus coagulans* in hot springs. *Journal of Applied Bacteriology* 37, 175-177.

Bergey, D.H., Harrison, F.C., Breed, R.S., Hammer, B.W. and Huntoon, F.M. (1925). "Bergey's Manual of Determinative Bacteriology", 1st edition. Baltimore: The Williams and Wilkins Co.

Berry, R.N. (1933). Some new heat resistant acid tolerant organisms causing spoilage in tomato juice. *Journal of Bacteriology* 25, 72-73.

Bonde, G.J. (1975). The genus *Bacillus*. *Danish Medical Bulletin* 22, 41-61.

Breed, R.S., Murray, E.G. and Parker-Hitchens, A. (1948). "Bergey's Manual of Determinative Bacteriology", 6th edition. London: Bailliere, Tindall and Cox.

Brock, T.D. (1978). Thermophilic Microorganisms and Life at High Temperatures. New York: Springer Verlag.

Brock, T.D. and Freeze, H. (1969). *Thermus aquaticus* gen.n. and sp.n. a non-sporulating extreme thermophile. *Journal of Bacteriology* 98, 289-297.

Cameron, E.J. and Esty, J.R. (1926). The examination of spoiled canned foods. 2. Classification of flat sour organisms from non-acid foods. *Journal of Infectious Diseases* 39, 89-105.

Cho, K.Y. and Salton, M.J.R. (1964). Fatty acid composition of the lipids of membranes of gram-positive bacteria and walls of gram-negative bacteria. *Biochemica et Biophysica Acta* 84, 773-775.

Cohn, F. (1876). Untersuchungen über Bakterien. IV. Beiträge zur Biologie der Bacillen. *Beiträge zur Biologie der Pflanzen* 2, 249-276.

Darland, G. and Brock, T.D. (1971). *Bacillus acidocaldarius* sp.nov., an acidophilic thermophilic spore-forming bacterium. *Journal of General Microbiology* 67, 9-15.

Daron, H.H. (1967). Occurrence of isocitrate lyase in a thermophilic *Bacillus* species. *Journal of Bacteriology* 101, 145-151.

Daron, H.H. (1970). Fatty acid composition of lipid extracts of a thermophilic *Bacillus* species. *Journal of Bacteriology* 101, 145-151.

De Rosa, M., Gambacorta, A. and Bu'lock, J.D. (1971-1973). An isolate of *Bacillus acidocaldarius* an acidophilic thermophile with unusual lipids. *Giornale Microbiologia* 19-21, 145-174.

Deranleau, D.A. and Zuber, H. (1977). Thermophilic aminopeptidase.

IV. Co-operative effects in ANS binding by the thermophilic amino-peptidase I from *B.stearothermophilus*. *International Journal of Peptide and Protein Research* **9**, 258-268.

Devanathan, T., Akagi, J.M. and Hersh, R.T. (1969). Ferredoxin from two thermophilic clostridia. *Journal of Biological Chemistry* **244**, 2845-2853.

Donk, R.J. (1920). A highly resistant thermophilic organism. *Journal of Bacteriology* **5**, 373-374.

Epstein, I. and Grossowicz, N. (1969). Prototrophic thermophilic *Bacillus*: isolation, properties and kinetics of growth. *Journal of Bacteriology* **99**, 414-417.

Esser, A.F. (1978). The influence of growth temperature and lipid state on the planer distribution of lipids and proteins in *Bacillus stearothermophilus* membranes. In "Biochemistry of Thermophily" (ed. S.M. Friedman), pp.45-60. New York, San Francisco, London: Academic Press.

Fields, M.L. (1970). The flat sour bacteria. *Advances in Food Research* **18**, 163-217.

Forrester, I.T. and Wicken, A.J. (1966). The chemical composition of the cell walls of some thermophilic bacilli. *Journal of General Microbiology* **42**, 147-154.

Frank, G., Haberstich, H.U., Schaer, H.P., Tratschin, J.D. and Zuber, H. (1975). Thermophilic and mesophilic enzymes from *B.caldotenax* and *B.stearothermophilus*, properties, relationships and formation. In "Enzymes and Proteins from Thermophilic Microorganisms" (ed. H. Zuber), pp.375-390. Basel and Stuttgart: Birkhauser Verlag.

Friedman, S.M. (1978). Studies on heat-stable ribosomes from thermophilic bacteria. In "Biochemistry of Thermophily" (ed. S.M. Friedman), pp.151-168. New York, San Francisco, London: Academic Press.

Friedman, S.M. and Mojica-a, T. (1978). Transformants of *Bacillus subtilis* capable of growth at elevated temperatures. In "Biochemistry of Thermophily" (ed. S.M. Friedman), pp.117-126. New York, San Francisco, London: Academic Press.

Friedman, S.M. and Weinsten, I.B. (1964). Lack of fidelity in the translation of synthetic polyribonucleotides. *Proceedings of the National Academy of Science USA* **52**, 988-996.

Galesloot, Th.E. and Labots, H. (1959). Thermophilic bacilli in milk. *Netherlands Milk and Dairy Journal* **13**, 155-179.

Gibson, T. and Gordon, R.E. (1974). *Bacillus*. In "Bergey's Manual of Determinative Bacteriology" (eds. R.E. Buchanan and N.E. Gibbons), pp.529-555. Baltimore: The Williams and Wilkins Co.

Gibson, T. and Topping, L.E. (1938). Further studies of the aerobic spore forming bacilli. *Society of Agricultural Bacteriology. Abstracts of Proceedings*, 43-44.

Golikowa, S.M. (1926). Zur Frage der Thermobiose. *Zentralblatt für Bakteriologie und Parasitenkunde II* **69**, 178-184.

Golovacheva, R.S., Egorova, L.A. and Loginova, L.G. (1965). Ecology and systematics of aerobic obligate thermophilic bacteria isolated from thermal localities on Mount Yangan-Tau and Kunashir isle of the Kuril chain. *Microbiology* (USSR) English edition **34**, 693-698.

Golovacheva, R.S., Loginova, L.G., Salikhov, T.A., Kolesmikov, A.A. and Zaitseva, G.N. (1975). A new thermophilic species *B.thermo-*

catenulatus nov. spec. *Microbiology* (USSR) English edition **44**, 230-233.

Gordon, R.E. (1977). Some taxonomic observations on the genus *Bacillus*. In "Biological Regulation of Vectors" (ed. J.D. Briggs), pp.67-82. Washington, D.C.: U.S. Department of Health, Education and Welfare.

Gordon, R.E., Haynes, W.C. and Pang, C.H-N. (1973). "The Genus *Bacillus*". Washington, D.C.: United States Department of Agriculture.

Gordon, R.E. and Smith, N.R. (1949). Aerobic spore forming bacteria capable of growth at high temperatures. *Journal of Bacteriology* **58**, 327-341.

Gorini, L. (1971). Ribosomal discrimination of tRNAs. *Nature, New Biology* **234**, 261-264.

Grinsted, E. and Clegg, L.F.L. (1955). Spore-forming organisms in commercial sterilised milk. *Journal of Dairy Research* **22**, 178-190.

Guicciardi, A., Biffi, M.R., Manachim, P.L., Craveri, A., Scolastico, C., Rindone, B. and Craveri, R. (1968). Novo termofilo del genere *Bacillus*. *Annali di Microbiologia ad Enzymologia* **18**, 191-205.

Gyllenberg, H. (1951). Studies on thermophilic bacteria of the genus *Bacillus* Cohn. *Acta Agraria Fennica, Helsinki* **73**, 1-88.

Hammer, B.W. (1915). Bacteriological studies on the coagulation of evaporated milk. *Iowa Agricultural Experiment Station Research Bulletin* **19**, 119-131.

Handley, P.S. (1977). Acid induced germination of *Bacillus acido-caldarius* spores. In "Spore Research, 1976", Vol. 2 (eds. A.N. Barker *et al.*), pp.735-751. London: Academic Press.

Hase, T., Ohmy, N., Matsubara, H., Mullingen, R.N., Rao, K.K. and Hall, D.O. (1976). Amino acid sequences of a 4-iron 4-sulphur ferredoxin from *B.stearothermophilus*. *Biochemical Journal* **159**, 55-63.

Heilbrun, L.V. (1924). The colloid chemistry of protoplasm. IV. The heat coagulation of protoplasm. *American Journal of Physiology* **69**, 190-199.

Heinen, W. (1971). Growth conditions and temperature-dependent substrate specificity of two extremely thermophilic bacteria. *Archiv für Mikrobiologie* **76**, 2-17.

Heinen, U.J. and Heinen, W. (1972). Characteristics and properties of a caldoactive bacterium producing extracellular enzymes and two related strains. *Archiv für Mikrobiologie* **82**, 1-23.

Heinen, W., Klein, H.P. and Vokman, C.M. (1970). Fatty acid composition of *Thermus aquaticus* at different growth temperatures. *Archiv für Mikrobiologie* **72**, 199-202.

Heinrich, M.R. (1976). "Extreme Environments: Mechanisms of Microbial Adaptation". New York: Academic Press.

Hocking, J.D. and Harris, J.I. (1976). Glyceraldehyde 3-phosphate dehydrogenase from an extreme thermophile, *Thermus aquaticus*. In "Enzymes and Proteins from Thermophilic Microorganisms" (ed. H. Zuber), pp.107-120. Basel: Birkhauser Verlag.

Hollaus, F. and Klaushofer, H. (1970). Taxonomische Untersuchungen an Hochthermophilen *Bacillus*-Stämmen aus Zuckerfabrikssaften. *Publication Faculty of Science, Purkyne University, Brno* **47**, 99-105.

Howell, N., Akagi, J.M. and Himes, R.H. (1969). Thermostability of glycolytic enzymes from thermophilic clostridia. *Canadian Journal of Microbiology* **15**, 461-464.

Hussong, R.V. and Hammer, B.W. (1928). A thermophile coagulating milk under practical conditions. *Journal of Bacteriology* **15**, 179-188.

Irwin, C.C., Akagi, J.M. and Himes, R.H. (1973). Ribosomes, polyribosomes and deoxyribonucleic acid from thermophilic, mesophilic and psychrophilic clostridia. *Journal of Bacteriology* **113**, 252-262.

Isono, K. (1970). Transformation of amylase producing ability in *Bacillus stearothermophilus*. *Japanese Journal of Genetics* **49**, 285-291.

Jacobsen, G. (1918). On factors influencing efficient pasteurisation. *Abstracts of Bacteriology* **2**, 215.

Jung, L., Jost, R., Stoll, E. and Zuber, H. (1974). Metabolic differences in *B.stearothermophilus* at 55° and 37°. *Archiv für Mikrobiologie* **95**, 125-138.

Klaushofer, H. and Hollaus, F. (1970). Zur Taxonomie der hochthermophilen, in Zuckerfabrikssaften vorkommenden, aeroben Sporenbildner. *Zeitschrift für die Zuckerindustrie* **20**, 465-470.

Klaushofer, H., Kunz, Ch., Bartelnus, W. and Schreyer, E. (1962). Schnell-Methode zur Keimziehung von *B.stearothermophilus* aus Zuckerfabrikssaften. *Zeitschrift für die Zuckerindustrie* **87**, 299-303.

Koffler, H. (1957). Protoplasmic differences between mesophiles and thermophiles. *Bacteriological Reviews* **21**, 227-240.

Koffler, H. and Gale, G.O. (1957). The relative thermostability of cytoplasmic proteins from thermophilic bacteria. *Archives of Biochemistry and Biophysics* **67**, 249-251.

Kushner, D.J. (1978). "Microbial Life in Extreme Environments". New York and London: Academic Press.

Laxa, O. (1900). Bakteriologische Studien über die Produkte des normalen Zuckerfabriksbetriebes. *Zentralblatt für Bakteriologie und Parasitenkunde II* **6**, 286-295.

Lee, A.G. (1977). Lipid phase transitions and phase diagrams. *Biochemica et Biophysica Acta* **472**, 237-281.

Lindsay, J.A. and Creaser, F.H. (1975). Enzyme thermostability is a transformable property between *Bacillus* spp. *Nature* **225**, 650-652.

Ljungdahl, L.G. (1979). Physiology of thermophilic bacteria. *Advances in Microbial Physiology* **19**, 149-243.

Ljungdahl, L. and Sherod, D. (1976). Proteins from thermophilic microorganisms. In "Extreme Environments: Mechanisms of Microbial Adaption" (ed. M.R. Heinrich), pp.147-188. New York and London: Academic Press.

Ljungdahl, L., Sherod, D., Moore, M.R. and Andreesen, J.R. (1976). Properties of enzymes from *Clostridium thermoaceticum* and *Clostridium formicoaceticum*. In "Enzymes and Proteins from Thermophilic Microorganisms" (ed. H. Zuber), pp.237-249. Basel and Stuttgart: Birkhauser Verlag.

Loginova, L.G. and Egorova, L.A. (1977). New Forms of Thermophilic Bacteria. *Nauka*, Moscow (in Russian).

Loginova, L.G., Khraptsova, G.I., Egorova, L.A. and Bogdanova, T.I. (1978). Acidophilic obligate thermophilic bacterium, *Bacillus acidocaldarius*, isolated from hot springs and soil of Kunashir

Island. *Microbiology* (USSR) English edition **47**, 771-775.

Loginova, L.G., Kosmachev, A.E., Golovacheva, R.S. and Seregina, L.M. (1962). A study of the thermophilic microflora of Mount Yangan-Tau in the Southern Urals. *Microbiology* (USSR) English edition **31**, 877-880.

Long, S.K. and Williams, O.B. (1959). Growth of obligate thermophiles at 37°C as a function of the cultural conditions employed. *Journal of Bacteriology* **77**, 545-547.

Long, S.K. and Williams, O.B. (1960). Factors affecting growth and spore formation of *B.stearothermophilus*. *Journal of Bacteriology* **79**, 625-628.

Lund, B.M. (1965). A comparison by the use of gel electrophoresis of soluble protein components and esterase enzymes of some Group D streptococci. *Journal of General Microbiology* **40**, 413-419.

MacFadyen, A. and Blaxall, F.R. (1894). Thermophilic bacteria. *Journal of Pathology and Bacteriology* **3**, 87-99.

Magsanaga, A. and Nosoh, Y. (1974). Conformational change with temperature and thermostability of glutamine synthetase from *Bacillus stearothermophilus*. *Biochemica et Biophysica Acta* **365**, 208-211.

Mandel, M. (1966). Deoxyribonucleic acid base composition in the genus *Pseudomonas*. *Journal of General Microbiology* **43**, 273-292.

Manning, G.B. and Campbell, L.L. (1961). Thermostable α-amylase of *Bacillus stearothermophilus*. 1. Crystallization and some general properties. *Journal of Biological Chemistry* **236**, 2952-2957.

Marmur, J. (1960). Thermal denaturation of deoxyribosenucleic acid isolated from a thermophile. *Biochemica et Biophysica Acta* **38**, 342-343.

McDonald, W.C. and Matney, S.T. (1963). Genetic transfer of the ability to grow at 55°C in *Bacillus subtilis*. *Journal of Bacteriology* **85**, 218-220.

McElhany, R.N. (1976). The biological significance of alterations in the fatty acid composition of microbial membrane lipids in response to changes in environmental temperatures. In "Extreme Environments: Mechanisms of Microbial Adaptation" (ed. M.R. Heinrich), pp. New York, San Francisco, London: Academic Press.

Militzer, W., Sonderegger, T.B., Tattle, L.C. and Georgi, C.E. (1949). Thermal enzymes. *Archives of Biochemistry* **24**, 75-82.

Minnikin, D.E., Abdolrahimzadeh, H. and Wolf, J. (1977). II. Taxonomic significance of polar lipids in some thermophilic members of *Bacillus*. In "Spore Research, 1976", Vol. 2 (eds. A.N. Barker *et al.*), pp.879-893. London: Academic Press.

Miquel, P. (1888). Monographie d'un bacille vivant au-dela de 70°C. *Annales Micrographic* **1**, 3-10.

Mishustin, E.N. (1950). Quoted by Golovacheva *et al.* [1965].

Norris, J.R. (1964). The classification of *Bacillus thuringiensis*. *Journal of Applied Bacteriology* **27**, 439-447.

Norris, J.R. and Burgess, H.D. (1963). Esterases of crystalliferous bacteria pathogenic for insects; epizootiological applications. *Journal of Insect Pathology* **5**, 460-466.

Norris, J.R. and Wolf, J. (1961). A study of the aerobic spore-forming bacteria. *Journal of Applied Bacteriology* **24**, 42-56.

Novitsky, T.J., Chan, M., Himes, R.H. and Akagi, J.M. (1974). Effect of temperature on the growth and cell wall chemistry of a facultative thermophilic *Bacillus*. *Journal of Bacteriology* **117**, 858-865.

O'Brien, W.E., Brewer, J.M. and Ljungdahl, L.G. (1976). Chemical, physical and enzymatic comparisons of formyltetrahydrofolate synthetases from thermophilic and mesophilic clostridia. In "Enzymes and Proteins from Thermophilic Microorganisms" (ed. H. Zuber), pp.249-262. Basel and Stuttgart: Birkhauser Verlag.

Olsen, E. (1944). Quoted by Gordon, Haynes and Pang [1973].

Oprescu, V. (1898). Studien über thermophile Bakterien. *Archiv für Hygiene und Bakteriologie* **33**, 164-186.

Oshima, T., Arakawa, H. and Baba, M. (1977). Biochemical studies on the thermophile *Bacillus acidocaldarius*. *Journal of Biochemistry* (Japan) **81**, 1107-1113.

Pace, B. and Campbell, L.L. (1967). Correlation of maximal growth temperature and ribosome heat stability. *Proceedings of the National Academy of Sciences, USA* **57**, 1110-1116.

Perutz, M.F. and Raidt, H. (1975). Deletion of α-globin genes in haemoglobin-H disease demonstrates multiple α-globin structural loci. *Nature, London* **225**, 256-259.

Pfueller, S.L. and Elliot, W.K. (1969). The extracellular α-amylase of *Bacillus stearothermophilus*. *Journal of Biological Chemistry* **244**, 48-54.

Prickett, P.S. (1928). Thermophilic and thermoduric microorganisms with special reference to species isolated from milk. Description of spore forming types. *New York State Agricultural Experimental Station Technical Bulletin* **147**.

Pringsheim, H. (1913). Über die Vergarung der Zellulose durch thermophile Bakterien. *Zentralblatt für Bakteriologie und Parasitenkunde II* **38**, 513-516.

Proom, H. and Knight, B.C.J.G. (1950). *Bacillus pantothenticus* (n. sp.). *Journal of General Microbiology* **4**, 539-541.

Ray, H.R., White, D.C. and Brock, T.D. (1971). Effect of temperature on the fatty acid composition of *Thermus aquaticus*. *Journal of Bacteriology* **106**, 25-30.

Renco, P. (1942). Ricerche su un fermento lattico sporigeno *(B. thermoacidicans)*. *Annale Microbiologia* **2**, 109-114.

Rice, A.C. and Pederson, C.S. (1954). Factors influencing growth of *B.coagulans* in canned tomato juice. II. *Food Research* **19**, 124-133.

Robinson, K. (1966). An examination of *Corynebacterium* sp. by gel electrophoresis. *Journal of Applied Bacteriology* **29**, 179-185.

Robson, E. and Pain, R.H. (1971). Analysis of the code relating sequence to conformation in proteins: Possible implications for the mechanism of formation of helical regions. *Journal of Molecular Biology* **58**, 237-259.

Sarles, W.B. and Hammer, B.W. (1932). Observations on *Bacillus coagulans*. *Journal of Bacteriology* **23**, 301-314.

Saunders, G.F. and Campbell, L.L. (1966). Ribonucleic acid and ribosomes of *Bacillus stearothermophilus*. *Journal of Bacteriology* **91**, 332-339.

Schenk, A. and Aragno, M. (1979). *Bacillus schlegelii*, a new species of thermophilic facultatively chemolithoautotrophic bacterium oxidising molecular oxygen. *Journal of General Microbiology* **115**, 333-341.

Schlanger, G. and Friedman, S.M. (1973). Ambiguity in a polypeptide synthesizing extract from *Saccharomyces cerevisiae*. *Journal of Bacteriology* **115**, 129-138.

Seki, T., Chung, C.-K., Mikami, H. and Oshima, V. (1978). Deoxyribonucleic acid homology and taxonomy of the genus *Bacillus*. *International Journal of Systematic Bacteriology* **28**, 182-189.

Sharp, R.J., Bingham, A.H.A., Comer, M.J. and Atkinson, A. (1979). Partial characterisation of a bacteriocin (thermocin) from *Bacillus stearothermophilus* RS93. *Journal of General Microbiology* **111**, 449-451.

Sharp, R.J., Bown, K.J. and Atkinson, A. (1980). Phenotypic and genotypic characterisation of some thermophilic species of *Bacillus*. *Journal of General Microbiology* **117**, 201-210.

Shen, P.Y., Coles, E., Foote, J.L. and Stenesh, J. (1970). Fatty acid distribution in mesophilic and thermophilic strains of the genus *Bacillus*. *Journal of Bacteriology* **103**, 479-481.

Shilo, M. (1979). "Thermophilic Microorganisms at High Temperature". Dahlem Konferenzen, Berlin. New York: Springer Verlag.

Sidler, W. and Zuber, H. (1977). The production of extracellular thermostable neutral proteinase and α-amylase by *Bacillus stearothermophilus*. *European Journal of Applied Microbiology* **4**, 255-266.

Singleton, R. (1976). A comparison of the amino acid compositions from thermophilic and non-thermophilic origins. In "Extreme Environments: Mechanisms of Microbial Adaption" (ed. M.R. Heinrich), pp.189-200. New York and London: Academic Press.

Singleton, R. and Amelunxen, R.E. (1973). Proteins from thermophilic microorganisms. *Bacteriological Reviews* **37**, 320-342.

Smith, N.R. (1948). Genus I *Bacillus* Cohn. In "Bergey's Manual of Determinative Bacteriology, Sixth Edition" (eds. R.S. Breed, E.G.D. Murray and A.P. Hitchens), pp.705-762. Baltimore: The Williams and Wilkins Co.

Smith, N.R. and Gordon, R.E. (1955a). Questionable adaptions of cultures to higher temperatures. *Journal of Bacteriology* **69**, 603-604.

Smith, N.R. and Gordon, R.E. (1955b). Questionable adaptions of cultures to higher temperatures. *Journal of Bacteriology* **70**, 488-489.

Smith, N.R., Gordon, R.E. and Clark, F.E. (1946). "Aerobic mesophilic spore-forming bacteria". *Miscellaneous Publication* **559**. Washington, D.C.: United States Department of Agriculture.

Smith, N.R., Gordon, R.E. and Clark, F.E. (1952). Aerobic spore-forming bacteria. *Agricultural Monograph* **16**. Washington, D.C.: United States Department of Agriculture.

Smith, D.G., Stein, W.H. and Moore, S. (1963). The sequence of amino acid residues in bovine pancreatic ribonuclease: Revisions and confirmations. *Journal of Biological Chemistry* **238**, 227-234.

Sokal, R. and Sneath, P.H.A. (1963). "Principles of Numerical Taxonomy". London: W.H. Freeman and Co.

Souza, K.A., Kostiw, L.L. and Tyson, B.J. (1974). Alterations in normal fatty acid composition in a temperature sensitive mutant of a thermophilic *Bacillus*. *Archiv für Mikrobiologie* **97**, 89-102.

Spackman, D.H., Stein, W.H. and Moore, S. (1960). The disulphide bonds of ribonuclease. *Journal of Biological Chemistry* **235**, 648-659.

Spizizen, J. (1969). Transformation of biochemically deficient strains of *Bacillus subtilis* by deoxyribonucleic acid. *Proceedings of the National Academy of Sciences, USA* **44**, 1072-1078.

Stahl, S. and Olsson, O. (1977). Temperature range variants of *Bacillus megaterium*. *Archiv für Mikrobiologie* **113**, 221-229.

Stark, E. and Tetrault, P. (1952). A determinative study of amylolytic, stearothermophilic bacteria isolated from soil. *Scientific Agriculture (Canada)* **32**, 81-92.

Stenesh, J. (1976). Information transfer in thermophilic bacteria. In "Extreme Environments: Mechanisms of Microbial Adaption" (ed. M.R. Heinrich), pp.85-101. New York, San Francisco, London: Academic Press.

Stenesh, J., Roe, B.A. and Snyder, T.L. (1968). Studies of deoxyribonucleic acid from mesophilic and thermophilic bacteria. *Biochemica et Biophysica Acta* **161**, 442-454.

Stenesh, J. and Roe, B.A. (1972). DNA polymerase from mesophilic and thermophilic bacteria. II. Temperature dependence of nearest neighbour frequencies of the product from the DNA polymerase reaction. *Biochemica et Biophysica Acta* **272**, 167-178.

Stern, R.M., Hegarty, C.P. and Williams, O.B. (1942). Detection of *B.thermoacidarans* (Berry) in tomato juice. *Food Research* **7**, 186-191.

Tajima, M., Urabe, I., Yutani, K. and Okada, H. (1976). Role of calcium ions in the thermostability of thermolysin and *Bacillus subtilis* var. *amylosacchariticus* neutral protease. *European Journal of Biochemistry* **64**, 243.

Tanaka, M., Haniu, M., Matsueda, G., Yasunobu, K.T., Himes, R.H., Akagi, J.M., Barnes, E.M. and Devanathan, T. (1971). The primary structure of the *Clostridium tartarivorum* ferredoxin, a heat stable ferredoxin. *Journal of Biological Chemistry* **246**, 3958-3960.

Taylor, R.B. (1953). A study of proteose peptone acid agar as a plating medium for the routine enumeration of *B.thermoacidarans* (Berry) in tomato juice. *Food Research* **18**, 516-521.

Thompson, G.A. and Nozawa, Y. (1977). *Tetrahymena*: A system for studying dynamic membrane alterations within the eukaryotic cell. *Biochemica et Biophysica Acta* **472**, 55-92.

Uchino, F. and Doi, S. (1967). Acido-thermophilic bacteria from thermal waters. *Agricultural and Biological Chemistry* **31**, 817-822.

Walker, P.D. (1959). A study of biochemical, physiological and serological properties of some strains of *B.stearothermophilus*. University of Leeds, Thesis.

Walker, P.D. and Wolf, J. (1961). Some properties of aerobic thermophiles growing at 65°C. *Journal of Applied Bacteriology* **24**, iv.

Walker, P.D. and Wolf, J. (1968). See Wolf, J. and Barker, A.N. [1968].

Walker, P.D. and Wolf, J. (1971). The taxonomy of *Bacillus stearothermophilus*. In "Spore Research, 1971" (eds. A.N. Barker, G.W. Gould and J. Wolf), pp.247-262. London: Academic Press.

Wedler, F.C. (1978). Properties and regulation of thermophilic glutamine synthetases. In "Biochemistry of Thermophily" (ed. S.M. Friedman), pp.325-345. New York, San Francisco, London: Academic Press.

Wedler, F.C. and Hoffman, F.M. (1974). Glutamine synthetase of

Bacillus stearothermophilus. II. Regulation and thermostability. *Biochemistry* **13**, 3215-3221.

Weerkamp, A. and Heinen, W. (1972). The effect of temperature on the fatty acid composition of the extreme thermophiles, *B.caldolyticus* and *B.caldotenax*. *Journal of Bacteriology* **109**, 443-446.

Weerkamp, A. and MacElroy, R.D. (1972). Lactate dehydrogenase from an extremely thermophilic *Bacillus*. *Archiv für Mikrobiologie* **85**, 113-122.

Welker, N.E. (1976). Microbial endurance and resistance to heat stress. In "The Survival of Vegetative Microbes" (eds. T.R.G. Gray and J.R. Postgate), pp.241-277. Cambridge, London, New York, Melbourne: Cambridge University Press.

Welker, N.E. (1978). Physiological and genetic factors affecting transfection and transformation in *Bacillus stearothermophilus*. In "Biochemistry of Thermophily" (ed. S.M. Friedman), pp.127-150. New York, San Francisco, London: Academic Press.

Welker, N.E. and Campbell, L.L. (1965). Induction and properties of a temperate bacteriophage from *Bacillus stearothermophilus*. *Journal of Bacteriology* **89**, 175-184.

Williams, R.A.D. (1975). Caldoactive and thermophilic bacteria and their thermostable proteins. *Science Progress, Oxford* **62**, 373-393.

Wisdon, C. and Welker, N.D. (1973). Membranes of *Bacillus stearothermophilus*: Factors affecting protoplast stability and thermostability of alkaline phosphatase and reduced nicotinamide adenine dinucleotide oxidase. *Journal of Bacteriology* **114**, 1336-1345.

Wolf, J. (1968). Quoted in Wolf and Barker [1968].

Wolf, J. and Barker, A.N. (1968). The genus *Bacillus*: Aids to the identification of its species. In "Identification Methods for Microbiologists. Part B" (eds. M. Gibbs and D.A. Shapton), pp.93-109. London: Academic Press.

Yutani, K. (1976). Role of calcium ions in the thermostability of α-amylase produced by *Bacillus stearothermophilus*. In "Enzymes and Proteins from Thermophilic Microorganisms" (ed. H. Zuber), pp.91-103. Basel and Stuttgart: Birkhauser Verlag.

Zeikus, J.G. (1979). Thermophilic bacteria: ecology, physiology and technology. *Enzyme Microbiology and Technology* **1**, 243-252.

Zuber, H. (1976). "Symposium on Enzymes and Proteins from Thermophilic Microorganisms". Basel: Birkhauser Verlag.

Zuber, H. (1978). Comparative studies of thermophilic and mesophilic enzymes: Objectives, problems, results. In "Biochemistry of Thermophily" (ed. S.M. Friedman), pp.267-287. New York, San Francisco, London: Academic Press.

Chapter 12

BACILLUS CEREUS AND OTHER *BACILLUS* SPECIES:
THEIR PART IN
FOOD POISONING AND OTHER CLINICAL INFECTIONS

R.J. GILBERT, P.C.B. TURNBULL, JENNIFER M. PARRY
and J.M. KRAMER

*Food Hygiene Laboratory, Central Public
Health Laboratory, Colindale Avenue, London, UK*

Introduction

Between 1906 and 1949 several accounts of food poisoning
associated with aerobic sporing rods were reported in the
European literature [Gilbert, 1979]. These reports shared
several features: they rarely presented a complete description of the implicated organism, but usually classified it as an 'anthracoid' bacillus or as a member of the
subtilis-mesentericus group, and they did not give the
number of these and other organisms in the incriminated
foods. Similarly, there were numerous reports during the
same period of a wide variety of other types of clinical
infection associated with *Bacillus* spp. other than *B.
anthracis*.

Since 1952, when the taxonomy of the genus was greatly
clarified by Smith *et al.* [1952], reports have continued
of infections in which *Bacillus* spp. featured significantly, particularly *B.cereus*, and also other species from
time to time (see pp.13,131,304, *et seq.*). There is now
little doubt that, on occasion, members of this genus can
be direct or opportunistic pathogens of man and animals.

In the same period, and particularly in the last 10
years, there have been a large number of well-documented
reports which have established *B.cereus* as a food poisoning
organism and the predominant interest in this organism over
the past decade has been from the standpoint of its role in
gastroenteritis. There have, in addition, been a relatively small number of reports in which the evidence has always
been less than complete but which, when taken together,
suggest that on occasion *B.subtilis* and *B.licheniformis* may
also be food poisoning agents.

As a result of the increased interest in *B.cereus* as a
food poisoning agent, this laboratory has become a centre
to which increasing numbers of *B.cereus* and other *Bacillus*
spp. isolated from food poisoning and non-gastrointestinal

situations are being sent.

This chapter reviews briefly the state of knowledge on those *Bacillus* spp. other than *B.anthracis* known to be or suspected of being pathogenic for man, and the features of the organisms taken advantage of during investigations of problems arising from foodborne or other infections. Since, in this context, *B.cereus* food poisoning is the situation most frequently encountered in the Food Hygiene Laboratory, most of the principles discussed are illustrated with reference to this organism.

Antigenic Constitution and Serology

Vegetative cell somatic, and flagellar and spore antigens are present in all strains of *B.cereus* and reviews on the serology of this and other *Bacillus* spp. have been published [Norris, 1962; Baba, 1969; Goepfert *et al.*, 1972]. Norris and Wolf [1961] studied the antigens of a number of *Bacillus* spp. and demonstrated that, whilst the spore antigen possessed the highest species specificity, the flagellar (H) antigen provided the highest degree of strain specificity. Using the H antigen of *B.cereus*, LeMille *et al.* [1969] reported 17 serotypes among 33 cultures, mainly isolates from insects.

In the Food Hygiene Laboratory, a serotyping scheme based on the flagellar antigen of *B.cereus* was developed by Taylor and Gilbert [1975] and Gilbert and Parry [1977] for the investigation of food poisoning outbreaks. Twenty-three strains of *B.cereus* isolated from foods and clinical specimens from outbreaks of food poisoning in Great Britain, the Netherlands, Canada and the USA, and one from a neonatal abscess were used to prepare agglutinating sera against the H antigen. Each of the sera agglutinates its homologous cell suspension to a titre within the range 1,280 - 20,480 and there are no significant cross-reactions between the sera.

The typing scheme has proved useful in the epidemiological investigation of food poisoning outbreaks [Gilbert and Parry, 1977; Gilbert, 1979] and also for non-food related clinical isolates [Fitzpatrick *et al.*, 1979]. For example, it has become apparent that *B.cereus* is the aetiological agent of 2 distinct types of food poisoning (Table 1). The first is similar in nature to *Staphylococcus aureus* food poisoning with symptoms predominantly of vomiting after an incubation period of 1 to 5 h and has almost always been associated with Chinese cooked rice dishes. The other is similar in nature to *Clostridium perfringens* food poisoning and is characterized mainly by diarrhoea after an incubation period of 8 to 16 h; a wide variety of foods have been implicated in reports of the diarrhoeal-type of *B. cereus* food poisoning.

Table 2 shows the distribution of serotypes of *B.cereus* from 107 incidents of food poisoning in 6 countries, asso-

TABLE 1

Food poisoning caused by Bacillus cereus, Clostridium perfringens
and Staphylococcus aureus: some clinical and epidemiological data*

	C.perfringens	B.cereus diarrhoeal-type+	B.cereus vomiting-type++	S.aureus
Incubation period (h)	8 - 22	8 - 16	1 - 5	2 - 6
Duration of illness (h)	12 - 24	12 - 24	6 - 24	6 - 24
Diarrhoea	Extremely common	Extremely common	Fairly common	Common
Vomiting	Rare	Occasional	Extremely common	Extremely common
Food most frequently implicated	Cooked meat and poultry	Meat products, soups, vegetables, puddings and sauces	Fried or boiled rice	Cooked meat and poultry and dairy products

*From Gilbert [1979].
+Outbreaks reported since 1950 in Norway, Denmark, Italy, the Netherlands, Hungary, Sweden, Poland, Rumania, the USSR, the USA, Germany, Canada and Great Britain.
++Outbreaks reported since 1971 in Great Britain, Canada, Australia, the Netherlands, Finland, the USA, Japan and India.

TABLE 2

Distribution of serotypes of Bacillus cereus *from
107 incidents of food poisoning (vomiting-type)*

Country		Source of strains	Serotype	Number of incidents
Great Britain	(90)		1	65
Netherlands	(6)		1 + others	10
		Food*, faeces or	3	3
Japan	(5)	vomitus,	4	2
India	(3)	or both food and clinical	5	4
		specimens	8	6
Australia	(2)		8 + others	4
Finland	(1)		12 or 12 and 19	3
			Not typable	10

*Fried or boiled rice (101), spaghetti (2), pasteurized cream (1),
vanilla slice (1), cooked vegetables (1) and omelette (1).

ciated with the vomiting-type syndrome. In 65 of the 107
incidents, foods, clinical specimens or both yielded serotype 1 only. Type 1 strains, together with other serotypes, were isolated in a further 10 episodes. The value
of the serotyping system lies primarily in making it
possible to relate isolates from the incriminated food to
those from clinical specimens submitted by the patients.
The fact that certain serotypes are more frequently implicated than others in outbreaks of food poisoning, as has
been illustrated with serotype 1, suggested the need for
further studies to clarify what is unusual about these
types.

In one such study (Table 3) a survey was carried out to
determine the frequency of serotype 1 strains in routine
samples of rice and other foods not associated with food
poisoning [Gilbert and Parry, 1977]. Among the observations made was that 23% of isolates from boiled and fried
rice were serotype 1 as compared with just 3% of isolates
from uncooked rice. In contrast, 15% of the cultures from
uncooked rice were serotype 17 as compared with only 1%
for cooked rice.

The implication from these findings that type 1 strains
might have a greater resistance to heat was subsequently
borne out in a study on the heat resistance of spores of
B.cereus strains from various sources [Parry and Gilbert,
1980]. Decimal reduction times (D) at 95°C for
10 strains of serotype 1 isolated from food or clinical
specimens from outbreaks of food poisoning were in the
range 22.4 to 36.2 mins while D values for 9 strains of
serotypes 3, 5, 8, 12, 17 and 18 isolated from samples of
uncooked rice lay only in the range 1.5 to 6.0 mins.

TABLE 3

Distribution of serotypes among 400 isolates of Bacillus cereus
*from various samples of food not associated with outbreaks of
food poisoning* * *and 100 isolates from faecal specimens of
healthy adults in the general population* +

Serotype	Number of isolates from				
	Uncooked rice	Boiled and fried rice	Milk and cream	Cooked meat and poultry	Human faeces
1	3	23	11	15	7
2	0	0	0	1	0
3	2	0	2	5	2
4	0	1	0	0	0
5	1	6	0	0	1
6	0	0	0	0	0
7	0	0	0	0	0
8	1	6	5	3	3
9	0	0	0	1	0
10	0	1	0	0	1
11	2	1	3	6	3
12	6	1	1	3	1
13	2	0	0	0	0
14	2	3	3	0	0
15	6	1	2	1	2
16	1	1	0	1	0
17	15	1	0	1	4
18	4	2	0	0	8
19	0	6	1	6	1
20	7	5	1	5	4
21	1	0	6	0	3
22	1	0	8	1	7
23	0	2	3	0	1
Not typable	46	40	54	51	52
Total	100	100	100	100	100

*From Gilbert and Parry [1977] and unpublished.
+From Ghosh [1978].

These results would account for the preponderance of type
1 in *B.cereus* food poisoning outbreaks in this and other
countries.

Other serotypes involved in vomiting-type incidents are
types 3, 4, 5, 8, 12 and 19. Only a few strains have been
received from diarrhoeal-type outbreaks; 2 incidents
yielded serotype 1 and 1 incident was associated with each
of types 2, 6, 8, 9, 10 and a mixture of 12 and an un-
typable strain. Thus types 1, 8 and 12 have been implicated in both vomiting and diarrhoeal outbreaks implying
that specific serotypes are not exclusively associated
with the 2 syndromes.

As far as isolates from non-gastrointestinal clinical infections are concerned, it was observed [Turnbull *et al.*, 1979a] that, of those that were typable (a little more than half the isolates) only 2, types 8 and 17, were at all commonly encountered in the Food Hygiene Laboratory.

In a survey on the prevalence of *B.cereus* in the faeces of healthy persons the organism was isolated from 100 of 711 (14%) of specimens: 15 of the 23 recognized serotypes were represented but 52% of the isolates were not typable [Ghosh, 1978; see Table 3]. The Taylor and Gilbert [1975] typing scheme has also been used with success by Terayama *et al.* [1978] in Japan in studies on the incidence of *B. cereus* in various foods: these workers have extended the scheme with additional serotypes.

It is of interest that of 21 *B.licheniformis* isolates received from various food poisoning and other clinical investigations, none were agglutinated with the *B.cereus* antisera. *B.subtilis* strains submitted to this laboratory from similar types of episodes have not been tested against the Food Hygiene Laboratory set of *B.cereus* antisera, but LeMille *et al.* [1969] found evidence of occasional cross-reactivity between *B.cereus* H antisera and *B.megaterium* antigens and between *B.subtilis* H antisera and *B.cereus* antigens. Taylor and Gilbert [1975] reported that the agglutinating antibodies in serum to *B.cereus* type 18 were completely absorbed by suspensions of *B.thuringiensis* serotype 4a,b and it has since been observed that a strain of *B.thuringiensis* type 5a,b agglutinated with antisera to *B.cereus* type 20. It should be noted, however, that strains interpreted in this laboratory as being *B.cereus* are not routinely examined for the parasporal bodies of *B.thuringiensis*.

Little success has been achieved in the serological identification of strains of *B.cereus* on the basis of somatic antigens. Studies with spore antigens have given better results [Norris and Wolf, 1961; Kim and Goepfert, 1971, 1972] but no definitive typing scheme has been developed.

Isolation and Identification Media and Tests

Identification of *Bacillus* spp. is carried out in this laboratory using the methods and interpretation of Cowan and Steel [1974]. A screening test is used for isolates with the typical colonial morphology of *B.cereus* on 5% horse-blood agar; if these are motile, lecithovitellin-positive and ferment glucose but not arabinose, xylose and mannitol in ammonium salt sugar bases, the identification of *B.cereus* is regarded as confirmed. Suspect *B.cereus* colonies giving variable or doubtful reactions and colonies of other *Bacillus* spp. are screened with tests 1 to 15 listed in Table 4. The table shows the results of these tests as applied to a large number of strains of *B.cereus* isolated from a wide variety of sources.

The existence of non-motile strains of *B.cereus* has been reported [Gordon *et al.*, 1973; Gilbert and Taylor, 1976] but such strains have not been encountered in this laboratory for more than 4 years; they would be regarded as possible *B.anthracis* and submitted to the Vaccine Research and Production Laboratory, PHLS Centre for Applied Microbiology and Research, Porton Down, Wiltshire, for specific identification.

TABLE 4

Biochemical characteristics of Bacillus cereus

Test	Number of strains tested	Result*	Percentage positive
1. Motility		+	97.9
2. Anaerobic growth		+	100
3. Citrate utilization (Christensen)		+	93.1
4. Gelatin hydrolysis		+	100
5. Casein hydrolysis	461	+	100
6. Potato starch hydrolysis		+	99.6
7. Indole		−	0
8. Voges-Proskauer		+	98.6
9. Nitrate reduction		+	93.7
10. Urease (Christensen)		−	11.9
11. Lecithovitellin (LV)		+	100
Ammonium salt sugar base			
12. Acid from: glucose		+	100
13. mannitol	2,275	−	0
14. xylose		−	0
15. arabinose		−	0
16. maltose		+	100
17. trehalose		+	100
18. glycerol		+	98.3
19. sucrose	113	+	89.8
20. lactose		−	8.5
21. dulcitol		−	0
22. inositol		−	0
23. sorbitol		−	0

*+, Positive in 80-100% of strains tested.
 −, Negative in 80-100% of strains tested.

Gilbert and Taylor [1976] reported that strains of *B. cereus* from the vomiting-type of food poisoning outbreaks did not ferment salicin but subsequent tests on a greater number of isolates have shown that some do slowly ferment this glucoside.

Details of the various distinguishing tests for *B.cereus*
and other *Bacillus* spp. can be found in various manuals and
texts such as Gordon *et al.* [1973], Murrell [1974], Cowan
and Steel [1974] and Gilbert [1979].

Bacillus Species as Pathogens

Reference has already been made to the fact that, from
time to time, *Bacillus* spp. other than *B.anthracis* feature
with apparent significance in food poisoning and other clinical infections. There is a wealth of literature now on
B.cereus as a food poisoning agent [Goepfert *et al.*, 1972;
Gilbert and Taylor, 1976; Terranova and Blake, 1978;
Gilbert, 1979] and as an organism capable of causing a
variety of severe infections [Goepfert *et al.*, 1972; Turnbull *et al.*, 1979a,b].

In addition to this, isolates of *B.subtilis*, *B.licheni-
formis* and, more rarely, other *Bacillus* spp. are received
periodically in this laboratory from investigations of
food poisoning and other clinical infections. *Bacillus*
spp. have long been thought of as capable of producing
"opportunistic" infections in already clinically compro-
mised hosts but, with the exception of *B.anthracis*, they
are usually regarded as non-pathogens or even just "contaminants" when isolated in the clinical laboratory. Despite
periodic reports indicating that *B.cereus* should be taken
more seriously, acceptance of even this organism's pathogenic potential has been slow to develop.

In general, the criteria which must be met before an
organism becomes accepted as capable of being pathogenic
are necessarily rigorous. It must be consistently isolated
from infections and in any one patient must be isolated on
several separate occasions. If other organisms are also
isolated from the infection(s) their importance and involvement must be ruled out. If the organism is present
only in small numbers it is unlikely to be of significance.
In clinical infections, the use of antimicrobial therapy
makes it difficult to meet these criteria and it is usually
only when the *Bacillus* is resistant to the initial treatment and fails to respond that it becomes noticed. This is
undoubtedly a major reason why *B.cereus*, normally a β-
lactamase producer and therefore resistant to penicillins,
has been found to be pathogenic [Turnbull *et al.*, 1979a].

In food poisoning, the ideal situation is to isolate the
organism in large numbers from both the implicated food,
which must have been held at refrigerated temperature
between the time of consumption and the time of examination, and from faecal or vomitus specimens submitted by the
patients. In practice, the actual food consumed is rarely
available for bacteriological examination or has not been
held refrigerated and patient specimens are usually submitted either not at all or long after the patient has recovered. Consequently the ideal conditions for implicating

an organism rarely occur.

As a result of the problems outlined the large majority
of investigations in which *Bacillus* spp. are isolated are
inconclusive and fail to reach the point of publication;
B.anthracis or *B.cereus* apart, those that do, have to date
always fallen short of the ideal criteria described above
in some respect or other.

Nevertheless, there are documented cases in which there
has been reasonable ground to implicate *B.subtilis* in food
poisoning [Tong *et al.*, 1962; Mortimer and Meers, 1975;
Winton and Sayers, 1975] and other clinical infections
such as the reports on septicaemia [Sathmary, 1958] and
respiratory infections [Ihde and Armstrong, 1973;
Pennington *et al.*, 1976] in cancer patients, postoperative
cellulitis [Behrend and Krouse, 1952], septicaemia in an
infant [Cox *et al.*, 1959] and respiratory afflictions in
otherwise healthy persons [Greenberg, 1970].

Similarly, there is an accumulating number of cultures
and reports of *B.licheniformis* submitted in association
with food poisoning [van Schothorst and Schipper, 1966;
Kampelmacher, 1969; Nielson and Pedersen, 1974; Jephcott
et al., 1977; Turnbull, 1978; Maddocks and Turnbull,
1978] and other infections such as bacteraemia and septicaemia [Amador *et al.*, 1976; Peloux *et al.*, 1976;
Fauchère *et al.*, 1977; Sugar *et al.*, 1977; Alix, 1978;
Turnbull, 1978], peritonitis [Sugar *et al.*, 1977;
Maddocks and Turnbull, 1978] and ophthalmitis [Anon, 1972].
Early evidence suggests that *B.licheniformis* food poisoning
is similar in nature to that of *C.perfringens* with fairly
mild diarrhoea or diarrhoea and vomiting after an incubation period of approximately 8 h. As with *B.cereus*, *B.
licheniformis* may also be of veterinary consequence on
occasion [Ryan, 1970; Maddocks and Turnbull, 1978].

Other *Bacillus* spp. reported to have caused or been
suspected of causing infections are *B.circulans* [meningi-
tis - Boyette and Rights, 1952], *B.macerans* [postoperative
wound infection - Ihde and Armstrong, 1973], *B.pumilus*
[rectal fistula - Melles *et al.*, 1969] and *B.sphaericus*
[meningitis, bacteraemia and endocarditis - Farrar, 1963;
meningitis and generalized Shwartzman reaction - Allen and
Wilkinson, 1969; and pulmonary pseudotumour - Isaacson *et
al.*, 1976]. Unidentified *Bacillus* spp. have been recorded
in association with conjunctivitis, skin, burn, wound,
bone and joint, urogenital, gall-bladder and other infections [Pearson, 1970]. Frequently, however, in these
types of infections, there are other underlying exacerba-
ting factors such as alcoholism, asthma, cancer or
diabetes.

In addition to these documented episodes, the Food
Hygiene Laboratory files for the last 3 years hold records
of 8 episodes of food poisoning or unexplained diarrhoea
in which *B.subtilis* was a suspect aetiological agent and
4 incidents in which *B.brevis* was isolated in large num-

bers either alone (2 incidents) or with *B.cereus* (2 incidents) from the implicated foods. In one incident it was also isolated with *B.cereus* from the patient's faeces and in a fifth incident it alone was isolated from the faeces. Incubation times in the incidents involving *B.brevis* ranged from 1 - 9½ h with symptoms of vomiting and diarrhoea.

In the *B.subtilis* incidents, symptoms were generally given as vomiting and diarrhoea although on 2 occasions vomiting only was specified and on 2 other occasions diarrhoea only; incubation times ranged from 1½ - 18 h. A similar variation in predominant symptoms and incubation periods is seen in the published reports referred to earlier. In a particularly interesting episode among the unreported records, a large number of elderly patients in a hospital suffered from intermittent and irregular diarrhoea; the only unusual faecal isolate common to faeces available for examination was *B.subtilis*. The source of the problem was not traced and the role of the *B.subtilis* was not conclusively established.

Considering each of the incidents involving *B.subtilis* and *B.brevis* alone, the epidemiological evidence for incriminating these organisms as "causative" is inadequate. Taken together, however, the pattern of their repeated occurrence in association with such incidents suggests a possible significant involvement. This applies also to the majority of the incidents associated with *B.licheniformis*. The Food Hygiene Laboratory is always on the alert for those organisms which are recurrently isolated in association with food poisoning.

Occasionally *Bacillus* spp. reach us from other interesting sources; samples of a medicated cream, for instance, being examined at St. Mary's Hospital Medical School, London, yielded *B.coagulans*, *B.firmus*, *B.licheniformis*, *B.megaterium*, *B.pantothenticus*, *B.polymyxa*, *B. pumilus* and *B.subtilis*, as well as *B.cereus*.

Pathogenicity and Toxins

The extracellular metabolites of various *Bacillus* spp. have been studied for many years for a wide variety of reasons. The developing interest in *B.cereus* as the agent of 2 types of food poisoning and other potentially serious infections has involved this and other laboratories in investigations into which of the many products that this organism is capable of elaborating [reviewed by Bonventre and Johnson, 1970 and Katsaras and Zeller, 1977] might be involved in its pathogenic activities in man. For many years it was thought that factors responsible for the organism's haemolytic, phospholipolytic and mouse lethal activities were the most likely contenders for this status. On the basis of work with whole cell cultures [Goepfert *et al.*, 1972] and cell free filtrates [Spira and Goepfert,

1972] in the ligated rabbit ileal loop model, the existence of a "diarrheagenic" enterotoxin, separable from the haemolytic and phospholipolytic factors but coincident with mouse lethality [Spira, 1974; Spira and Goepfert, 1975], was recognized. Subsequently it was found that this toxin could be more conveniently tested for with a skin test in rabbits [Glatz and Goepfert, 1973; Turnbull *et al.*, 1979b] and that it was a logarithmic growth metabolite produced to widely variable degrees by different *B.cereus* strains [Turnbull *et al.*, 1979b]. When produced strongly this toxin was capable of causing necrosis in skin or intestinal mucosa and it was proposed that it was the pyogenic and pyrogenic factor in non-gastrointestinal *B.cereus* infections as well as being the agent responsible for the diarrhoeal-type of *B.cereus* food poisoning [Turnbull *et al.*, 1979a,b].

Firm evidence that a second unrelated toxin was responsible for emesis in vomiting-type *B.cereus* food poisoning was first presented by Melling *et al.* [1976] on the basis of monkey feeding tests.

The 2 toxins have now been partially characterized [Melling and Capel, 1978; Kramer *et al.*, 1978; Turnbull *et al.*, 1979b - see also Table 5] and shown to be separable from 2 haemolysins and the egg yolk turbidity factor. Table 6 represents an attempt to bring some order to the

TABLE 5

Stability to various agents of the emetic toxin produced by B.cereus *strain 4810/72 and the diarrhoeal toxin produced by strains 4433/73 and 671/78*

Treatment		Activity remaining	
		Emetic toxin*	Diarrhoeal toxin**
4°C	14 days	+	+
4°C	2 months	+	NT
35°C	60 mins	NT	+
45°C	30 mins	NT	reduced
50°C	20 mins	NT	reduced
55°C	20 mins	NT	-
80°C	10 mins	+	-
115°C	10 mins	+	NT
126°C	90 mins	+	NT
pH 2	2 h	+	reduced
pH 4	2 h	+	reduced
pH 11	2 h	+	reduced
Trypsin		+	-
Pepsin		+	NT
Pronase		NT	-

*Data of Melling and Capel [1978]. **Data of Turnbull *et al.* [1979b]; J.M. Kramer and P.C.B. Turnbull (unpublished).
+, No diminution of activity; -, elimination of activity; NT, not tested.

TABLE 6

The toxins of Bacillus cereus − *current status*

	Toxin	References	Properties
(a)	Emetic	1,2,8	Low M.W. (<5,000) enterotoxin, not formed above 35°C. Stable to 126°C x 1.5 h, pH 2 x 2 h, pH 11 x 2 h, trypsin and pepsin digestion. May be associated with sporulation.
(b)	Diarrheagenic toxin Fluid accumulation factor Vascular permeability factor Dermonecrotic toxin Intestino-necrotic toxin Mouse lethal factor 1	3,4,5, 6,7,8, 13	Relatively unstable enterotoxigenic protein, M.W. about 50,000, pI 4.85. Susceptible to trypsin and pronase digestion. Antigenic. Activity may involve stimulation of adenylate cyclase-cAMP system. Probable role in non-gastrointestinal infections.
(c)	Cereolysin Haemolysin I	7,9,10, 11,12,13	Thiol-activated cytolytic toxin related to streptolysin O and clostridial θ toxin. Thermolabile antigenic protein M.W. 49-59,000, pI 6.3-6.7. Inactivated by cholesterol. Lethal to mice.
(d)	Secondary haemolysin Haemolysin II	7,8,11, 12	Thermolabile antigenic protein, M.W. 29-34,000, pI 4.92. Susceptible to pronase, pepsin and trypsin. *In vitro* activity unaffected by thiols, cholesterol and anti-streptolysin O. *In vivo* toxicity not yet established.
(e)	Phospholipase C Lecithinase Egg yolk turbidity factor	13,14	Relatively stable cytotoxic metallo-enzyme (Zn²⁺ion requirement). Resistant to trypsin. M.W. 23-29,000. pI 6.5-8.1 − possibly two enzymes, including 'phosphatasemia factor'. Neither lethal nor dermonecrotic in rabbits.
(f)	Mouse lethal factor 2	7,8	Unstable factor, closely related to, or identical with, cereolysin by M.W. and pI. Similarly inactivated by normal serum.

(g) Toxin of Ezepchuk and Fluer   15   Thermolabile antigenic protein, M.W. 57,000. Inactivated by 60°C x 20 mins. Lethal to mice and rabbits. I/V injection induces emesis in cats. Possibly related to (b).

1, Melling et al. [1976]; 2, Melling and Capel [1978]; 3, Turnbull et al. [1979a]; 4, Spira and Goepfert [1975]; 5, Turnbull [1976]; 6, Turnbull et al. [1977]; 7, Kramer et al. [1978]; 8, Turnbull et al. [1979b]; 9, Bernheimer and Grushoff [1967]; 10, Cowell et al. [1976]; 11, Coolbaugh and Williams [1978]; 12, Kramer and Turnbull (in preparation); 13, Johnson and Bonventre [1967]; 14, Möllby [1978]; 15, Ezepchuk and Fluer [1973].

confusion of names and toxic activities that have become
associated with *B.cereus* and which, from time to time,
have been incriminated in both human and animal infections.
The precise relationship of the toxin of Ezepchuk and
Fluer [1973] to the others tested remains uncertain; however, in our hands, their strain produced the vascular
permeability/dermonecrotic toxin very strongly and it is
suspected that this is where the identity lies. The
levels of haemolysins, phospholipase and mouse lethal
toxins produced by isolates from non-gastrointestinal infections with *B.cereus* did not appear to be correlated
with the recorded severity of the infections [Turnbull *et
al.*, 1979a].

So far the toxin tests used in the study of *B.cereus*
pathogenesis have failed to reveal an ability among the
B.subtilis and *B.licheniformis* isolates from food poisoning
and other infections (see p.304) to elaborate vascular
permeability, necrotic or mouse lethal toxins and a model
to explain and support the suggestion that they are potential pathogens has yet to be found.

References

Alix, P. (1978). Les infections disséminées à *Bacillus* à propos de
    deux observations. Thèse pour le Doctorat en Médecine Université
    de Bretagne Occidentale, Brest, France.
Allen, B.T. and Wilkinson, H.A. III. (1969). A case of meningitis and
    generalized Shwartzman reaction caused by *Bacillus sphaericus*.
    Johns Hopkins Medical Journal **125**, 8-13.
Amador, A., Garcia, J.F., Estan, R. and Perales, J. (1976).
    Bacteremia por *Bacillus licheniformis*. *Medicina Clinica* **67**, 535-
    536.
Anon. (1972). *Bacillus licheniformis* infection. *Communicable Disease
    Report* **No. 1.** Communicable Disease Surveillance Centre, Public
    Health Laboratory Service.
Baba, T. (1969). Analytical serology of Bacillaceae. In "Analytical
    Serology of Microorganisms" (ed. J.B.G. Kwapinski), Vol. 2, pp.609-
    642. New York: J. Wiley and Sons Inc.
Behrend, M. and Krouse, T.B. (1952). Postoperative bacterial syner-
    gistic cellulitis of abdominal wall: fatality following
    herniorrhaphy. *Journal of the American Medical Association* **149**,
    1122-1124.
Bernheimer, A.W. and Grushoff, P. (1967). Cereolysin: production,
    purification and partial characterization. *Journal of General
    Microbiology* **46**, 143-150.
Bonventre, P.F. and Johnson, C.E. (1970). *Bacillus cereus* toxin. In
    "Microbial Toxins", Vol. III (eds. T.C. Montie, S. Kadis and S.J.
    Ajl), pp.415-435. New York: Academic Press.
Boyette, D.P. and Rights, F.L. (1952). Heretofore undescribed aerobic
    spore bearing bacillus in child with meningitis. *Journal of the
    American Medical Association* **148**, 1223-1224.
Coolbaugh, J.C. and Williams, R.P. (1978). Production and character-
    ization of two haemolysins of *Bacillus cereus*. *Canadian Journal of*

Microbiology **24**, 1289-1295.

Cowan, S.T. and Steel, K.J. (1974). "Manual for the Identification of Medical Bacteria", Second edition. Cambridge: Cambridge University Press.

Cowell, J.L., Grushoff-Kosyk, P.S. and Bernheimer, A.W. (1976). Purification of cereolysin and the electrophoretic separation of the active (reduced) and inactive (oxidized) forms of the purified toxin. *Infection and Immunity* **14**, 144-154.

Cox, R., Sockwell, G. and Landers, B. (1959). *Bacillus subtilis* septicemia. Report of a case and review of the literature. *New England Journal of Medicine* **261**, 894-896.

Ezepchuk, Y.V. and Fluer, F.S. (1973). Der enterotoxische Effekt. *Moderne Medizin* **3**, 20-25.

Farrar, W.E. (1963). Serious infections due to "non-pathogenic" organisms of the genus *Bacillus*. Review of their status as pathogens. *American Journal of Medicine* **34**, 134-141.

Fauchère, J.L., Berche, P., Ganeval, D., Daniel, F., Bournerias, F., Daoulas-Lebourdelles, F. and Vernon, M. (1977). Une septicémie à *Bacillus licheniformis*. *Médecine et Maladies Infectieuses* **7**, 191-195.

Fitzpatrick, D.J., Turnbull, P.C.B., Keane, C.T. and English, L.F. (1979). Two gas gangrene-like infections due to *Bacillus cereus*. *British Journal of Surgery* **66**, 577-579.

Ghosh, A.C. (1978). Prevalence of *Bacillus cereus* in the faeces of healthy adults. *Journal of Hygiene (Cambridge)* **80**, 233-236.

Gilbert, R.J. and Taylor, A.J. (1976). *Bacillus cereus* food poisoning. In "Microbiology in Agriculture, Fisheries and Food", Society for Applied Bacteriology, Symposium Series No.4 (eds. F.A. Skinner and J.C. Carr), pp.197-213. London: Academic Press.

Gilbert, R.J. and Parry, J.M. (1977). Serotypes of *Bacillus cereus* from outbreaks of food poisoning and from routine foods. *Journal of Hygiene (Cambridge)* **78**, 69-74.

Gilbert, R.J. (1979). *Bacillus cereus* gastroenteritis. In "Foodborne Infections and Intoxications", Second edition (eds. H. Riemann and F.L. Bryan), pp.495-518. New York: Academic Press.

Glatz, B.A. and Goepfert, J.M. (1973). Extracellular factor synthesized by *Bacillus cereus* which evokes a dermal reaction in guinea pigs. *Infection and Immunity* **8**, 25-29.

Goepfert, J.M., Spira, W.M. and Kim, H.U. (1972). *Bacillus cereus:* food poisoning organism. A review. *Journal of Milk and Food Technology* **35**, 213-227.

Gordon, R.E., Haynes, W.C. and Pang, C.H.-N. (1973). "The Genus *Bacillus*". Washington, D.C.: United States Department of Agriculture.

Greenberg, M., Milne, J.F. and Watt, A. (1970). Survey of workers exposed to dusts containing derivatives of *Bacillus subtilis*. *British Medical Journal* **ii**, 629-633.

Ihde, D.C. and Armstrong, D. (1973). Clinical spectrum of infection due to *Bacillus* species. *American Journal of Medicine* **55**, 839-845.

Isaacson, P., Jacobs, P.H., Mackenzie, A.M.R. and Mathews, A.W. (1976). Pseudotumour of the lung caused by infection with *Bacillus sphaericus*. *Journal of Clinical Pathology* **29**, 806-811.

Jephcott, A.E., Barton, B.W., Gilbert, R.J. and Shearer, C.W. (1977).

An unusual outbreak of food-poisoning associated with meals-on-wheels. *Lancet* **ii**, 129-130.

Johnson, C.E. and Bonventre, P.F. (1967). Lethal toxin of *Bacillus cereus*. 1. Relationships and nature of toxin, hemolysin and phospholipase. *Journal of Bacteriology* **94**, 306-316.

Kampelmacher, E.H. (1969). In "The Microbiology of Dried Foods" (eds. E.H. Kampelmacher, M. Ingram and D.A.A. Mossel). Proceedings of the Sixth International Symposium on Food Microbiology, Bilthoven, Netherlands. International Association of Microbiological Societies.

Katsaras, K. and Zeller, U.P. (1977). Nachweis von *Bacillus cereus*-Toxinen. *Zentralblatt für Bakteriologie, Parasitenkunde, Infektionskrankheiten und Hygiene. I. Abteilung: Originale* **238**, 255-262.

Kim, H.U. and Goepfert, J.M. (1971). Occurrence of *Bacillus cereus* in selected dry food products. *Journal of Milk and Food Technology* **34**, 12-15.

Kim, H.U. and Goepfert, J.M. (1972). Efficacy of a fluorescent-antibody procedure for identifying *Bacillus cereus* in foods. *Applied Microbiology* **24**, 708-713.

Kramer, J.M., Turnbull, P.C.B., Jørgensen, K., Parry, J.M. and Gilbert, R.J. (1978). Separation of exponential growth exotoxins of *Bacillus cereus* and their preliminary characterisation (Abstract). *Journal of Applied Bacteriology* **45**, xix.

Kramer, J.M. and Turnbull, P.C.B. Properties of a novel *Bacillus cereus* haemolytic toxin isolated by dextran gel isoelectric focusing (in preparation).

LeMille, F., de Barjac, H. and Bonnefoi, A. (1969). Essai sur la classification biochimique de 97 bacillus du groupe I. Appartenant à 9 espèces differente. *Annales de l'Institut Pasteur* **117**, 31-38.

Maddocks, A.C. and Turnbull, P.C.B. (1978). Diarrhoea associated with *Bacillus licheniformis*. *Communicable Disease Report* **No.48**. Communicable Disease Surveillance Centre, Public Health Laboratory Service.

Melles, Z., Nikodémusz, I. and Abel, A. (1969). Die pathogene Wirkung aerober sporenbildender Bakterien. *Zentralblatt für Bakteriologie, Parasitenkunde, Infektionskrankheiten und Hygiene, I. Abteilung: Originale* **212**, 174-176.

Melling, J., Capel, B.J., Turnbull, P.C.B. and Gilbert, R.J. (1976). Identification of a novel enterotoxigenic activity associated with *Bacillus cereus*. *Journal of Clinical Pathology* **29**, 938-940.

Melling, J. and Capel, B.J. (1978). Characteristics of *Bacillus cereus* emetic toxin. *FEMS Microbiology Letters* **4**, 133-135.

Möllby, R. (1978). Bacterial phospholipases. In "Bacterial Toxins and Cell Membranes" (eds. J. Jeljaszewicz and T. Wadström), pp.367-424. London: Academic Press.

Mortimer, P.R. and Meers, P.D. (1975). Two food poisoning incidents, possibly associated with a *Bacillus* species. *Communicable Disease Report* **No.30**. Communicable Disease Surveillance Centre, Public Health Laboratory Service.

Murrell, W.G. (1974). *Bacillus cereus*. In "Food-borne microorganisms of public health significance", Vol. 1. 15.1-15.13. Joint AIFST/CSIRO/UNSW Publication, Australia.

Nielsen, S.F. and Pedersen, H.O. (1974). Studier over bakteriefore-komsten i røraeg. *Dansk Veterinärtidsskrift* **57**, 756-759.

Norris, J.R. and Wolf, J. (1961). A study of the antigens of the aerobic spore-forming bacteria. *Journal of Applied Bacteriology* **24**, 42-56.

Norris, J.R. (1962). Bacterial spore antigens: a review. *Journal of General Microbiology* **28**, 393-408.

Parry, J.M. and Gilbert, R.J. (1980). Studies on the heat resistance of *Bacillus cereus* spores and growth of the organism in boiled rice. *Journal of Hygiene (Cambridge)* **84**, 77-82.

Pearson, H.E. (1970). Human infections caused by organisms of the *Bacillus* species. *American Journal of Clinical Pathology* **53**, 506-515.

Peloux, Y., Charrel-Taranger, C. and Govin, F. (1976). Nouvelle affection opportuniste à *Bacillus*: un cas de bactériémie à *Bacillus licheniformis*. *Pathologie Biologique* **24**, 97-98.

Pennington, J.E., Gibbons, N.D., Strobeck, J.E., Simpson, G.L. and Myerowitz, R.L. (1976). *Bacillus* species infection in patients with hematologic neoplasia. *Journal of the American Medical Association* **235**, 1473-1474.

Ryan, A.J. (1970). Abortion in cattle associated with *Bacillus licheniformis*. *Veterinary Record* **86**, 650-651.

Sathmary, M.E. (1958). *Bacillus subtilis* septicemia and generalized aspergillosis in a patient with acute myeloblastic leukemia. *New York State Journal of Medicine* **58**, 1870-1876.

Schothorst, M. van and Schipper, K. (1966). Een voedselvergiftinging waarschijnlijk veroorzaakt door *Bacillus licheniformis*. *Verslagen en Madedelingen betreffende de Volbsgezondheid*, 3038-3039. Rijks Instituut voor de Volksgezondheid, Bilthoven.

Smith, N.R., Gordon, R.E. and Clark, F.E. (1952). Aerobic Spore-Forming Bacteria. Agriculture Monograph No. 16. Washington, D.C.: United States Department of Agriculture.

Spira, W.M. and Goepfert, J.M. (1972). *Bacillus cereus*-induced fluid accumulation in rabbit ileal loops. *Applied Microbiology* **24**, 341-348.

Spira, W.M. (1974). Purification and characterization of *Bacillus cereus* enterotoxin. Ph.D. Thesis. University of Wisconsin, USA.

Spira, W.M. and Goepfert, J.M. (1975). Biological characteristics of an enterotoxin produced by *Bacillus cereus*. *Canadian Journal of Microbiology* **21**, 1236-1246.

Sugar, A.M. and McCloskey, R.V. (1977). *Bacillus licheniformis* sepsis. *Journal of the American Medical Association* **238**, 1180-1181.

Taylor, A.J. and Gilbert, R.J. (1975). *Bacillus cereus* food poisoning: a provisional serotyping scheme. *Journal of Medical Microbiology* **8**, 543-550.

Terayama, T., Shingaki, M., Yamada, S., Ushioda, H., Igarashi, H., Sakai, S. and Zen-Yoji, H. (1978). Incidence of *Bacillus cereus* in commercial foods and serological typing of isolates. *Journal of Food Hygienic Society of Japan* **19**, 98-104.

Terranova, W. and Blake, P.A. (1978). *Bacillus cereus* food poisoning. *New England Journal of Medicine* **298**, 143-144.

Tong, J.L., Engle, H.M., Cullyford, J.S., Shimp, D.J. and Love, C.E. (1962). Investigation of an outbreak of food poisoning traced to turkey meat. *American Journal of Rublic Health* **52**, 976-990.

Turnbull, P.C.B. (1976). Studies on the production of enterotoxins by *Bacillus cereus*. *Journal of Clinical Pathology* **29**, 941-948.

Turnbull, P.C.B., Nottingham, J.F. and Ghosh, A.C. (1977). A severe necrotic enterotoxin produced by certain food, food poisoning and other clinical isolates of *Bacillus cereus*. *British Journal of Experimental Pathology* **58**, 273-280.

Turnbull, P.C.B. (1978). *Bacillus licheniformis*. A possible food poisoning agent and clinical pathogen. *Communicable Disease Report* **No.18**. Communicable Disease Surveillance Centre, Public Health Laboratory Service.

Turnbull, P.C.B., Jørgensen, K., Kramer, J.M., Gilbert, R.J. and Parry, J.M. (1979a). Severe clinical conditions associated with *Bacillus cereus* and the apparent involvement of exotoxins. *Journal of Clinical Pathology* **32**, 289-293.

Turnbull, P.C.B., Kramer, J.M., Jørgensen, K., Gilbert, R.J. and Melling, J. (1979b). Properties and production characteristics of vomiting, diarrheal and necrotizing toxins of *Bacillus cereus*. *American Journal of Clinical Nutrition* **32**, 219-228.

Winton, F.W. and Sayers, J.O. (1975). An outbreak of food poisoning caused by *Bacillus subtilis*? *Communicable Diseases Scotland Report* **No.46**.

Chapter 13

THE TAXONOMY OF SOME NITROGEN-FIXING *BACILLUS* SPECIES WITH SPECIAL REFERENCE TO NITROGEN FIXATION

MURIEL E. RHODES-ROBERTS

Department of Botany and Microbiology, University College of Wales, Aberystwyth, UK

Introduction

The dominance of *Ammophila arenaria* L. (marram grass) in nutrient-poor, dry and saline mobile sand dunes presents many biological problems, not least of which are the sources of nitrogen for such prolific plant growth. It is likely that free-living nitrogen-fixing bacteria, able to endure intermittent salinity and drought, are important components of such an environment. There is a vast lite-rature on the possible role of free-living nitrogen-fixing azotobacters, cyanobacteria and anaerobic clostridia in nitrogen-poor soils, but very little information concerning nitrogen-fixing *Bacillus* spp. The latter have only been found in low numbers even when highly selective methods for their enrichment, namely heat treatment of samples followed by aerobic or anaerobic cultivation in or on nitrogen-poor media, have been employed [Moore and Becking, 1963; Chang and Knowles, 1964; Moore, 1966; Meiklejohn and Weir, 1968; Bland, 1968; Line and Loutit, 1971; Knowles *et al.*, 1974]. Their nitrogen contribution was calculated by Bland [1968] to be less than that derived from rainfall. Alexander [1964] stated that *Bacillus* spp. were not found in soils of high salinity, but they are now known to exist in salt marshes [Turner and Jervis, 1968]. More relevant perhaps to their nitrogen-fixing activity *in situ* is the fact that the pH value at a root surface is claimed to be <5.0, which would preclude active growth of many bacteria. Webley *et al.* [1952] studied the bacteria in the vicinity of marram grass roots and in root-free sand of Aberdeen-shire dunes with a pH value of 6.68-7.8 and a minimal salt concentration of 1.66% in the free water fraction. They found that only 1.7% of a random selection of isolates from this nitrogen-deficient sand were aerobic endospore-forming rods which were equally abundant in the root-free sand. $^{15}N_2$ fixation in the damper dune slacks at Blakeney Point, Norfolk, with a 2.6% (w/v) salinity and pH 8.6 was cer-tainly established by Stewart [1965], but he concluded that this was primarily a light-dependent phenomenon and there-

fore due to photosynthetic cyanobacteria, although the
possibility of some fixation by non-photosynthetic bacteria
was not excluded.

An early claim by Bredemann [1908] for nitrogen fixation
by some strains of *B.asterosporus* is probably valid because
this is now regarded as a synonym for *B.polymyxa*, a taxon
known to contain nitrogen-fixing strains [Gibson and Gordon,
1974]. In 1958 Hino and Wilson confirmed by $^{15}N_2$ fixation
methods that *Bacillus* strain Hino isolated from Japanese
soil was a true fixer, although the fixation was inhibited
by as little as 1.0% (v/v) oxygen in the gas phase over the
nitrogen-poor medium. This *Bacillus* was fully described;
it is of interest that it was originally thought to be a
Clostridium sp. on the basis of anaerobic growth and nitrogen fixation, and acid and gas formation from several fermentable carbohydrates, together with *Clostridium*-like
sporangia containing swollen endospores. Hino and Wilson
[1958] recognized that it was in fact a facultative anaerobe with many properties similar to *B.polymyxa*, namely,
it was a biotin-requiring Voges-Proskauer (VP)-positive
organism, feebly proteolytic, able to grow only very
slightly at 42°C, showing poor growth in the absence of
glucose, which, as well as fructose, galactose, mannose,
mannitol, sorbitol and sucrose, was fermented with acid and
gas formation. However, no acid or gas was produced from
glycerol, arabinose or lactose, neither was nitrate reduced; it was emphasized that these characters were atypical for *B.polymyxa sensu* Smith *et al.* [1952]. This quali-
fication was commendable because later studies [Oulette *et
al.*, 1969] revealed that the Hino *Bacillus* had 51.5 moles
%G+C in its DNA, a value clearly nearer the range of 49.0
to 51.0 shown by these authors for 5 strains of *B.macerans*,
compared with the 43.2 to 45.6 moles %G+C for 5 strains of
B.polymyxa. Hill [1966] had reported similar values, namely 50.0 to 54.0% for *B.macerans* and 44.0 to 48.0% for *B.
polymyxa*.

Nitrogen fixation by 17 isolates of *B.polymyxa*, all producing acid and gas from glycerol, arabinose and lactose,
was tested by Grau and Wilson [1962] using labelled nitrogen, and again fixation occurred only anaerobically and as
little as 1.0% (v/v) oxygen was inhibitory for the 15
strains which were positive fixers: 2 strains of *B.poly-
myxa* and the 2 *B.macerans* strains tested were never induced
to fix nitrogen, despite attention to several environmental
factors such as adequate supplies of iron and molybdenum to
support growth on nitrogen. These authors differentiated
the 17 strains of *B.polymyxa* from their 2 strains of *B.
macerans* by the characters listed in Table 1. They were
not able to correlate nitrogen fixation with any other
property, including the possession of hydrogenase, which
was active in all organisms grown on sucrose (with acid and
gas formation), and confirmed by the rapid reduction of
methylene blue by both species.

TABLE 1

Differentiation of B.polymyxa *and* B.macerans
[*Grau and Wilson, 1962*]

Property	*B.polymyxa*	*B.macerans*
Acid and gas from		
glycerol	+	-
arabinose	+	-
lactose	+	-
Growth at 43°C	-	+
Voges Proskauer reaction	+	-
Growth requirements		
biotin	+	+
thiamine	-	+

Nitrogen fixation by both whole cells and cell extracts for 12 out of the 13 isolates of *B.macerans* tested, and for 15 of the 17 isolates of *B.polymyxa* was confirmed by Witz *et al.* [1967] by estimating nitrogen fixation in liquid cultures by both Kjeldahl and $^{15}N_2$ analytical techniques. In 1971 it was claimed that *B.circulans* was also capable of $^{15}N_2$ fixation [Line and Loutit, 1971]. They screened several hundred isolates from acid (pH 4.5 to 5.8) New Zealand soils for their ability to reduce acetylene, a rapid and sensitive assay method for the estimation of nitrogen fixation [Dilworth, 1966]. Eventually 5 facultatively anaerobic bacterial species were obtained which were unquestionably nitrogen-fixers; these included *B.polymyxa* and *B.circulans*-like organisms. $^{15}N_2$ fixation was confirmed for these 2 species and 0.480 Atoms% $^{15}N_2$ excess was found for *B.polymyxa* and 0.324 Atoms% $^{15}N_2$ excess for *B. circulans*. That they were both relatively inefficient nitrogen fixers was confirmed by the figures of only 5.3 mg nitrogen fixed per gram glucose utilized for *B.polymyxa* and 2.3 mg for *B.circulans*, compared with 8.0 mg for 1 strain of *Clostridium butyricum*. These 2 *Bacillus* species were fully described: the properties of the *B.polymyxa* strains were in full accord with the description in "Bergey's Manual of Determinative Bacteriology" [Smith and Gordon, 1957] and a bacteriocin produced by the isolates was active towards the type strain *B.polymyxa* NCTC 10343. The *B. circulans* isolates occurred more frequently, but at that time such strains had not been reported to fix nitrogen; other aberrancies compared with the current "Bergey Manual" description for *B.circulans* were that the spore size varied from 0.7 to 1.4 x 1.0 to 2.4 μm, and that its position in the rod was either terminal or sub-terminal, and generally caused swelling of the rod; but both swollen and non-swollen spores had been noted from blood agar cultures which, again uncharacteristically, showed hydrolysis.

Another aberrancy was lack of milk coagulation, but the
authors concluded that it could not be decided whether
these organisms merited an alternative nomenclature until
the whole *B.circulans* group, 'the most difficult to charac-
terize of all the species of the genus' [Smith *et al.*,
1952], and still a 'markedly heterogeneous collection'
with 'an ill-defined boundary around the species' [Gibson
and Gordon, 1974], had been further studied.

Ecologically it is noteworthy that these 2 species were
isolated from New Zealand soils with pH values of 4.5, 5.4
and 5.8, although the growth range of *B.polymyxa* is given
as 5.2 to 9.2 and that for *B.circulans* as 5.5 to 7.8.

Isolation of Nitrogen-Fixing *Bacillus*
Species from Ynyslas Sand Dunes

Since 1962 many samples of local Ynyslas dune sand (containing about 0.007% (w/w) organic N) from various depths,
and from the rhizosphere and rhizoplane of *Ammophila arenaria* have been plated on a variety of both nitrogen-poor
and fully nutrient media. The 'nitrogen-free' glucose-
salts agar, pH 7.1, devised by Brown *et al.* [1962] for
quantitative estimation of *Azotobacter* spp. has been extensively used, supplemented with yeast extract (0.05% w/v) to
provide growth factors for *Bacillus* spp. The nitrogen content of this medium was 0.0024 to 0.0046% (w/w). Pour-
and spread-plates were incubated both aerobically and anaerobically, but *Bacillus* strains were never conspicuous
on the poor medium, although numerous species, especially
B.cereus and *B.subtilis*, were evident on the glucose
nutrient agar plates after either aerobic or anaerobic incubation. It was concluded therefore that nitrogen-fixing
Bacillus spp. were not present in large numbers. Furthermore, significantly fewer *Bacillus* spp. occurred in the
root area, a result in good agreement with those of
Vágnerova *et al.* [1960] who recorded the amino-acid requiring *B.circulans, B.brevis, B.laterosporus, B.pumilus*
and *B.sphaericus* on the roots, in contrast to *B.firmus,
B.licheniformis, B.megaterium* and *B.subtilis* in root-free
soil.

Ynyslas sand suspensions were also incubated anaerobically in the media of Chang and Knowles [1964] and Ross
[1958] for the detection of nitrogen-fixing *Clostridium
butyricum*. These media are suitable for the enrichment of
facultatively anaerobic *B.polymyxa* and *B.macerans*, but
such were not detected, again confirming their low incidence in the dunes.

During the period 1964-9 further studies on the nutrition of *Ammophila arenaria* were carried out in the laboratory and greenhouse [Wahab, 1969]. This involved the
growth of seedlings in dune sand in the greenhouse with
different controlled nitrogen nutrition regimes. One set
of potted plants showed a definite increase in nitrogen

levels over 6 months, in the absence of any added combined
nitrogen. From this sand Wahab obtained (on the nitrogen-
poor glucose medium of Brown *et al.* [1962] plus 0.05% (w/w)
yeast extract, incubated aerobically for 5 days at 25°C) a
few colonies of *Azotobacter chroococcum*, a few colonies of
Klebsiella and *Enterobacter* spp., and an occasional
Bacillus isolate, all of which were usually capable of
acetylene reduction in pure culture. About 15 isolates of
Bacillus were thus obtained and partially characterized by
Wahab [1969]. They were remarkably similar and the late
Dr. T. Gibson examined these cultures and confirmed that
they were not typical *B.polymyxa* or *B.macerans*, *B.licheni-
formis* or *B.pumilus* [personal communication, 1969].

Concurrent acetylene-reduction studies by the author,
however, revealed a disconcerting presence of this
Bacillus in several 'sterile medium' controls. The
Bacillus was traced to the equipment used by both workers,
so it is therefore possible that the *Bacillus* isolated by
Wahab [1969] was a laboratory contaminant that may have
originated from the cylinder of argon or acetylene then in
use. Further attempts were therefore made by both Wahab
and R. Sylvester-Bradley to reisolate this *Bacillus* directly from Ynyslas dune sand, using the highly selective procedures of heating followed by aerobic or anaerobic cultivation in or on nitrogen-poor media. Even so, only 2
acetylene-reducing *Bacillus* isolates, G8 from the aerobic
series and Sp4 from the anaerobic series, were isolated by
Wahab. These were not the same as the earlier isolates.
The 11 *Bacillus* strains of R. Sylvester-Bradley contained
only 3 acetylene-reducing strains, D+ (now lost), R10 and
D10, and these all proved to be *B.polymyxa* (see below).
The properties of all these *Bacillus* spp. have been re-
examined together with those isolated by Wahab [1969] and
are described later.

*Acetylene reduction studies of the Ynyslas
isolates of* Bacillus

Twelve isolates of the *Bacillus* studied by Wahab [1969,
1975] reduced acetylene both aerobically and anaerobically,
but he found that acetylene reduction was much greater in
an argon (A) 90:acetylene (C_2H_2) 10% (v/v) gas mixture compared with an aerobic gas mixture of A78:O_2 22% (v/v) with
a 10.0% (v/v) displacement with acetylene. This was confirmed by determining the increase in the nitrogen content
of 5-day liquid cultures in nitrogen-poor media by Kjeldahl
analyses using the semi-micro procedure of Wilson and
Knight [1952]. Strain A33 showed a net increase in N content of 6.85 mg N ml^{-1} culture aerobically compared with
9.8 mg anaerobically; for strain A122 the increments were
5.84 mg compared with 7.2 mg ml^{-1} respectively [Wahab,
1975].

These results were confirmed by the author in some

cases, but far too often the results for acetylene reduction by a given purified isolate under closely standardized
conditions were disconcertingly variable. Therefore, many
modifications concerning the growth and acetylene-reducing
conditions of the isolate under test were made. The growth
medium of Brown *et al*. [1962] supplemented by yeast extract
(0.05 or even 0.01% w/v) was always suitable for the growth
of *Azotobacter* spp. and the *Bacillus* strains, but only the
former consistently reduced acetylene. Therefore, the
growth of the isolates in nitrogen-poor liquid media was
checked by determining the nitrogen gains by microKjeldahl
analyses (see Table 2). It is obvious that nitrogen fixation had occurred in all cultures, including the type
strain of *B.polymyxa* NCTC 10343 which had never shown any
acetylene-reducing activity. Thus, it was concluded that
suitable conditions for demonstrating acetylene reduction
had not been found.

The *Azotobacter* medium of Brown *et al*. [1962] was relatively poorly buffered with 0.8 g K_2HPO_4 1^{-1}, and the
initial pH of 7.1 fell to between 4.9 to 5.15 after 3 days'
growth at 25°C of all the *Bacillus* and *Azotobacter* cultures. The medium was therefore modified by dissolving the
ingredients in 0.1M sodium phosphate buffer solution to
give a final pH value of 7.6. After 2 days' growth the pH
values for the same 18 isolates were between 6.8 to 7.05
and acetylene reduction was always markedly positive.
Line and Loutit [1971] likewise found that less concentrated buffer solutions did not maintain sufficiently high pH
values for acetylene reduction, and particular difficulty
was experienced with *Clostridium butyricum* and their *B.
circulans*-like isolates; their search for non-inhibitory
buffers capable of maintaining adequately high pH values
proved unsuccessful. Whether the inhibition of acetylene
reduction was a straightforward effect of pH, or whether
there was a phosphate deficiency in the medium, or both,
is still unknown.

Further isolates of Bacillus *from Ynyslas sand dunes*

In 1971 a further 10 strains of *Bacillus* were isolated
(the RSB isolates) directly from Ynyslas dune sand and the
rhizosphere of *Ammophila arenaria*. These were taxonomically different from those of Wahab [1969]. RSB strains
D10 and R10 were typical *B.polymyxa* and consistently reduced acetylene anaerobically. Isolates RSB6 and RSB7
seemed to grow equally well on nitrogen-poor media but
never reduced acetylene. When 50 ml volumes of liquid
cultures were analysed for their net gains in nitrogen by
the microKjeldahl methodology described above, it was evident that the total nitrogen content of the cultures of
the acetylene-reducing strains was not always greater than
that for strains RSB6 and RSB7 (Table 2). The accuracy of
the microKjeldahl analytical technique used was confirmed

by a 91.3% recovery of ammonia from bovine serum albumin, and a 93.1% recovery from the standard solutions of ammonium sulphate. This strongly suggested that conditions were not optimal for the demonstration of acetylene reduction by isolates other than *Azotobacter* and some strains of *B.polymyxa*.

The addition of yeast extract (0.05%, w/v) had a markedly stimulating effect on the growth of the *Bacillus* and *Azotobacter* cultures, but it did not elicit ethylene formation by strains of RSB6 and 7, or *B.polymyxa* NCTC 10343 (the type strain) or *B.polymyxa* 22V (a laboratory strain originally from the Wellcome Research Laboratories via B.C.J.G. Knight). Furthermore, although the nitrogen-poor phosphate medium adjusted to pH 7.6 always markedly enhanced ethylene formation by all the Wahab isolates, it failed to promote acetylene reduction by the above 4 *Bacillus* isolates (see Table 2). Again, when these 4 isolates were grown anaerobically (under nitrogen or in a 95 N_2:5O_2% v/v gas mixture) on either the pH 7.1 or the pH 7.6 medium, they still failed to reduce acetylene although they all accumulated nitrogen in nitrogen-poor liquid media, often to levels greater than those found in the acetylene-reducing cultures of *B.polymyxa* (Table 2). One must conclude that the correct conditions for the demonstration of acetylene reduction have not yet been achieved for these isolates. The effect of the addition of a suitably chelated solution of trace elements merits investigation, especially because (see below) the growth of the *Bacillus* isolates was markedly enhanced after such supplements, and the need for adequate amounts of available iron and molybdenum for the growth of *B.polymyxa* on nitrogen, but not on ammonium-nitrogen, was clearly shown by Grau and Wilson [1962]. It has been amply confirmed since at the molecular level that all bacterial nitrogenases are enzymes containing iron and molybdenum [Postgate, 1978].

Description of the Nitrogen-Fixing Isolates of *Bacillus*

The Wahab isolates

The 12 isolates of *Bacillus*, A11, A17, A17a, A28, A33, A35/1, A35/2, A36, A122, BP, 2a2c/1 and 2a2c/2 [Wahab, 1969], which may have originated from the sand used in the greenhouse pot experiments planted with seedlings of *Ammophila arenaria*, or may have been a laboratory contaminant, were shown to be capable of $N_2(C_2H_2)$-fixation, especially under anaerobic conditions and when grown on a nitrogen-poor medium with yeast extract (0.05%, w/v) at a pH of 7.6, achieved by the use of 0.1M phosphate buffer. They were also capable of fixation in the presence of oxygen (20.0%, v/v), when tested on the medium buffered to pH 7.6. On less well buffered media at pH 7.1 acetylene reduction was very sporadic.

TABLE 2

Acetylene reduction and nitrogen accumulation by Bacillus strains and Azotobacter chroococcum

Strain	Acetylene reduction anaerobically at pH 7.1 (9 tests)	Acetylene reduction anaerobically at pH 7.6 (3 tests)	µg N mg⁻¹ dry weight of cells
RSB6	-	-	22.558
RSB7	-	-	22.681
RSB D10 (*B.polymyxa*)	+	++	29.727
RSB R10 (*B.polymyxa*)	+	++	7.883
Wahab *Bacillus* A11	+/-	+++	.
Wahab *Bacillus* A17a	+/-	+++	
Wahab *Bacillus* A35/1	+/-	+++	.
Wahab *Bacillus* A36	+/-	+++	11.274
Wahab *Bacillus* A122	+/-	+++	10.361
Wahab *Bacillus* 2a2c/1	+/-	+++	21.66
Bacillus Sp.4	++	++	20.974
B.polymyxa NCTC 10343 (type strain)	-	-	46.787
B.polymyxa 22V	-		32.64
Azotobacter chroococcum Y1a	+++	n.g	38.407
Azotobacter chroococcum Y3a	+++	n.g	24.514
Medium control	-	-	
Control (culture boiled after inoculation)	.	.	0.007

$N \mg^{-1}$

-, no ethylene formation; +/-, ethylene production sometimes slightly positive and sometimes negative;
++, good ethylene production; +++, abundant ethylene reduction; n.g, no growth; ., test not done.

The following description of the isolates is based upon cultures incubated aerobically at 30°C unless otherwise stated. The 12 isolates were remarkably uniform in their properties, which perhaps favours the possibility that they were derived from the same laboratory contaminant. All were evenly-staining Gram-positive rods 2.5 to 5.0 x 0.8 to 1.0 μm, with rounded ends and no inclusions of poly-β-hydroxybutyrate; some short chains or filament formation were occasionally observed. Terminal or subterminal spores formed readily within 2 days on glucose nutrient agar; they were ovoids measuring 1.8 to 2.3 x 1.2 μm, and spores of 3 isolates were kindly examined under the electron microscope by Dr. P.D. Walker and reported to be smooth-walled, resembling *B.subtilis* or *B.pumilus* rather than ridged like the spores of *B.polymyxa* or *B.macerans*. Endo-spores caused slight to definite swelling of the sporangium at maturity, and sporangium remnants adhered to the liberated spores and stained Gram-positive (see Fig. 1).

Cells grown on nutrient agar slopes with added distilled water and incubated for 2 days at 25°C were actively motile and had 2 to 10 peritrichous flagella, as revealed by the silver-plating method of Rhodes [1958]. Capsules were often seen (Fig. 1) on such preparations even though no carbohydrate was present in the growth medium.

Growth was very sparse in nutrient broth and barely visible on nutrient agar even after prolonged incubation: 2-day old colonies were <2 mm diameter, smooth, shining, but barely visible. Glucose (1.0%, w/v) greatly enhanced growth in either liquid or on solid media, with the formation of white-cream, smooth or matt surfaced colonies, opaque and with entire edges, with a diameter of 2 to 3 mm after 2 days. Markedly pleomorphic colonial morphology developed later (see Fig. 2). Yeast extract (0.05%, w/v) slightly enhanced the rate of growth in nutrient broth, but not the ultimate turbidity, which was very sparse unless a

---

Footnote to Table 2 (opposite)

Cultures were grown aerobically on the medium of Brown *et al.* [1962] with glucose (0.5%, w/v) and yeast extract (0.05%, w/v) adjusted to pH 7.1, or in a similar medium made with 0.1M phosphate buffer to give a final pH of 7.6, for 2 days at 25°C and then incubated in the an-aerobic gas mixture of A90:C_2H_2 10 (v/v). Net nitrogen accumulation was determined in 50 ml volumes of the liquid glucose yeast extract medium of Brown *et al.* [1962]: cells from cultures grown aerobically for 8 days at 25°C were lyophilized and their nitrogen content deter-mined by microKjeldahl analysis. Bovine serum albumin and ammonium sulphate solutions were employed as standards and the microtitrations of the liberated ammonia were triplicated; the average of the 3 values are given, and the values were always within ± 10.0%.

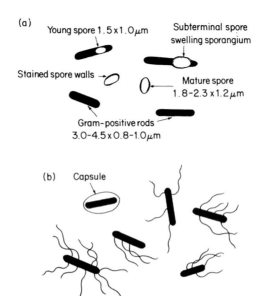

Fig. 1 The Wahab *Bacillus* strain A36: (a) Gram-stained cells from glucose nutrient agar incubated aerobically for 2 days at 30°C, showing young and mature spores; (b) cells from nutrient agar slopes plus distilled water incubated aerobically for 2 days at 25°C and silverplated by the staining method of Rhodes [1958] to show capsules and flagella.

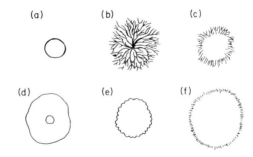

Fig. 2 The Wahab *Bacillus* strain A36: showing the pleomorphic morphology of colonies grown aerobically for 7 days at 30°C on nutrient agar with 1.0% (w/v) starch. (a) 2 mm diameter - flat, white, shining, opaque; (b) 4-7 mm diameter - flat, white, dull, rough, ridged, rhizoid, opaque; (c) 4 mm diameter - low convex, white, smooth, shining, opaque centre, rough rhizoid edge; (d) 4 mm diameter - very thin, white, smooth, shining, transparent; (e) 3 mm diameter - flat, white, shining, smooth, crenate edge, translucent; (f) 5 mm diameter - flat, white, shining, smooth, slightly rough margin, opaque

utilizable carbohydrate was present. No growth occurred
on classic potato slopes resting on a glycerol-soaked wad
of cotton wool unless the inoculum was in contact with the
glycerol, when moderately thick, white shiny growth was
visible after about 5 days. Growth on the nitrogen-poor
glucose salts medium of Brown *et al.* [1962] supplemented
with yeast extract (0.05%, w/v) was markedly better and
faster to develop than growth on nutrient agar, and this
feature was almost diagnostic for this organism. All isolates were also able to grow well anaerobically on either
glucose nutrient agar or the nitrogen-poor glucose medium.
They were catalase-positive facultative anaerobes, and
weakly oxidase-positive, but growth was so sparse on glu-
cose-free media that Kovac's oxidase test was not strictly
applicable. All isolates grew at 40°C, a few sparsely at
41°C but none grew at 42°C. Very slight slimy growth
occurred in glucose nutrient broth with either 1.0 or 2.0%
(w/v) NaCl, but this could easily be overlooked; no
growth occurred in the 3.0% (w/v) salt broths.

Biochemical properties Carbohydrate utilization was
tested using the Hugh and Leifson O/F method, as well as
classic sugar broths: acid was rapidly formed from glycerol, glucose, mannitol, and more slowly (requiring 5 days)
from salicin. Gas was never detected, and xylose, arabinose and lactose were not utilized and such media failed
to support visible growth. Nutrient agar supplemented
with starch (1.0%, w/v) was hydrolysed only to the dextrin
stage (red with iodine solution); pleomorphic colonial
morphology was particularly evident on this medium (see
Fig. 2), but the isolates were in fact pure. Isolates
were Voges-Proskauer-positive and methyl red-negative.
They grew very sparsely in peptone water, so the standard
test for indole was invalid. Likewise they were unable to
grow in the presence of urea (2.0%, w/v) in Christensen's
medium. Poor growth occurred on gelatin (1.0%, w/v) in
nutrient agar, with 4 to 7 mm hydrolysis after 7 days.
H_2S formation from cysteine was feeble. No hydrolysis
occurred on skim milk agar (25.0%, w/v) or in litmus milk
because growth was so sparse; very slight acidity or no
change was the typical litmus milk reaction. Moderate
growth occurred on human blood (10.0%, v/v) agar plates,
but there was no hydrolysis. All isolates failed to grow
on egg yolk agar, or on the test margarine agar stained
with Victoria Blue for the detection of lipolysis. Growth
on the pectin agar of Hankin *et al.* [1971] was also too
sparse for the degradation of pectin to be apparent. Nitrate was not reduced in the classic peptone-nitrate water,
but growth was again sparse; heavier cell suspensions
grown on glucose nutrient agar failed to reduce nitrate
when the microtest method of Brough [1950] was applied, and
no growth was obtained in the denitrification medium of
Stanier *et al.* [1966]. No growth occurred in Koser's

citrate solution.

The G+C value of the DNA of isolate A36 was 34.7 ± 0.3 moles % (T_m 83.6), De Ley [personal communication] and this is in the range shown by *B.alvei, B.anthracis* or *B.cereus* rather than that of *B.polymyxa* whose range is 43 to 51% [Gibson and Gordon, 1974; Priest, Chapter 3]. De Ley [personal communication] reported that in his hands 12 strains of *B.polymyxa* showed an average value of 45 moles %G+C in their DNA, and so the Wahab *Bacillus* was indeed not closely related to this species, nor to *B.macerans* with 49 to 51% G+C values [Gibson and Gordon, 1974].

Wahab strains G8 and Sp.4

Because of the controversial origin of the previous isolates, Wahab attempted to re-isolate similar organisms directly from the rhizosphere of *Ammophila arenaria* growing in the Ynyslas sand dunes. Sand suspensions were heated at 80°C for 10 min and then cultured on the nitrogen-poor glucose salts yeast extract agar as before. Isolate G8 was obtained from the plates incubated aerobically and Sp.4 from the anaerobically incubated plates. No other colonies of *Bacillus* or *Clostridium* were found.

The *Bacillus* strain G8 never reduced acetylene anaerobically or aerobically; it was an obligate aerobe, and grew only moderately well after 10 successive sub-cultures on the nitrogen-poor medium incubated aerobically. It was markedly different from Wahab's previous isolates and all its properties were in good agreement with the description of *B.megaterium* [Gibson and Gordon, 1974].

Bacillus Sp.4 on the other hand, isolated from the plates incubated anaerobically, consistently reduced acetylene anaerobically. Nevertheless, this facultative anaerobe was markedly different from the Wahab [1969] isolates of *Bacillus*, and this is clearly shown in Table 3. The properties of the fully-described isolates of nitrogen-fixing *B.polymyxa* and *B.circulans*-like organisms of Line and Loutit [1971] are included for relevant comparisons (rather than the criteria of Gordon *et al.* [1973] or Gibson and Gordon [1974] which do not emphasize nitrogen fixation). Table 3 shows that many of the properties of Sp.4, such as rod and spore morphology, temperature tolerance, growth on potato with maceration and gas formation, pectinolysis and nitrate reduction, resemble those for *B.polymyxa*. However, this strain differed markedly from the 2 *B.polymyxa* isolates NCTC 10343 and the Reading University stock strain 22V (origin unknown) used here; Sp.4 grew much more vigorously on a wide range of media, tolerated NaCl (3.0%, w/v), and also grew rapidly from minimal inocula in 3 chemically defined glucose-ammonia or glucose-citrate media. Thus, neither biotin nor thiamine were necessary for growth. The carbohydrate utilization pattern, acid only being formed from glucose, sucrose and lactose, al-

though both acid and gas were produced from mannitol, salicin, arabinose and xylose, was noteworthy, suggesting affinities with *B.circulans* (*cf*. Line and Loutit, 1971; Gordon
et al., 1973], although no haemolysis occurred on blood
agar (*cf*. *B.circulans*, Line and Loutit).

Altogether, Sp.4 was capable of vigorous growth on many
diagnostic media, and rapidly digested casein (unlike *B.
macerans*) and gelatin. Atypical lack of gas formation from
sugars has been noted for 1 strain of *B.macerans*, and anaerogenic strains are not uncommon; Knowles *et al.* [1974]
isolated low numbers of facultatively anaerobic nitrogen-
fixing *Bacillus* spp. from pulp and paper mill effluents:
these were very similar to *B.polymyxa* except for their inability to produce gas from glucose, xylose, mannitol,
rhamnose and sucrose, and the authors concluded that they
were anaerogenic strains of *B.polymyxa*. Anaerogenic
strains of many Gram-negative Enterobacteriaceae are not
uncommon either, and so Sp.4 seems to be another atypical
isolate of *B.polymyxa*, or, a vigorously-growing and growth-
factor-independent strain of *B.circulans*, which remains a
complex 'exhibiting variation in several directions' to
such an extent that one may add Sp.4 as a further variant.
Until further isolates are more fully characterized no
definitive conclusions are possible. The comments of
Gordon *et al.* [1973] drawing attention to the alkali-tole-
rant isolates of *B.circulans* described by Chislett and
Kushner [1961] are noteworthy, Sp.4 having been isolated
from sand dunes with pH values of 7.5 to 8.4. A strain
fully conforming to the description of *B.circulans* Jordan
was isolated from cultures of *B.cereus* by Chislett and
Kushner [1961] during their alkali-tolerance training experiments; this strain was capable of active growth at pH
11, even after intermittent growth at pH 7.4. The authors
believed that it was not a variant of *B.cereus*, but that it
was a contaminant revealed after enrichment under the highly alkaline conditions. They further examined a collection
of 26 strains of *B.circulans* from different sources, but
none of these strains grew at pH 10.7; Gordon *et al.*
commented that the alcalophilic strains of *B.circulans* resembled *B.alcalophilus* Vedder 1934.

The RSB isolates

The 13 *Bacillus* strains isolated by R. Sylvester-Bradley
in 1971 (unpublished) from the rhizosphere of *Ammophila
arenaria* and root-free dune sand were fully characterized
for comparative purposes, but only 4 strains merit discussion here; all 4, RSB6, RSB7, RSB D10 and RSB R10, continued to grow well aerobically on nitrogen-poor media,
with definite increments of combined nitrogen occurring in
liquid culture as determined by microKjeldahl analyses (see
Table 2). Nitrogen fixation (acetylene reduction) has,
however, never been shown for isolates RSB6 and 7, although
acetylene reduction anaerobically was consistently positive

TABLE 3

A comparison of the facultatively anaerobic Bacillus spp. studied here, including the type strain B.polymyxa NCTC 10343, with the isolates of B.polymyxa and B.circulans-like strains described by Line and Loutit [1971]

Character	Wahab Bacillus (12 strains)	Wahab Sp.4	B.circulans [Line and Loutit, 1971]	B.polymyxa [Line and Loutit, 1971]	B.polymyxa NCTC 10343	RSB isolates 6 and 7	RSB isolates D10 and R10
Acetylene reduction	+	+	+	+	-	-	+
Gram reaction	+	+	-	+V	+	+	+
Cell length (µm)	2.5-5.0	3.5-6.0	1.7-4.8	2.0-7.0	3.0-8.0	2.5-8.0	3.0-4.5
Cell width (µm)	0.8-1.0	0.7-0.9	0.5-0.7	0.6-1.0	0.8	1.3-1.5	0.7-0.9
Capsules	+V	+	.	.	+	+++	+V
Motility	+	+	+	+	+	+	+
Flagella	peri 1-10	peri 2-10	peri	peri	peri 4-8	peri 2-10	peri 3-10
Endospores							
length (µm)	1.8-2.3	2.0-2.5	1.0-2.4	1.5-2.5	1.5-2.5	1.5	1.5-2.5
width (µm)	1.1-1.3	1.2-1.5	0.7-1.4	1.2-1.5	1.1-1.5	0.8-1.0	1.0-1.5
position	T/ST	T/ST	T/ST	T/ST	T/ST	ST	T/ST
swelling the rod	+	+	+ or -	+	+	-	+
thick-walled	+	+	.	+	+	+	+
Growth on nutrient agar	(+)	+++	+++	(+)	(+)	+++	(+)
Glucose enhanced growth	+++	-	.	.	+++	-	+++
Yeast extract enhanced growth	(+)	-	.	.	+++	-	+++
Growth on nitrogen-poor media + yeast extract	+++	+++	+++	+++	+++	+++	+++
Facultative anaerobe	+	+	+	+	+	-	+
Catalase	(+)	+	+	+	(+)	+++	(+)
Oxidase	(+)	+++	-	-	+	+++	+
Growth at 40°C	+	+	-	-	+	+	+
Growth at 42°C	-	-	-	-	+	-	-
Growth in 2% (w/v) salt broth	(+)	+	-	-	+	+	+
Growth in 3% (w/v) salt broth	-	+	-	.	-	+	-
Growth on potato slope	(+)	+++	.	.	(+)	+++	+++
maceration and gas	-	+++	.	.	+++	-	+++
Starch hydrolysis	Dex	+++	+	+	Dex	++	+++

Character					
Carbohydrate utilization					
glucose	A	AG	-	AG	AG
sucrose	A	AG	-	AG	AG
lactose	-	AG	-	AG	AG
mannitol	A	AG	-	AG	AG
salicin	A	AG	-	AG	AG
arabinose	-	AG	-	AG	AG
xylose	-	AG	-	AG	AG
Voges-Proskauer	-	+	+	+	+
Methyl red	-	-	-	+	+
Indole	-	+	.	+	+
Growth in litmus milk	ACDR	AC	ADR or AR	ARG	ACRDG
Casein hydrolysis	(+)	+	+ or -	+	+
Gelatin hydrolysis	+	+++	+ or -	+++	+++
Growth on pectin agar	-	+	-	+	+
pectin hydrolysis	-	+++	-	+++	+++
Growth on egg yolk agar	-	+++	+	(+)	(+)
lecithinase production	-	-	-	-	-
Growth on human blood agar	(+)	(+)	+++	(+)	(+)
α-haemolysis	(+)	-	+	-	-
H$_2$S formation from cysteine	-	+	.	+	+
Nitrate reduction to nitrite (3 methods)	-	+	+	+[1]	+[1]
Growth in ammonium-glucose defined medium	-	+	-	-	-
Growth in ammonium-citrate defined medium	-	+	+	+	+
Biotin requirement	+/-	+	-	-	-
Thiamin requirement	+/-	-	-	+	-

Note: Further details of the methods used are given in the section describing the Wahab *Bacillus* in detail.

., not recorded; -, negative reaction; (+), slight, slow, or weak reaction; +, ++ and +++, moderate, good and abundantly positive reactions or growth; +V, Gram-positive to variable reaction; peri, peritrichous flagella; T/ST, terminal or subterminal spores; Dex, dextrin formation only; A, acid; AG, acid and gas; under litmus milk reactions A, acid; C, clot; D, digestion; R, reduction; G, gas; 1, addition of yeast extract (0.05%, w/v) to the peptone-nitrate water essential for nitrate reduction.

for isolates D10 and R10 (Table 2).

Some properties of these 4 isolates are given in Table 3. The 2 acetylene-reducing strains D10 and R10 were closely similar to the type strain *B.polymyxa* NCTC 10343 and also resembled the laboratory strain 22V of *B.polymyxa*, although neither of these reduced acetylene.

The 2 isolates RSB6 and 7, both obligate aerobes, resembled each other closely, except that the growth on nutrient agar of RSB6 was pale orange and showed clumps of crystals in the opaque, rough-surfaced growth, whereas RSB7 colonies were smooth, cream-white and opaque. Cytologically both were large Gram-positive rods 4.5 to 8.0 x 1.3 to 1.5 µm in size and conspicuously capsulated (although peritrichously flagellate under suitable conditions). Neither acid or gas production was evident in any classic sugar broths, although marked acidity developed in litmus milk. RSB6 was proteolytic towards both casein and gelatin whereas RSB7 was not. Marked α-haemolysis and green-brown discolouration with ammoniacal odours were marked on human blood agar. Growth on the nitrogen-poor agar was much more abundant at pH 7.6 than at 7.1 and yeast extract supplementation was not essential; this was confirmed by abundant growth in glucose-ammonium or citrate-ammonium salts media in the absence of biotin or thiamine. Overall the characters were in accord with the current description of *B.megaterium* [Gibson and Gordon, 1974].

Supplementary studies concerning growth factor requirements

The need for great technical care when attempting to elucidate growth factor requirements for growth in minimal media, as emphasized by Owens and Keddie [1969] was fully appreciated. Nevertheless, non-reproducible results were often obtained during attempts to determine the need for supplements of biotin and/or thiamine in various defined glucose-ammonium-salts media, including the glucose (3.0%, w/v) medium recommended by Knight and Proom [1950] for the growth of *Bacillus* spp.

The defined vitamin-free basal test medium recommended by Stanier *et al.* [1966] for use with pseudomonads was tested; this basal medium contains a heavily chelated solution of many trace metals and mineral salts in 0.035M phosphate buffer at pH 6.8, with 1.0 g l^{-1} ammonium sulphate. With either glucose or sucrose carbon sources, this medium supported very good growth of the *Bacillus* strains of Wahab. The possibility that the other defined glucose-ammonia test media, for example those of Koser or Knight and Proom, might be deficient in essential trace elements was therefore ascertained by adding to all the 'no growth' cultures which had been incubated for 7 days at 30°C, 0.1 ml of Hunter's vitamin-free mineral base (a component of the medium of Stanier *et al.* containing the

chelated metal and salt mixture) to each 5 ml volume of
liquid culture. These were reincubated without further
inoculation. In every case, for each of the 12 Wahab isolates under test, growth then occurred even in the minus-
biotin and minus-thiamine media. When only 0.1 ml of the
Hunter's solution was added to the 5 ml cultures, growth
occurred variously after 2 to 14 days reincubation; a
supplement of 0.2 ml per 5 ml culture resulted in abundant
growth within 2 days.

Thus, there was overwhelming evidence that an adequate
amount of chelated salts and trace metals substituted for
vitamin requirements for the *Bacillus* strains isolated by
Wahab [1969]. It also had a similar 'sparing effect' for
the 2 test strains of *B.polymyxa*, a preliminary finding
which merits further investigation. It also explains the
more consistent acetylene reduction experienced by Wahab,
who used the Stanier-type medium rather than the defined
medium of Brown *et al*. [1962] in his later studies.

Conclusions

Optimal conditions for the determination of either
growth requirements or acetylene reduction by the *Bacillus*
spp. studied here have not been found; pH values, phosphate levels, the buffering capacity of the medium and an
adequate supply of chelated trace metals and salts have all
been shown to exert a pronounced effect. A vitamin-sparing
action by inorganic growth factor supplements has been
found.

Many of the nitrogen-fixing *Bacillus* spp. and certainly
the 4 isolates of *B.polymyxa* studied here often showed very
poor aerobic growth on nutrient agar and many of the standard diagnostic media; this was not always remedied by the
addition of yeast extract, and a trace element deficiency
or an unfavourable pH value may have been responsible for
this, and account for some of the unsatisfactory diagnostic
tests for the identification of *B.circulans* and *B.macerans*,
'a difficult group' [Gordon *et al*., 1973; Gibson and
Gordon, 1974; Bonde, 1975, personal communication, 1978;
Seki *et al*., 1978]. The anomalous properties of the
acetylene-reducing sand dune strain Sp.4 isolated here were
such that it remains uncertain whether it is an atypical
B.polymyxa or yet another *B.circulans*-type organism.
Studies of further isolates and G+C analyses of their DNA
might be helpful, but recent DNA-DNA homology studies reported by Seki *et al*. [1978] are not promising; their 3
isolates of *B.circulans*, including the type or neotype
strain ATCC 4513 (and also ATCC 9966, and ATCC 7049, which
latter was, like the type strain, supposedly derived from
the W.W. Ford strain 26) showed only 2 to 3% homology
between themselves (!); the same value was obtained when
B.circulans ATCC 4513 DNA was matched with *B.polymyxa* ATCC
842 (the neotype strain). No strains of *B.macerans* were

included in this or previous studies by Seki and co-
workers.

Two typical *B.polymyxa* isolates, RSB D10 and RSB R10,
which consistently reduced acetylene anaerobically, have
been isolated from Ynyslas dunes. Two further *Bacillus*
isolates, RSB6 and RSB7 never reduced acetylene despite
numerous variations in the test conditions, but they
showed substantial gains in nitrogen content in liquid
culture (sometimes indeed greater than those shown by the
acetylene-reducing isolates of *B.polymyxa*, see Table 2).
RSB6 and 7 were identified as isolates of *B.megaterium*,
and although *B.megaterium* has never been claimed to fix
nitrogen in pure culture, it has been interestingly implicated in mixed culture with *Rhodopseudomonas palustris*
wherein nitrogen fixation was greater than that shown by
R.palustris alone: *B.mesentericus*, on the other hand, pre-
vented the growth of *Azotobacter* due to toxic secretions
(see Moore, 1966, for an excellent review of non-symbiotic
nitrogen fixation in soil and soil-plant systems). There
is one report of a deliberate inoculation with proven
nitrogen-fixing cultures of *B.polymyxa*, of plants (maize,
wheat and tomato) grown in sterile nitrogen-deficient soil
[Rovira, 1963]. This did not result in a significant increase in the nitrogen content of any host plant.

Studies here have certainly reinforced the need for
great care when assessing either growth factor requirements
or acetylene-reducing ability of *Bacillus* isolates. The
author believes that the 12 isolates of *Bacillus* studied
by Wahab [1969] probably originated in laboratory equipment
rather than in the sand dunes. They were remarkably
homogeneous, not particularly salt-tolerant, and could not
be re-isolated from the dunes despite the availability of
highly selective techniques for this purpose. Nevertheless, it is clear that the verbal opinion of the late Dr.
T. Gibson that the nitrogen-fixing strains of *Bacillus*
isolated by Wahab belong to neither *B.polymyxa* nor *B.mace-
rans* has been amply confirmed. It is believed that these
strains represent a new species, which in view of the involvement of Dr. T. Gibson, should be named after him when
the necessary fuller studies have been performed. Isolate
A36 with 34.7 ± 0.3 moles %G+C (T_m 83.6) in its DNA is
suggested here as a suitable reference strain; it has been
deposited in the National Collection of Industrial
Bacteria, NCIB 11495.

The low G+C value for the DNA is of especial interest,
being more reminiscent of the 23 to 43% range for *Clostridium*, although the value for nitrogen-fixing *C.butyricum*
is 27 to 28%. The overall similarities between some of the
nitrogen-fixing *Bacillus* spp., especially *B.polymyxa*, and
some clostridia merit speculation; as discussed, they have
often been isolated by the same selective procedures, they
frequently actively ferment many carbohydrates with abundant acid and gas formation, they form endospores with

markedly swollen sporangia, oxygen (1.0%, v/v) markedly
inhibits nitrogen fixation by *B.polymyxa* and may well in-
hibit other properties because *B.polymyxa* and the strains
isolated by Wahab grew very poorly on a wide range of
diagnostic media incubated aerobically. Furthermore, when
Couchot and Maier [1974] tested the ability of 16 strains
of 11 facultatively anaerobic species of *Bacillus* to sporulate strictly anaerobically, only the single strain of *B.
macerans* ATCC 8244, and both strains of *B.polymyxa* (ATCC
12056 and UWO 320) readily did so. Thus, one might
question whether there is a meaningful boundary between
facultatively anaerobic *Bacillus* spp. which grow poorly
aerobically in the absence of fermentable carbohydrate,
and oxygen-tolerant *Clostridium*, or whether indeed 'Nature
knows no sharp distinctions'.

Acknowledgements

I am grateful to Professor P.D. Walker for electron
microscope examination of the spores of 3 strains of the
Bacillus isolated by Wahab, and to Professor J. De Ley for
determining the G+C value of the DNA of strain A36. The
excellent technical assistance throughout of Mrs. G. Hall
has been indispensable; the advice and encouragement of
the late Dr. T. Gibson in this unfamiliar taxonomic
terrain is likewise sincerely acknowledged.

References

Alexander, M. (1964). Biochemical ecology of soil microorganisms.
Annual Review of Microbiology **18**, 217-252.
Bland, B.F. (1968). Nitrogen contribution from the soil for herbage
growth. *Plant and Soil* **28**, 217-225.
Bonde, G.J. (1975). The genus *Bacillus*: an experiment with cluster
analysis. *Danish Medical Bulletin* **22**, 41-61.
Bredemann, G. (1908). Untersuchungen über die Variation und das
Stickstoff-Bindungsvermögen des *Bacillus asterosporus* A.M., ausge-
führt an 27 Stämmen verschiedener Herkunft. *Zentralblatt für
Bakteriologie, Parasitenkunde und Infektionskrankheiten Abteilung
II* **22**, 44-89.
Brough, F. (1950). Microtechnique for nitrate reduction. *Journal of
Bacteriology* **60**, 365-366.
Brown, M.E., Burlingham, S.K. and Jackson, R.M. (1962). Studies on
Azotobacter species in soil. I. Comparison of media and techniques
for counting *Azotobacter* in soil. *Plant and Soil* **17**, 309-319.
Chang, P.-C. and Knowles, R. (1964). Non-symbiotic nitrogen fixation
in some Quebec soils. *Canadian Journal of Microbiology* **11**, 29-38.
Chislett, M.E. and Kushner, D.J. (1961). A strain of *Bacillus circulans* capable of growing under highly alkaline conditions. *Journal
of General Microbiology* **24**, 187-190.
Couchot, K.R. and Maier, S. (1974). Anaerobic sporulation in facultatively anaerobic species of the genus *Bacillus*. *Canadian Journal
of Microbiology* **20**, 1291-1296.

Dilworth, M.J. (1966). Acetylene reduction by nitrogen-fixing preparations from *Clostridium pasteurianum*. *Biochimica et Biophysica
Acta* **127**, 285-294.

Gibson, T. and Gordon, R.E. (1974). *Bacillus*. In "Bergey's Manual of
Determinative Bacteriology" (eds. R.E. Buchanan and N.E. Gibbons),
pp.529-550. Baltimore: The Williams and Wilkins Co.

Gordon, R.E., Haynes, W.C. and Pang, C.H.-N. (1973). "The Genus
Bacillus". Washington, D.C.: United States Department of Agriculture.

Grau, F.H. and Wilson, P.W. (1962). Physiology of nitrogen fixation
by *Bacillus polymyxa*. *Journal of Bacteriology* **83**, 490-496.

Hankin, L., Zucker, M. and Sands, D.C. (1971). Improved solid medium
for the detection and enumeration of pectolytic bacteria. *Applied
Microbiology* **22**, 205-209.

Hill, L.R. (1966). An index to deoxyribonucleic acid base compositions of bacterial species. *Journal of General Microbiology* **44**,
419-447.

Hino, S. and Wilson, P.W. (1958). Nitrogen fixation by a facultative
Bacillus. *Journal of Bacteriology* **75**, 403-408.

Knight, B.C.J.G. and Proom, H. (1950). A comparative survey of the
nutrition and physiology of mesophilic species in the genus
Bacillus. *Journal of General Microbiology* **4**, 508-538.

Knowles, R., Neufeld, R. and Simpson, S. (1974). Acetylene reduction
(nitrogen fixation) by pulp and paper mill effluents and by *Klebsiella* isolated from effluents and environmental situations.
Applied Microbiology **28**, 608-613.

Line, M.A. and Loutit, M.W. (1971). Non-symbiotic nitrogen-fixing
organisms from some New Zealand tussock-grassland soils. *Journal
of General Microbiology* **66**, 309-318.

Meiklejohn, J. and Weir, J.B. (1968). Nitrogen-fixers - pseudomonads
and other bacteria - from Rhodesian soils. *Journal of General
Microbiology* **50**, 487-496.

Moore, A.W. (1966). Non-symbiotic nitrogen fixation in soil and soil-
plant systems. *Soils and Fertilizers* **29**, 113-128.

Moore, A.W. and Becking, J.H. (1963). Nitrogen fixation by *Bacillus*
strains isolated from Nigerian soils. *Nature* **198**, 915-916.

Oulette, C.A., Burris, R.H. and Wilson, P.W. (1969). Deoxyribonucleic
acid base composition of species of *Klebsiella*, *Azotobacter* and
Bacillus. *Antonie van Leeuwenhoek Journal of Microbiology and
Serology* **35**, 275-286.

Owens, J.D. and Keddie, R.M. (1969). The nitrogen nutrition of soil
and herbage coryneform bacteria. *Journal of Applied Bacteriology*
32, 338-347.

Postgate, J.R. (1978). Biological nitrogen fixation. In "Companion
to Microbiology" (eds. A.T. Bull and P.M. Meadow), pp.343-361.
London: Longman, London Group Ltd.

Rhodes, M.E. (1958). The cytology of *Pseudomonas* spp. as revealed by
a silver-plating staining method. *Journal of General Microbiology*
18, 639-648.

Ross, D.J. (1958). Influence of media on the counts of *Clostridium
butyricum* in soils. *Nature* **181**, 1142-1143.

Rovira, A.D. (1963). Microbial inoculation of plants. I. Establishment of free-living nitrogen-fixing bacteria in the rhizosphere and

their effects on maize, tomato, and wheat. *Plant and Soil* **19**, 304-314.

Seki, T., Chung, C.-K., Mikami, H. and Oshima, Y. (1978). Deoxyribo-nucleic acid homology and taxonomy of the genus *Bacillus*. *International Journal of Systematic Bacteriology* **28**, 182-189.

Smith, N.R. and Gordon, R.E. (1957). *Bacillus*. In "Bergey's Manual of Determinative Bacteriology' (eds. R.S. Breed, E.G.D. Murray and N.R. Smith), pp.613-634. London: Baillière, Tindall and Cox Ltd.

Smith, N.R., Gordon, R.E. and Clark, F.E. (1952). "Aerobic Spore-forming Bacteria". United States Department of Agriculture Monograph No. 16, Washington, D.C.

Stanier, R.Y., Palleroni, N.J. and Doudoroff, M. (1966). The aerobic pseudomonads: a taxonomic study. *Journal of General Microbiology* **43**, 159-271.

Stewart, W.D.P. (1965). Nitrogen turnover in marine and brackish habitats. I. Nitrogen fixation. *Annals of Botany, New Series* **29**, 229-239.

Turner, M. and Jervis, D.I. (1968). The distribution of pigmented *Bacillus* species in saltmarsh and other saline and non-saline soils. *Nova Hedwigia* **16**, 293-298.

Vágnerova, K., Macura, J. and Catska, V. (1960). Rhizosphere flora of wheat. I. Composition and properties of bacterial flora during the first stages of wheat growth. *Folia Microbiologica* **5**, 298-310.

Wahab, A.M. Abdel (1969). "The role of micro-organisms in the nitrogen nutrition of *Ammophila arenaria*". Ph.D. thesis. University of Wales, Aberystwyth.

Wahab, A.M. Abdel (1975). Nitrogen fixation by *Bacillus* strains isolated from the rhizosphere of *Ammophila arenaria*. *Plant and Soil* **42**, 703-708.

Webley, D.M., Eastwood, D.J. and Gimmingham, C.H. (1952). Development of a soil microflora in relation to plant succession on sand-dunes, including the 'rhizosphere' flora associated with colonizing species. *Journal of Ecology* **40**, 168-178.

Wilson, P.W. and Knight, S.G. (1952). "Experiments in Bacterial Physiology". Minneapolis: Burgess Publishing Co.

Witz, D.F., Detroy, R.W. and Wilson, P.W. (1967). Nitrogen fixation by growing cells and cell-free extracts of the Bacillaceae. *Archiv für Mikrobiologie* **55**, 369-381.

Chapter 14

SPOROSARCINA AND SPOROLACTOBACILLUS

J.R. NORRIS

*Agricultural Research Council, Meat Research Institute,
Langford, Bristol, UK**

Introduction

So much of our understanding of the structure, composition,
formation and behaviour of endospores is derived from
studies of *Bacillus* and *Clostridium* species that microbiologists are apt to dismiss similar resistant bodies observed in other genera as being of 'uncertain status' or
as representing some other 'poorly understood' phenomena.
It is not unusual to find workers involved with *Bacillus*
spores apparently avoiding such organisms as *Sporosarcina*
and *Sporolactobacillus*. Perhaps the traditional microbiologist feels uncomfortable at the idea of a motile,
spore-forming coccus or a motile, microaerophilic spore-
forming lactobacillus. The object of this chapter is to
review our knowledge of these 2 organisms and discuss
their relationship to the genus *Bacillus*.
  The Eighth Edition of "Bergey's Manual of Determinative
Bacteriology [Buchanan and Gibbons, 1974] includes both
genera in the Family Bacillaceae.
I.   Cells rod-shaped
      A.   Aerobic or facultative, catalase usually
            produced
                              Genus I *Bacillus*
      B.   Microaerophilic, catalase not produced
                              Genus II *Sporolactobacillus*
      C.   Anaerobic
            1. Sulphate not reduced to sulphide
                              Genus III *Clostridium*
            2. Sulphate reduced to sulphide
                              Genus IV *Desulfotomaculum*
II.   Cells spherical, in packets
                              Genus V *Sporosarcina*

*For present address see List of Contributors.

Endospores are the definitive characteristic of all of these genera and the Manual lists refractivity, reduced stainability, resistance to heat and other destructive agents, possession of dipicolinic acid to 5-15% of the dry weight, a peptidoglycan cortex and an outer spore coat as the criteria by which endospores differ from vegetative cells. It is important to bear these criteria in mind when considering the homology of the spores of the 'unusual' genera.

Sporosarcina

History

The earliest descriptions of motile, packet-forming cocci producing heat resistant bodies were those of Beijerinck [1901] and Ellis [1903]. Beijerinck found that the heat resistant bodies would withstand exposure to 90°C for 10 mins but, strangely, did not believe that typical endospores were formed. His isolate decomposed urea and he named it *Planosarcina ureae*. Löhnis [1911] changed the name to *Sarcina ureae* and that name was retained right up to the Seventh Edition of "Bergey's Manual" [Breed *et al.*, 1957] which classified it as a member of the Micrococcaceae.

Prior to Beijerinck's description motile sarcinae were described by Maurea [1892] and by Sames [1898] but spore-formation was not mentioned [Hucker and Thatcher, 1928] and spore-forming but apparently non-motile sarcinae were described by Hauser [1887] and again by Lehmann and Neumann [1927].

These and other early records are discussed in detail by Gibson [1935] who isolated several strains of *Sarcina ureae* from soil. Gibson clearly stated that in his opinion the heat-resistant bodies which survived 100°C for up to 5 mins were true endospores. A further interesting isolate was that of Wood [1946] who obtained a motile, spore-forming coccus from sea water which could survive 99.5°C for 3 mins.

After the study of Gibson [1935] attention focused largely on the nature of the spores and composition of the organism with few workers examining more than small numbers of strains and none attempting to isolate the organism on any scale until Pregerson [1973; Pregerson and Spotts, 1973] who examined 198 soils from various parts of the world and obtained 61 isolates. They were able, for the first time, to carry out a substantial study of strains in parallel and form some idea of the extent of variation within the species.

Isolation and cultivation

Isolation of *Sporosarcina ureae* from soil is not diffi-cult. Beijerinck [1901] used beef broth containing 10%

(w/v) urea as a selective medium and Gibson [1935] also incorporated 10% (w/v) urea in a meat extract-peptone agar
medium, isolating several strains by direct plating of soil
dilutions.

Pregerson [1973], having shown the advantageous effect
of high pH and finding that urea at 10% (w/v) was somewhat
inhibitory, used Tryptic soy-yeast agar the pH of which was
adjusted to 8.5 before autoclaving to which filter steri-
lized urea was added to give a final concentration of 1%
(w/v). An incubation temperature of 22°C proved suitable
for isolation, colonies of *Sporosarcina ureae* being readily
distinguished by their characteristic yellow pigment and
uniform surface granularity after 3 days' incubation.
Attempts at selective enrichment offered little or no advantage.

Once isolated in pure culture *Sporosarcina ureae* is a
robust organism growing well on Tryptic soy-yeast extract
agar (pH about 7) at 30°C and remaining viable for months
at refrigerator temperatures.

Distribution

With very few exceptions all isolates of *Sporosarcina
ureae* have derived from soils. Lehmann and Neumann [1927]
[Gibson, 1935] described a urea-decomposing, spore-forming
sarcina which they isolated from the human respiratory
tract and Sames' [1898] isolation of a motile urea-decomposing sarcina was from liquid manure. These organisms are
no longer available and the only indisputable *Sporosarcina
ureae* not associated directly with soil appears to be that
of Wood [1946] which came from surface sea water off the
coast of Australia. Wood, however, found the general flora
of his samples to be similar to that of soil so the
organism could well have originated from soil.

All other isolates have come from soils of various types
and from many parts of the world. Usually isolates have
been made from soils which have received liberal applications of farm manure or from land used for animal grazing
and Pregerson [1973] showed a specific relationship between
the presence of *Sporosarcina ureae* and soil contamination
by urine of man or dog. She was unsuccessful in attempts
to isolate the organism from zoo compound soils heavily
contaminated with faecal matter or from 'undisturbed' soils
from deserts, fields, mountains or other relatively un-
populated areas.

The role played by urea in the ecology of *Sporosarcina
ureae* remains to be fully elucidated. The organism is
isolated most easily from soils relatively rich in urea but
is unable to use the compound as a source of carbon or
energy for growth. Most strains, however, break it down
readily and use the ammonia formed as a nitrogen source and
Pregerson [1973] suggests that resistance to urea concentrations over 3% (w/v) and the enhancement of growth seen
at moderately elevated pH when nutrients are in limited
supply possibly confer a selective advantage on the

organism in organic-rich soils.

Gibson [1935] calculated that his soils contained 1 to
2 x 10⁴ cells per g dry weight and Pregerson [1973] detected similar populations.

Clearly *Sporosarcina ureae* is primarily an inhabitant
of urea-containing urban soils but some facets of its
ecology remain to be studied.

Morphology

Cells   There is no doubt at all that the bacterium is a
typical sarcina. The cocci vary in diameter from 1.0 to
2.5 μm and are characteristically arranged in packets of 4
or 8 cells. Electron microscopy [Mazanec *et al.*, 1965]
clearly shows the division walls forming at right angles
to one another and Thompson and Leadbetter [1962] were unable to find any evidence of rod-like forms (Figs. 1a and
b).

Motility   Wider-ranging studies than have been so far reported are necessary before the correlation between spore
formation and motility in sarcinae can be fully determined.
Motile cocci not producing spores have been described, for
example by Hucker and Thatcher [1928], and Pijper *et al.*
[1955] reported that a motile, non-spore-forming sarcina
(*Sarcina agilis*) did not cross react serologically with
Sarcina ureae. Observations of spore formation in non-
motile sarcinae are rare and for the most part occur in
the very early literature.

There are technical difficulties. Motility is often
sluggish and appears to be highly dependent on the use of
suitable growth media [Pregerson, 1973]. One strain referred to the author as being non-motile proved to be quite
vigorously motile when grown on plate-count agar (Oxoid).
Gibson [1974], in describing the genus *Sporosarcina* for the
Eighth Edition of "Bergey's Manual of Determinative Bacteriology" [Buchanan and Gibbons, 1974], states that strains
may be motile or non-motile but MacDonald and MacDonald
[1962] found motility to be a feature of the strains they
studied, as did Pregerson [1973].

The vast majority of strains of spore-forming, urea-
decomposing sarcinae have been shown to be motile when
carefully examined and electron microscopy readily shows
the presence of typical flagella. There are discrepancies
over the type of flagellation. Gibson refers to several
'randomly spaced' flagella on each cell whilst Kocur and
Martinec [1963] state that each cell possesses a single
long flagellum. In the author's experience the flagella
are very easily detached from the cells during preparation
for electron microscopy and the heavy clumping of the
cells makes micrographs difficult to interpret (Fig. 2).

Spores   The nature of the spores of *Sporosarcina ureae* has

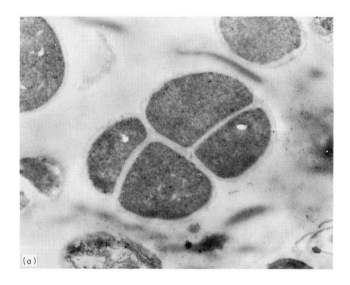

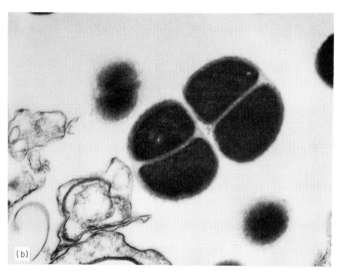

Fig. 1 Electromicrograph of a thin section of recently divided cells of *Sporosarcina ureae* x 28,000 (a) and x 25,200 (b).

attracted much attention in recent years as knowledge of the endospores of *Bacillus* and *Clostridium* has developed. Early observations established that the bodies which could be seen under the microscope, and which resembled spores in appearance and staining reactions, showed the same order of heat resistance as did the spores of the better known endospore-forming bacteria. The first detailed study of heat resistance was that of Gibson [1935] who showed that spores of *Sarcina ureae* could withstand boiling for up to 5 mins and that vegetative cells resisted 75°C for 15 mins. MacDonald and MacDonald [1962] working with a different strain recorded a lower resistance of the vegetative cells (50°C for 10 mins) but also detected some spores surviving boiling water for 10 mins.

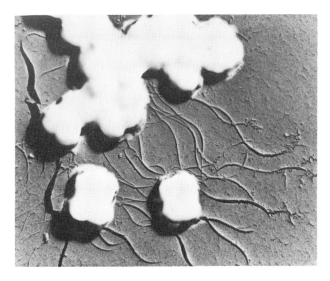

Fig. 2 Electromicrograph of a shadowed preparation of packets of cells and flagella of *Sporosarcina ureae* x 5,600.

Spore formation is generally poor on media of relatively high pH, particularly when cultures are incubated at temperatures in excess of 25°C. Gibson [1935] obtained good yields of spores using a peptone-meat extract agar with added ammonium chloride at pH 6.8 and incubating below 25°C. MacDonald and MacDonald [1962] modified the sporulation medium G of Stewart and Halvorson [1953] by adding peptone and malt extract to obtain sporulation approaching 100% and Pregerson [1973] has confirmed the need for a mixture of mineral ions for good spore production. Under suitable conditions most strains produce spores freely, the majority of cells showing a fully refractile, central spherical spore after 5 days at 22°C, but there is apparently wide variation from strain to strain in the nature of the nutritional requirements for maximal sporulation.

Until 1965 observations of spores were restricted to light microscopy which indicated that they were spherical, highly refractile and measured 0.8 to 1.2 μm in diameter. Mazanec *et al*. [1965] published electron micrographs which showed the spores to be slightly elliptical in shape, measuring 0.6 by 0.8 µm and having a structure which was indistinguishable from that of normal endospores. The spore coat consisted of 5 layers totalling 24 nm and closely resembled that of *Bacillus cereus*. A typical cortex 60 to 90 nm thick was readily visible. Silva *et al*. [1973] used more sophisticated fixation techniques and showed typical endospore structure in thin sections of *Sporosarcina ureae* spores (Fig. 3).

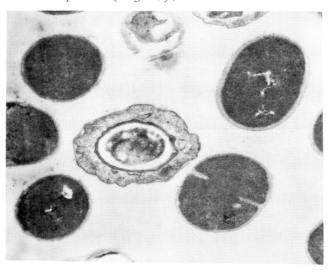

Fig. 3 Electronmicrograph of a thin section of vegetative cells and a spore of *Sporosarcina ureae* x 19,600.

An important piece of evidence for the homology with *Bacillus* spores was the demonstration by Thompson and Leadbetter [1962, 1963] of dipicolinic acid to a level of 4 to 7% of the dry weight in spores of *Sporosarcina ureae*. On germination the spore swells and takes on the characteristics of a vegetative cell [Thompson and Leadbetter, 1963].

Vegetative cells From the earliest observations the shape and mode of division of the vegetative cells of *Sporosarcina ureae* have been clearly described as characteristic of the packet-forming sarcinae and electronmicrographs such as those published by Mazanec *et al*. [1965] and by Silva *et al*. [1973] have amply confirmed the spherical shape of the free-lying cell, normally modified by interfacial flattening in the doublets, tetrads and packets of 6 or 8 cells that make up the majority of units in a culture, and by the occurrence of division walls at right angles to one another (Figs. 1a and b).

The cell wall consists of 2 layers of low electron density separated by a more dense layer which sometimes appears double [Mazanec *et al*., 1965], the wall being some

24 nm thick. By using fixation conditions more appropriate to bacteria, Silva *et al.* [1973] demonstrated meso-
somes within vegetative cells and showed that the fine
structure of membranes and surface structures were similar
to those of *Bacillus* species rather than to those of
Sarcina.

A number of Gram-positive bacilli including *Bacillus
pasteurii* and *B.sphaericus* are unusual in that they contain L-lysine in the tetrapeptide of the peptidoglycan of
the cell wall but *meso*-diaminopimelic acid instead of L-
lysine in the peptidoglycan of the spore cortex. Linnett
et al. [1974] demonstrated that *Sporosarcina ureae* also
shows this situation having a γ-D-glutamylglycine cross
link between the L-lysine and D-alanine residues of adja-
cent peptides in the vegetative cell wall and *meso*- di-
aminopimelic acid in the spore cortex.

Classification

The obvious taxonomic problems of *Sporosarcina ureae*
have attracted several microbiologists who have applied a
wide range of techniques in attempts to elucidate the
relationships of the organism with other species.

Cell wall analysis Diaminopimelic acid and sugars found
in *Bacillus* cell walls are absent from the walls of *Sporosarcina ureae* which contain principally alanine, glutamic
acid, lysine and glycine, a composition similar to that of
Gram-positive cocci [Cummins and Harris, 1956; Salton,
1964].

DNA composition Guanine + cytosine contents (G+C%) of the
DNA of *Sporosarcina ureae* have been variously determined
as, for example, 38 to 40% [MacDonald, 1962], 43% [Auletta
and Kennedy, 1966] and 39.2 to 44% [Boháček *et al.*, 1968].
These figures are informative since they indicate that
Sporosarcina ureae is not closely related to either the
anaerobic sarcinae (G+C% = 29 to 30) or the aerobic sarcinae (G+C% = 68 to 73). The G+C% of *Sporosarcina ureae*
is similar to the figures for some members of the genus
Bacillus, B.pasteurii for instance having a G+C% value of
42.5 [Boháček *et al.*, 1968].

Only 1 study has employed nucleic acid pairing techniques. Herndon and Bott [1969] hybridized RNA from
Sporosarcina ureae with DNA from members of the Bacillaceae
and of the Micrococcaceae and obtained results indicating
a closer relationship to the aerobic spore-formers than to
the cocci.

Serology Using agglutination reactions with vegetative
cells and gel diffusion analysis of antigens solubilized
by ultrasonic irradiation, MacDonald and MacDonald [1962]
failed to demonstrate cross reactions between *Sporosarcina*

ureae and *Sarcina flava* or *B.subtilis*. There was evidence
from agglutination reactions that strains of *Sporosarcina
ureae* could be sub-divided into several antigenic groups.
There were slight cross agglutination reactions between
Sporosarcina ureae and *B.pasteurii* and between *Sporosarcina ureae* and *B.sphaericus* which MacDonald [1962] regarded as significant.

Enzyme control Jensen and Stenmark [1970] characterized
the control mechanism of 3-deoxy-D-arabinoheptulosonate
7-phosphate synthetase and showed that control in *Sporosarcina ureae* resembled that in *Bacillus* but differed from
that in the genus *Micrococcus*.

16S ribosomal RNA Pechman *et al.* [1976] applied a relatively little used technique to examine the taxonomic relationships of *Sporosarcina ureae*. Using T_1 ribonuclease
to break down ^{32}P-labelled 16S ribosomal RNA the authors
constructed catalogues of the oligonucleotides so produced
and then examined the catalogues from different organisms
for the presence of identical oligonucleotides.
   Sporosarcina ureae shared few common oligonucleotides
with *Micrococcus luteus*. On the other hand, all but 1 of
the sequences (hexamer and larger) common to the catalogues of *B.subtilis*, *B.stearothermophilus* and *B.pasteurii*
were also present in *Sporosarcina ureae*. Furthermore, the
relationship of *Sporosarcina ureae* was closer to *B.
pasteurii* than to the other *Bacillus* species.
   Developing this approach Fox *et al.* [1977] used computer analysis to produce a similarity matrix of the relationships between *Bacillus* species, *Sporosarcina ureae*
and *Sporolactobacillus inulinus*. *B.subtilis* and *B.pumilus*
clustered together as did *B.cereus* and *B.megaterium*. *B.
stearothermophilus* did not cluster with any of the other
species but *Sporosarcina ureae* clustered with *B.pasteurii*.
Sporolactobacillus inulinus was more distantly related to
this whole group of organisms and *Escherichia coli* more
distantly still.

Biochemical characteristics Uniformity of biochemical
reactions is the main feature of the strains of *Sporosarcina ureae* which have been studied by various authors
[MacDonald and MacDonald, 1962; Kocur and Martinec, 1963;
Pregerson, 1973]. Nitrate reduction, urease and catalase
production are uniformly positive and gelatin hydrolysis,
indole and H_2S production and the Voges-Proskauer reaction
negative. Carbohydrates in general are not utilized for
growth nor is glucose but most strains use glutamate or
acetate as carbon and energy sources.
   Pregerson [1973] demonstrated differences in nutritional
requirements and was able to assign her strains to 4
groups; one growing on a minimal acetate medium without
added growth factors, a large group with a spectrum of

definable growth factors all members of which required
biotin either alone or in combination with niacin and/or
thiamine and some members of the group requiring aspartate,
a third group with complex growth factor requirements
satisfied by low concentrations of yeast extract and
casamino acids and a small group of strains which were
nutritionally fastidious.

Requirement for biotin and/or thiamine and ability to
utilize ammonium salts as nitrogen source are nutritional
characteristics often encountered in the genus *Bacillus*
[Knight and Proom, 1950; Proom and Knight, 1955]. *B.
pasteurii* shows some nutritional properties not generally
encountered in the genus but which it shares with *Sporosarcina*. Knight and Proom [1950] showed that its nutritional requirements were more heterogeneous than those of
most other *Bacillus* species and that some strains had an
unusual requirement for niacin. Perhaps most striking is
the inability of both *B.pasteurii* and *Sporosarcina ureae*
to utilize glucose, a character which is rare among aerobic
spore-formers.

Nomenclature and taxonomic status

All subsequent studies have served to confirm
Beijerinck's original description of a motile, packet-
forming coccus producing heat resistant bodies and decomposing urea [Beijerinck, 1901]. Although Beijerinck did
not describe spores as such, Gibson [1935] was of the
opinion that typical endospores were produced and again
subsequent work has demonstrated the soundness of this
view.

In the light of Gibson's verdict that the spores of
Sporosarcina ureae 'cannot be recognized as anything other
than true endospores', it is surprising that the literature
contains so many later allusions to abnormality. The
earlier editions of "Bergey's Manual of Determinative
Bacteriology" state 'typical endospores absent' and this is
only slowly modified in later editions to 'atypical endospores present' (Sixth Edition) and 'endospores of an unusual type are produced' (Seventh Edition). Thompson and
Leadbetter [1963] point out that what deserves emphasis is
'not that the spores are "atypical", which is contrary to
fact, but that in the group of spherical bacteria typical
spore formation is unknown except in the case of *S.ureae*'.

Of the taxonomic criteria applied, only cell wall composition suggests a close relationship of *Sporosarcina
ureae* to other cocci. DNA composition, RNA/DNA pairing
studies, serology and biochemical characteristics all
suggest a closer relationship to the genus *Bacillus* in
general and within the aerobic spore-formers to *B.
pasteurii* in particular. This is recognized in the Eighth
Edition of "Bergey's Manual of Determinative Bacteriology"
in which *Sporosarcina* is given generic rank in the
Bacillaceae.

Reluctance to accept the fact of endospore formation has been paralleled by indecision about the correct name for the organism. As early as 1909 Orla-Jensen proposed that Beijerinck's *Planosarcina ureae* should be placed in a new genus - *Sporosarcina* and this was also recognized by Kluyver and van Niel [1936]. Nevertheless, even as late as the Seventh Edition of "Bergey's Manual of Determinative Bacteriology" published in 1957 [Breed *et al.*, 1957] the organism was still referred to as *Sarcina ureae*. It was left to Gibson when writing for the Eighth Edition of the "Manual" to bring nomenclature and observed fact together giving the correct designation as *Sporosarcina* Kluyver and van Niel, and placing the genus firmly in the Bacillaceae with the type species *Sporosarcina ureae* (Beijerinck) Kluyver and van Niel [Buchanan and Gibbons, 1974].

Perhaps 2 issues remain to be resolved; the relationships of the various urea-splitting, motile, non-spore-forming sarcinae reported in the literature and the detailed status of *Sporosarcina* within the Bacillaceae.

When Gibson [1935] studied cultures of Beijerinck's original isolate of *Planosarcina ureae* he experienced initial difficulty in persuading them to produce spores and several workers have commented on the reluctance of some strains of *Sporosarcina ureae* to sporulate. Indeed, production of spores in a coccus might easily pass unnoticed, especially if sporulation were suppressed by cultivation on alkaline media. It could well be that other motile, or indeed non-motile, sarcinae would form spores under suitable conditions and there is a need for a thorough comparative study of new and existing isolates of these interesting organisms along the lines of Pregerson's work with *Sporosarcina ureae* [Pregerson, 1973].

Pregerson [1973] suggests that selective pressures in a common, enriched soil environment may have resulted in evolutionary convergence of *Sporosarcina ureae* and *B. pasteurii* as evidenced by their possession of a unique constellation of mutual properties and indications of a general restrictive type of metabolism. This is certainly consistent with the generally accepted view that vegetative cell type is of fundamental importance in defining widely different taxa but the other evidence discussed above argues against this, suggesting divergence from a common prototype. Particularly interesting here are the studies of Pechman *et al.* [1976] and Fox *et al.* [1977] involving analysis of 16S ribosomal RNA which indicate that *Sporosarcina ureae* is more closely related to *B. pasteurii* than are several other species of *Bacillus*. These authors believe that *Sporosarcina ureae* should not only be classified as a member of the genus *Bacillus* but should be placed in a subgroup containing *B. pasteurii*. The idea of sub-dividing the genus *Bacillus* is not by any means a new one; the DNAs of *Bacillus* species exhibit a very wide range

of G+C% which many microbiologists would consider far too
wide to justify grouping them all into one taxon of generic
rank (see Priest, Chapter 3). There is a clear need for
comparative studies of *Sporosarcina* and a full range of
aerobic spore-forming bacteria involving a variety of techniques to examine the genetic information present in the
cells at different levels of its expression; at the level
of the genome, in relation to the structure of protein
molecules, in terms of the chemical composition of the
cells, and at the morphological and behavioural level.

Perhaps the most important issue is the concept of
classifying together bacteria whose cell type differs as
fundamentally as a rod and a coccus. This is the problem
which has caused so much difficulty for the traditional
microbiologist faced with the reality of *Sporosarcina*
ureae. Evidence seems to be accumulating that this bacterium should be thought of as a circular *Bacillus* rather
than as a spore-forming coccus; the implications of that
concept for bacterial taxonomy are both fascinating and
profound.

Sporolactobacillus

History

In contrast to *Sporosarcina*, *Sporolactobacillus* was relatively recently described and has, as yet, been little
studied. Nevertheless, such studies as are published have
been competently executed and cultures are readily available.

The borderline between *Lactobacillus* and *Bacillus* has
received considerable attention and numerous 'aberrant'
types of bacteria sharing characteristics of both genera
have been described [Kitahara and Suzuki, 1963; Nakayama
and Yanoshi, 1967a, b; Davis, 1964; Gemmell and Hodgkiss,
1964; Nonomura *et al.*, 1965].

Lactobacillus is defined as being Gram-positive, non-
motile, non-sporulating, catalase negative and microaerophilic. The rod shaped bacteria produce lactic acid from
glucose either by homo- or hetero-fermentative pathways.
Nakayama studied a range of intermediate forms which produced spores in sugar-deficient media, were motile, catalase positive and produced L(+)-lactic acid, classifying
them with *B.coagulans* [Suzuki and Kitahara, 1964].

Other workers [Harrison and Hansen, 1950; Deibel and
Niven, 1958] described organisms closely resembling homo-
fermentative *Lactobacillus* except for motility by peritrichous flagella in young cultures. The lactic acid produced was L(+) or DL. They were catalase negative.

Kitahara and Suzuki [1963] isolated from chicken feed a
catalase negative, spore-forming bacterium which was
motile by peritrichous flagella, microaerophilic and
showed a typical homo-fermentative metabolism producing

D(-)-lactic acid. They created a new subgenus within the Lactobacillaceae to accommodate this unusual organism calling it *Sporolactobacillus inulinus* on account of its ability to ferment inulin.

Nakayama and Yanoshi [1967b] isolated 7 strains of catalase negative, spore-bearing, lactic acid producing bacteria which were motile and produced DL-lactic acid, from the rhizosphere of wild plants. These authors considered their isolates to be further members of *Sporolactobacillus* which they suggested merited generic rank within the Bacillaceae. Nakayama subsequently isolated further organisms which differed in the type of lactic acid produced and named his strains *Sporolactobacillus laevus* and *Sporolactobacillus racemicus* [Uchida and Mogi, 1973].

Isolation and cultivation

Sporolactobacillus grows reasonably well on media of the type used for lactobacilli. Kitahara and Suzuki [1963] in their original isolation used a glucose, yeast extract, peptone medium solidified with agar and obtained pin-point colonies. These colonies were more distinct in plate cultures when calcium carbonate was present in the medium since they were surrounded by clear zones formed by the lactic acid produced. Nakayama and Yanoshi [1967b] used a similar medium and also obtained small colonies of their rhizosphere isolates.

Sporolactobacillus inulinus is microaerophilic, producing large colonies in a band below the surface of semi-solid agar shake cultures but the soil species are more robustly facultatively anaerobic. *Sporolactobacillus inulinus* has an optimal growth temperature about 37°C, the soil species growing best about 30°C.

Isolation of these rather feebly growing bacteria from their natural environment is complicated by the presence of clostridia and aerobic spore-formers. The most effective method involves enrichment by inoculating a pasteurized (80°C for 20 mins) specimen into a fluid medium containing glucose and allowing clostridia to develop by anaerobic incubation. These prevent the growth of aerobic spore-formers but do not prevent development of the lactic acid bacteria. After a period of a few days' incubation cultures showing a pH of less than 4.0 are streaked onto a glucose agar medium containing a little calcium carbonate. The agar surface is covered with polyvinylidenechloride film, again to prevent growth of aerobic spore-formers and after incubation, pin-point colonies surrounded by clear zones are selected.

Distribution

The original isolate of *Sporolactobacillus inulinus* [Kitahara and Suzuki, 1963] was from chicken feed. Subse-

quent isolates of this species and of other sporolactobacilli have been made from the rhizospheres of a variety
of wild plants [Nakayama and Yanoshi, 1967b]. Using similar isolation techniques these authors [Nakayama and
Yanoshi, 1967a] also isolated spore-forming lactic acid-
producing bacteria which were catalase positive, *B.laevo-
lacticus* and *B.racemilacticus*, from the same source material.

Nakayama and Yanoshi [1967a] have pointed out that lactic acid bacteria are found mainly in milk products,
pickles, fermented mashes and similar materials but that
these are features of the environment associated with
human activities and are unlikely to be the primary habitats. They argue that soil, and more specifically the
rhizosphere, will provide a habitat with the nutrients
necessary for growth of lactic acid bacteria and that this
situation with its localized nutrient concentrations and
tendency to undergo drying and heating by sunlight impin-
ging on the soil surface, would favour motile, spore-
forming lactic acid bacteria. The logic is interesting
but more extensive ecological studies are obviously indicated.

Morphology

Cells The cells are straight Gram-positive rods measuring
0.7 to 0.8 μm by 3 to 5 μm. They differ from typical
lactobacilli only in being motile with peritrichous flagella which vary from many to few or even 1 in some
strains.

Spores Cultures do not spore freely. Kitahara and Lai
[1967] obtained improved sporulation of *Sporolactobacillus
inulinus* up to 1% of the vegetative cell population on a
carefully designed medium containing yeast extract, meat
extract, α-methylglucoside, calcium carbonate and a source
of manganese ion and a further improvement to 10% by incubating in a carbon dioxide atmosphere. Nakayama and
Yanoshi [1967b] found starch to stimulate sporulation in
their soil isolates and glucose to be effective in some
strains.

In all species the spores are ellipsoidal, terminal or
subterminal and the sporangia are distinctly swollen. Heat
resistance is of the order of 80°C for 10 mins. Dipicolinic acid is present in the spores of *Sporolactobacillus
inulinus* to a concentration of 5% [Kitahara and Lai, 1967].

Electronmicroscopy has been little used but indicates
a typical endospore structure for the spores of *Sporolactobacillus inulinus* although fine detail has not, so
far, been revealed [Kitahara and Lai, 1967].

Classification

The so-called 'intermediate' forms of lactobacilli which resemble aerobic spore-forming bacteria by producing spores and exhibiting motility include both catalase positive and catalase negative organisms and they have attracted some attention, particularly from Japanese workers.

DNA composition Two determinations of G+C% for the DNA of *Sporolactobacillus inulinus* are available and unfortunately they differ substantially; Suzuki and Kitahara [1964] reported a value of approximately 39% whilst Miller *et al.* [1970] found a value of 47.3%. Both groups agree in finding a wide range of values across the genus *Lactobacillus*. The Japanese authors found *Lactobacillus leichmannii* to be the nearest species in terms of G+C% and Miller *et al.*'s value most resembled that of *Lactobacillus casei*.

Dellaghio *et al.* [1975] performed an extensive series of DNA homology tests on lactobacilli and concluded from the low homologies recorded that *Sporolactobacillus inulinus* was genetically distinct from the other species studied.

Miller *et al.* [1971] could find no evidence from DNA homology for a relationship between *Sporolactobacillus inulinus* and *B.coagulans* or *Lactobacillus plantarum* in spite of their having similar G+C%s.

Cellular fatty acids There are basic differences between the cellular fatty acid spectra of *Bacillus* and *Lactobacillus* [Uchida and Mogi, 1973]. *Bacillus* cells contain predominantly saturated fatty acids with odd numbers of carbon atoms, iso- and anteiso-branched C_{15} and C_{17} and smaller amounts of iso-C_{16} are also present. Lactobacilli, by contrast, contain even numbered saturated straight chain acids (C_{16} predominating), even numbered, straight chain unsaturated acids (C_{16} and C_{18} predominating) and C_{17} and C_{19}(lactobacillic)-cyclopropane acids.

Uchida and Mogi [1973] found a sharp distinction in fatty acid pattern when they examined a range of '*Bacillus-Lactobacillus* intermediates'. *B.subtilis*, *B.coagulans*, *B.racemilacticus*, *B.laevolacticus*, *B.myxolactis*, *Sporolactobacillus inulinus*, *Sp.laevus* and *Sp.racemicus* all showed the *Bacillus* pattern of fatty acids. *Lactobacillus yamanashiensis* (motile), *L.plantarum* and *L.casei* showed the typical *Lactobacillus* spectrum.

Isoprenoid quinones Collins and Jones [1979] determined the isoprenoid quinone composition as a guide to the classification of *Sporolactobacillus* and representative *Bacillus*, *Lactobacillus* and intermediate organisms. Similar menaquinone patterns were seen in *Bacillus*, in *B.dextrolacticus*, *B.laevolacticus*, *B.myxolactis*, *B.race-*

milacticus, and all 3 species of *Sporolactobacillus*. They were not, however, detected in *Lactobacillus plantarum* or in several other lactobacilli examined.

Cell wall composition   Like *Bacillus*, but unlike typical *Lactobacillus*, *Sporolactobacillus* cell walls contain di-aminopimelic acid.

16S ribosomal RNA   Fox *et al*. [1977] in their studies of 16S ribosomal RNA oligonucleotides included *Sporolacto-bacillus inulinus* and showed that it was only distantly related to *Bacillus* species and to *Sporosarcina ureae*.

Nomenclature and taxonomic status

When *Sporolactobacillus inulinus* was first described by Kitahara and Suzuki [1963] the authors recognized immediately the similarity to some motile *Lactobacillus* strains and named their new isolate *Sporolactobacillus (Lacto-bacillus) inulinus* proposing a new subgenus of the Lacto-bacillaceae to accommodate it.

In doing so they were influenced by the treatment of *Sporosarcina* (see *Sarcina ureae*) in the Seventh Edition of "Bergey's Manual of Determinative Bacteriology" [Breed *et al.*, 1957] and they inadvertently reduced *Lactobacillus* to subgenus rank.   This presumably was a *lapsus calami* for *Lactobacillus (Sporolactobacillus) inulinus*.

The recognition of generic rank as *Sporolactobacillus inulinus* is due to Kitahara and Lai [1967] and Kitahara and Toyota [1972] transferred the genus to Bacillaceae.

In the Eighth Edition of the "Manual" Kitahara [1974] has *Sporolactobacillus inulinus* (Kitahara and Suzuki) Kitahara and Lai, 1967.

The other species, *Sporolactobacillus laevus* and *Sporo-lactobacillus racemicus*, were first named by Nakayama in a little known publication [Uchida and Mogi, 1973].   They must await further study before their characteristics and classification can be verified or reassessed.

Kitahara and Suzuki [1963] were well aware of the existence of a spectrum of organisms showing affinities with both lactobacilli and the spore-forming rods, and discussed at some length the taxonomic implications of the possession of a homolactic fermentative metabolism, spores and flagella in the absence of catalase.   They concluded that their new isolate was intermediate in nature between typical *Lactobacillus* and the anaerobic, catalase negative genus *Clostridium*.

Nakayama and Yanoshi [1967a] demonstrated the existence of spore-forming catalase positive lactic acid bacteria assigning them to the genus *Bacillus* and pointing out the similarity to *B.coagulans* and other lactic acid-producing bacilli.   In discussing their catalase negative isolates, the same authors [Nakayama and Yanoshi, 1967b] point out

the similarity of *B.coagulans* to *Lactobacillus thermophilus*
and of *Sporolactobacillus inulinus* to *Lactobacillus leichmannii* if spore formation and motility are ignored. They
also comment that spore formation is difficult to demonstrate in their catalase negative strains and suggest close
similarities to several non-spore-forming, motile, lactobacilli, especially *Lactobacillus plantarum*.

Uchida and Mogi [1973] take the analysis further, pointing out that *Lactobacillus thermophilus* is, in fact, identical with *B.coagulans* and that the various motile lactobacilli, including *Sporolactobacillus*, most closely resemble *Lactobacillus plantarum* with its diaminopimelic acid
in the cell wall and its distinctive serological behaviour
[Sharp, 1955].

It is clearly possible to identify a series of organisms
which range from the typical spore-forming, non-lactic
acid-producing, motile, catalase positive *Bacillus* to the
lactic acid-producing, non-motile, non-spore-forming,
catalase negative *Lactobacillus* (Table 1) and there seems
little to support a closer relationship with *Clostridium*.

Conclusions

Both *Sporosarcina* and *Sporolactobacillus* undoubtedly
produce typical bacterial endospores and in both cases
there are good arguments based on a variety of considerations to suggest taxonomic links with the genus *Bacillus*.
Sporosarcina has a number of features in common with *B.
pasteurii* and there are indications that these may reflect
a closer relationship than mere convergence resulting from
the habitation of the same environmental niche.

Both *Bacillus* and *Lactobacillus* are large genera whose
G+C% values range from the low 30s to the 50s casting
serious doubt on their validity as taxa at the generic
level. The borderline between them is certainly ill-
defined and demands further study.

Perhaps the most important conclusion to be drawn is
that although microbiologists may find *Sporosarcina* and
Sporolactobacillus uncomfortable they really do exist,
cultures are freely available for study, and they should
be taken into account in any comprehensive taxonomic study
of the genus *Bacillus*.

Acknowledgements

I am grateful to C.A. Voyle for the electron micrographs and to Bernadine Pregerson for providing me with a
copy of her informative thesis on *Sporosarcina ureae*.

TABLE 1

Some characteristics of Bacillus-Lactobacillus *intermediates*
(after Uchida and Mogi, 1973)

Taxonomic characteristic	Typical *Bacillus*	*Bacillus coagulans*	*Bacillus racemilacticus*	*Sporolactobacillus inulinus*	*Lactobacillus yamanashiensis*	*Lactobacillus plantarum*	Typical *Lactobacillus*
Lactic acid fermentation	-	+	+	+	+	+	+
Nitrate reduction	+	±	-	-	-	-	-
Catalase production	+	+	+	-	-	-	-
Spore formation	+	+	+	+	-	-	-
Motility	+	+	+	+	+	-	-
Diaminopimelic acid in cell wall	+	+	+	+	+	+	-
G+C% in DNA	33.50	45.4 46.9		47.3 39.3		44.5 42.9	33-54
Type of fatty acid spectrum	B	B	B	B	L	L	L

B, *Bacillus* type; L, *Lactobacillus* type   (see text)

References

Auletta, A.E. and Kennedy, E.R. (1966). Deoxyribonucleic acid base composition of some members of the Micrococcaceae. *Journal of Bacteriology* **92**, 28-34.

Beijerinck, M.W. (1901). Anhäufungsversuche mit Ureumbakterien, Ureumspaltung durch Urease und durch Katabolismus. *Zentralblatt für Bakteriologie, Parasitenkunde, Infektionskrankheiten und Hygiene*, II. Abt. **7**, 33-61.

Boháček, J., Kocur, M. and Martinec, T. (1968). Deoxyribonucleic acid base composition of *Sporosarcina ureae*. *Archiv für Mikrobiologie* **64**, 23-28.

Breed, R.S., Murray, E.G.D. and Smith, N.R. (eds.) (1957). "Bergey's Manual of Determinative Bacteriology, 7th Edition". London: Baillière, Tindal and Cox Ltd.

Buchanan, R.E. and Gibbons, N.E. (eds.) (1974). "Bergey's Manual of Determinative Bacteriology, 8th Edition". Baltimore: The Williams and Wilkins Co.

Collins, M.D. and Jones, D. (1979). Isoprenoid quinone composition as a guide to the classification of *Sporolactobacillus* and possibly related bacteria. *Journal of Applied Bacteriology* **47**, 293-297.

Cummins, C.S. and Harris, H. (1956). The relationships between certain members of the *Staphylococcus-Micrococcus* group as shown by their cell wall composition. *International Bulletin of Bacteriological Nomenclature and Taxonomy* **6**, 111-119.

Davis, G.H.G. (1964). Notes on the phylogenetic background to *Lactobacillus* taxonomy. *Journal of General Microbiology* **34**, 177-184.

Deibel, R.H. and Niven, C.F. (1958). Microbiology of meat curing. I. The occurrence and significance of a motile micro-organism of the genus *Lactobacillus* in ham curing brines. *Applied Microbiology* **6**, 323-327.

Dellaghio, F., Bottazzi, V. and Vescova, M. (1975). Deoxyribonucleic acid homology among *Lactobacillus* species of the subgenus *Streptobacterium* Orla-Jensen. *International Journal of Systematic Bacteriology* **25**, 160-172.

Ellis, D. (1903). Untersuchungen über *Sarcina, Streptococcus* und *Spirillum. Zentralblatt für Bakteriologie, Parasitenkunde, Infektionskrankheiten und Hygiene I Abt.* **33**, 1-17, 81-96, 161-166.

Fox, G.E., Pechman, K.R. and Woese, C.R. (1977). Comparative cataloging of 16*S* ribosomal ribonucleic acid: molecular approach to procaryotic systematics. *International Journal of Systematic Bacteriology* **27**, 44-57.

Gemmell, M. and Hodgkiss, W. (1964). The physiological characters and flagellar arrangement of motile homofermentative lactobacilli. *Journal of General Microbiology* **35**, 519-526.

Gibson, T. (1935). An investigation of *Sarcina ureae*, a spore forming motile coccus. *Archiv für Mikrobiologie* **6**, 73-78.

Gibson, T. (1974). *Sporosarcina*. In "Bergey's Manual of Determinative Bacteriology" (eds. R.E. Buchanan and N.E. Gibbons), pp.573-574. Baltimore: The Williams and Wilkins Co.

Harrison, A.P. and Hansen, P.A. (1950). A motile *Lactobacillus* from the cecal feces of turkeys. *Journal of Bacteriology* **59**, 444-446.

Hauser, G. (1887). Ueber Lungensarcine. *Deutsches Archiv für klinische Medizin* **42**, 146-158.

Herndon, S.E. and Bott, K.F. (1969). Genetic relationship between *Sarcina ureae* and members of the genus *Bacillus*. *Journal of Bacteriology* **97**, 6-12.

Hucker, G. and Thatcher, L. (1928). Studies on the *Coccaceae*. X. The motility of certain cocci. *Technical Bulletin No.136*. Geneva, New York: New York State Agricultural Experimental Station.

Jensen, R.A. and Stenmark, S.L. (1970). Comparative allostery of 3-deoxy-D-arabino-heptulosonate-7-phosphate synthetase as a molecular basis for classification: two cases in point. *Journal of Bacteriology* **101**, 763-769.

Kitahara, K. (1974). *Sporolactobacillus*. In "Bergey's Manual of Determinative Bacteriology" (eds. R.E. Buchanan and N.E. Gibbons), pp.550-551. Baltimore: The Williams and Wilkins Co.

Kitahara, K. and Suzuki, J. (1963). *Sporolactobacillus* nov. subgen. *Journal of General and Applied Microbiology* **9**, 59-71.

Kitahara, K. and Lai, C.-L. (1967). On the spore formation of *Sporolactobacillus inulinus*. *Journal of General and Applied Microbiology* **13**, 197-203.

Kitahara, K. and Toyota, T. (1972). Auto-spheroplastization and cell-permeation in *Sporolactobacillus inulinus*. *Journal of General and

Applied Microbiology **18**, 99-107.

Kluyver, A.J. and van Niel, C.B. (1936). Prospect for natural system of classification of bacteria. *Zentralblatt für Bakteriologie, Parasitenkunde, Infektionskrankheiten und Hygiene, II Abt.* **94**, 369-403.

Knight, B.C.J.G. and Proom, H. (1950). A comparative survey of the nutrition and physiology of mesophilic species in the genus *Bacillus*. *Journal of General Microbiology* **4**, 508-538.

Kocur, M. and Martinec, T. (1963). The taxonomic status of *Sporosarcina ureae* (Beijerinck) Orla-Jensen. *International Bulletin of Bacteriological Nomenclature and Taxonomy* **13**, 201-209.

Lehmann, K.B. and Neumann, R.O. (1927). Bakteriologie insbesondere Bakteriologische Diagnostik. II. Allgemeine und spezielle Bakteriologie. 7 Auflage.

Linnett, P.E., Roberts, R.J. and Strominger, J.L. (1974). Biosynthesis and cross-linking of the γ-glutamylglycine-containing peptidoglycan of vegetative cells of *Sporosarcina ureae*. *Journal of Biological Chemistry* **249**, 2497-2506.

Löhnis, F. (1911). *Landwirtschaftlichbakteriologisches Prakticum. Berlin.* 1-156.

MacDonald, R.E. (1962). Some further observations on the physiology and taxonomic position of *Planosarcina ureae* Beijerinck. *Bacteriological Proceedings* **G24**, 40.

MacDonald, R.E. and MacDonald, S.W. (1962). The physiology and natural relationships of the motile sporeforming sarcinae. *Canadian Journal of Microbiology* **8**, 795-808.

Maurea, G. (1892). Ueber eine bewegliche Sarcine. *Zentralblatt für Bakteriologie, Parasitenkunde, Infektionskrankheiten und Hygiene I Abt.* **11**, 228-231.

Mazanec, K., Kocur, M. and Martinec, T. (1965). Electron microscopy of ultrathin sections of *Sporosarcina ureae*. *Journal of Bacteriology* **90**, 808-816.

Miller, A., Sandine, W.E. and Elliker, P.R. (1970). Deoxyribonucleic acid base composition of lactobacilli determined by thermal denaturation. *Journal of Bacteriology* **102**, 278-280.

Miller, A., Sandine, W.E. and Elliker, P.R. (1971). Deoxyribonucleic acid homology in the genus *Lactobacillus*. *Canadian Journal of Microbiology* **17**, 625-630.

Nakayama, O. and Yanoshi, M. (1967a). Spore-bearing lactic acid bacteria isolated from rhizosphere. I. Taxonomic studies on *Bacillus laevolacticus* nov. sp. and *Bacillus racemilacticus* nov. sp. *Journal of General and Applied Microbiology* **13**, 139-153.

Nakayama, O. and Yanoshi, M. (1967b). Spore-bearing lactic acid bacteria isolated from rhizosphere. II. Taxonomic studies on the catalase-negative strains. *Journal of General and Applied Microbiology* **13**, 155-165.

Nonomura, H., Yamazaki, T. and Ohara, Y. (1965). Die Apfelsäure-Milchsäure-Bakterien, welche aus japanischen Weinen isoliert wurden. *Mitteilungen der Hoeheren Bundeslehr- und Versuchsanstalt fuer Wein-, Obst- und Gartenbau-Klosterneuburg* **15A**, 241-254.

Orla-Jensen, Ø. (1909). Die Hauptlinien des natürlichen Bakteriensystems. *Zentralblatt für Bakteriologie, Parasitenkunde, Infektionskrankheiten und Hygiene I Abt.* **22**, 305-346.

Pechman, K.J., Lewis, B.J. and Woese, C.R. (1976). Phylogenetic status of *Sporosarcina ureae*. *International Journal of Systematic Bacteriology* 26, 305-310.

Pijper, A., Crocker, C.G. and Savage, N. (1955). Sarcinae: motility, kind of flagella and specific agglutination. *Journal of Bacteriology* 69, 151-158.

Pregerson, B. (1973). "The Distribution and Physiology of *Sporosarcina ureae*". M.Sc. Thesis, California State University at Northridge.

Pregerson, B. and Spotts, C. (1973). Distribution and nutritional properties of *Sporosarcina ureae*. *Abstracts of the Annual Meeting of the American Society for Microbiology* G117, 45.

Proom, H. and Knight, B.C.J.G. (1955). The minimal nutritional requirements of some species in the genus *Bacillus*. *Journal of General Microbiology* 13, 474-480.

Salton, M.R.J. (1964). "The Bacterial Cell Wall". New York: Elsevier Publishing Co., Inc.

Sames, T. (1898). Eine bewegliche Sarcine. *Zentralblatt für Bakteriologie, Parasitenkunde, Infektionskrankheiten und Hygiene II Abt.* 4, 664-669.

Sharp, M.E. (1955). A serological classification of lactobacilli. *Journal of General Microbiology* 12, 107-122.

Silva, M.T., Lima, M.P., Fonseca, A.F. and Sousa, J.C.F. (1973). The fine structure of *Sporosarcina ureae* as related to its taxonomic position. *Journal of Submicroscopic Cytology* 5, 7-22.

Stewart, B.T. and Halvorson, H.O. (1953). Studies on the spores of aerobic bacteria. I. The occurrence of alanine racemase. *Journal of Bacteriology* 65, 160-166.

Suzuki, J. and Kitahara, K. (1964). Base compositions of deoxyribonucleic acid in *Sporolactobacillus inulinus* and other lactic acid bacteria. *Journal of General and Applied Microbiology* 10, 305-311.

Thompson, R.S. and Leadbetter, E.R. (1962). Endospores of *Sarcina ureae*. *Bacteriological Proceedings* G61, 49.

Thompson, R.S. and Leadbetter, E.R. (1963). On the isolation of dipicolinic acid from endospores of *Sarcina ureae*. *Archiv für Mikrobiologie* 45, 27-32.

Uchida, K. and Mogi, K. (1973). Cellular fatty acid spectra of *Sporolactobacillus* and some other *Bacillus-Lactobacillus* intermediates as a guide to their taxonomy. *Journal of General and Applied Microbiology* 19, 129-140.

Wood, E.J.F. (1946). The isolation of *Sarcina ureae* (Beijerinck) Löhnis from sea water. *Journal of Bacteriology* 51, 287-289.

ORGANISMS INDEX

SUBJECT INDEX